Networks: From Biology to Theory

Jianfeng Feng, Jürgen Jost and
Minping Qian (Eds.)

Networks: From Biology to Theory

Jianfeng Feng, PhD
Computer Science Department
Warwick University, Coventry, UK

Jürgen Jost, PhD
Max Planck Institute for Mathematics in the Sciences
Leipzig, Germany

Minping Qian, PhD
Centre for Theoretical Biology
Peking University, Beijing, PR China

British Library Cataloguing in Publication Data
A catalogue record for this book is available from the British Library

Library of Congress Control Number: 2006928726

ISBN-10: 1-84628-485-6 e-ISBN-10: 1-84628-780-4
ISBN-13: 978-1-84628-485-4 e-ISBN-13: 978-1-84628-780-0

Printed on acid-free paper

9 8 7 6 5 4 3 2 1

Springer Science+Business Media
springer.com

Contents

List of Contributors

Basim Al-Shaikhli
Physics Department
Neurophysics Group
Philipps University
35032 Marburg
Germany
Basim.Al-Shaikhli@physik.uni-marburg.de

Fatihcan M. Atay
Max Planck Institute for Mathematics in the Sciences
04103 Leipzig
Germany
atay@member.ams.org

Andreas Bruns
Physics Department
Neurophysics Group
Philipps University
35032 Marburg
Germany
Andreas.Bruns@physik.uni-marburg.de

Hilary Buxton
Department of Informatics
University of Sussex
Brighton BN1 9QH

United Kingdom
hilaryb@central.susx.ac.uk

Minghua Deng
LMAM
School of Mathematical Sciences and Center for Theoretical Biology
Beijing University
Beijing 100871
People's Republic of China
dengmh@pku.edu.cn

Andreas Dress
Max Planck Institute for Mathematics in the Sciences
04103 Leipzig
Germany
(present address:
CAS-MPG Partner Institute for Computational Biology
Shanghai Institutes for Biological Sciences
Chinese Academy of Sciences
200031 Shanghai
People's Republic of China)
andreas@picb.ac.cn

Reinhard Eckhorn
Physics Department
Neurophysics Group
Philipps University
35032 Marburg
Germany
Reinhard.Eckhorn@physik.uni-marburg.de

Jianfeng Feng
Department of Mathematics
Hunan Normal University
Changsha 410081
People's Republic of China
Department of Computer Science and Mathematics
Warwick University
Coventry CV4 7AL
United Kingdom
Jianfeng.Feng@warwick.ac.uk

Xiucheng Feng
Center for Theoretical Biology and School of Mathematical Sciences
Beijing University
Beijing 100871
People's Republic of China
xcfeng@ctb.pku.edu.cn

Andreas Gabriel
Physics Department
Neurophysics Group
Philipps University
35032 Marburg
Germany
Andreas.Gabriel@physik.uni-marburg.de

Alexander Gail
Physics Department
Neurophysics Group
Philipps University
35032 Marburg
Germany
(present address:
German Primate Center
Cognitive Neuroscience Laboratory
37037 Göttingen
Germany)
alexander.gail@dpz.gwdg.de

Benoit Gaillard
Department of Informatics
Sussex University
Brighton BN1 9QH
United Kingdom
B.Gaillard@sussex.ac.uk

Mengwen Jia
MOE Key Laboratory of Bioinformatics
Department of Automation
Tsinghua University
Beijing 100084

People's Republic of China
mwjia@mail.tsinghua.edu.cn

Jürgen Jost
Max Planck Institute for Mathematics in the Sciences
04103 Leipzig
Germany
jost@mis.mpg.de

Fangting Li
Department of Physics and Center for Theoretical Biology
Beijing University
Beijing 100871
People's Republic of China

Tao Long
Center for Theoretical Biology
Beijing University
Beijing 100871
People's Republic of China
Longtao@ctb.pku.edu.cn

Liaofu Luo
Laboratory of Theoretical Biophysics
Faculty of Science and Technology
Inner Mongolia University
Hohhot 010021
People's Republic of China
lfluo@mail.imu.edu.cn

Ying Lu
Center for Theoretical Biology
Beijing University
Beijing 100871
People's Republic of China
Lying@ctb.pku.edu.cn

Axel Mosig
Bioinformatics Group
Department of Computer Science

Interdisciplinary Center for Bioinformatics
University of Leipzig
04107 Leipzig
Germany
(present address:
CAS-MPG Partner Institute for Computational Biology
Shanghai Institutes for Biological Sciences
Chinese Academy of Sciences
200031 Shanghai
People's Republic of China)
axel.mosig@gmail.com

Jörg Ontrup
Neuroinformatics Group
Faculty of Technology
Bielefeld University
33501 Bielefeld
Germany
jontrup@techfak.uni-bielefeld.de

Qi Ouyang
Department of Physics and Center for Theoretical Biology
Beijing University
Beijing 100871
People's Republic of China
qi@pku.edu.cn

Sonja J. Prohaska
Bioinformatics Group
Department of Computer Science
Interdisciplinary Center for Bioinformatics
University of Leipzig
04107 Leipzig
Germany
sonja@bioinf.uni-leipzig.de

Minping Qian
Center for Theoretical Biology and School of Mathematical Sciences
Beijing University
Beijing 100871

People's Republic of China
qianmp@math.pku.edu.cn

Helge Ritter
Neuroinformatics Group
Faculty of Technology
Bielefeld University
33501 Bielefeld
Germany
helge@techfak.uni-bielefeld.de

Mirko Saam
Physics Department
Neurophysics Group
Philipps University
35032 Marburg
Mirko.Saam@physik.uni-marburg.de

Peter F. Stadler
Bioinformatics Group
Department of Computer Science
Interdisciplinary Center for Bioinformatics
University of Leipzig
04107 Leipzig
Germany
Institute for Theoretical Chemistry
1090 Wien
Austria
Santa Fe Institute
Santa Fe, NM 87501
U.S.A.
studla@bioinf.uni-leipzig.de

Jochen Steil
Neuroinformatics Group
Faculty of Technology
Bielefeld University
33501 Bielefeld
Germany
jsteil@techfak.uni-bielefeld.de

Fengzhu Sun
Molecular and Computational Biology Program
Department of Biological Sciences
University of Southern California
Los Angeles, CA 90089
U.S.A.
fsun@hto.usc.edu

Chao Tang
California Institute for Quantitative Biomedical Research (QB3)
Department of Pharmaceutical Sciences
University of California
San Francisco, CA 94143-2540
U.S.A.
tangc@pharmacy.ucsf.edu

Henry C. Tuckwell
Max Planck Institute for Mathematics in the Sciences
04103 Leipzig
Germany
tuckwell@mis.mpg.de

Lin Wan
Center for Theoretical Biology and School of Mathematical Sciences
Beijing University
Beijing 100871
People's Republic of China
wanlin@ctb.pku.edu.cn

Sebastian Weng
Neuroinformatics Group
Faculty of Technology
Bielefeld University
33501 Bielefeld
Germany
sweng@techfak.uni-bielefeld.de

Huan Yu
School of Mathematical Sciences and Center for Theoretical Biology
Beijing University

Beijing 100871
People's Republic of China

Yuping Zhang
Center for Theoretical Biology and School of Mathematical Sciences
Beijing University
Beijing 100871
People's Republic of China
zhangyp@ctb.pku.edu.cn

Introduction

It seems certain that, to truly understand the staggeringly complex phenomena observed in contemporary biological experiments, both at molecular and system levels, we need to seek inspiration from mathematics, physics, and computer science. Equally, challenging problems for applied mathematics, physics, and computer science come from real applications.

Let us first look at the nervous system as an example. A rat brain contains about 10^7 nerve cells (neurons; the human brain contains about 10^{12} neurons), each making up to 10,000 connections (synapses) with other neurons. This wiring is not preprogrammed in detail; instead, the brain is the product of "bootstrap" instructions for developmental assembly and neuronal differentiation, and the precise wiring is the complex product of chance, experience-driven assembly, and broad preprogrammed principles. The process of developmental assembly is robust; for all the differences between your genes and mine and your experience and mine, our brains work pretty much the same. The final product is incredibly robust. Deletion of individual genes often has no apparent phenotypic consequence; early damage to large areas of the brain is often compatible with apparently normal function; and in the fully developed adult brain, neurodegenerative diseases such as Parkinson's or Alzheimer's are asymptomatic until neural loss in the affected areas is of the order of 70% cell death.

Yet individual neurons and synapses are, in general, noisy and unreliable. Neurons are certainly complex, and this is often (mis)taken for sophistication. However, much complexity arises from the difficulties of creating a living cell from proteins, and much apparent complexity may reflect imprecision of specification (so-called superfluous gene expression). There are only about 30,000 genomic domains in the human DNA, about one third of which appear to be concerned purely with development. Many of them are homologous with genes in simple organisms such as *Drosophila*, and many code for "housekeeping" genes, concerned with basic cellular regulation. Strictly, a gene encodes a protein (or in some cases many proteins), and includes rules that specify in what circumstances and in what amounts this protein is made. What that protein does depends on the environment (networks) in which it is expressed; expression of some proteins in a neuron will change the information processing capability of that neuron by altering its electrophysiological phenotype (for instance, by enabling it to discharge action potentials in a particular pattern), by altering its secretory phenotype (by making it secrete excitatory or inhibitory substances with a particular time-scale of action), by altering its morphological or anatomical phenotype (by giving it a single long process perhaps that goes from A to B, or many short processes), or by altering its receptive phenotype (by specifying receptors for particular molecules). The functional consequences of the expression of a particular protein depend on which neurons (and which circumstances) that protein is expressed in. The outcome of development is a brain that comprises many phenotypically distinct populations; in the rat hypothalamus, for example, we can estimate that each consists of a few thousand neurons. Each of these populations is variable—individual neurons of a population are not clones of each other but have considerable heterogeneity—and these populations are massively (and quasi-randomly) interconnected. Individual neurons are both unreliable and apparently quite loosely specified. However, the emergent structure has remarkable information storage capabilities and information processing capacity, and these are robust and reproducible between individual organisms.

We may know in considerable detail how individual genes, proteins, and neurons work, but how this relates to "higher level" behaviour is poorly understood. For example, in neuroscience, this is often referred to as the "problem of neural coding." Many paradigms that underlie our current thinking about neural function arose from analogies with computational systems where classical approaches take the single neuron as the atomic element of parallel computation so that complex processing is the result of the activity of large assemblies of these elements. This paradigm sets a hard limit to the processing power of the conventional machine learning approach, so perhaps it is not surprising that they have not led us to understand how the brain can do anything that conventional digital computers cannot do well.

An architecture with such self-evident information processing and storage capacity, so robustly specified by so few encoded rules, that operates with high reliability despite the fragility and imperfection of its individual components, and is so robust against changing environment and experience and against damage warrants close study, in particular, from the system level (interacting networks) point of view. But if biology is to be the inspiration, we must look at the biological facts themselves and not at interpretations of biology based on preconceptions that are themselves based on a rather superficial understanding of existing, mathematical, physical, and computational principles. This means that we need a dialogue between biologists and theoreticians, and shared problems of deep interest and concern across this cultural divide.

Understanding the nature and limits of the strategies employed by neural systems to represent, process, and transmit sensory information to higher-level areas that make behavioural decisions is fundamental to learning how brains work, and to developing novel computation. In general, neurons respond to sensory or synaptic inputs by generating a train of stereotypical responses called action potentials or spikes. Deciphering the encoding process that transforms continuous, analog signals (photon fluxes, acoustic vibrations, chemical concentrations, etc.) or outputs from other neurons into spike trains is essential to understanding neural information processing, since often the nature of representation determines the nature of computation that is possible. Researchers, however, remain divided on the issue of the neural code used to represent information. Information as used by the brain is likely to be highly distributed given the massively parallel network architecture, but our thinking, and especially that of biologists, is generally based on intuition about serial and single-unit processing strategies. On the one hand, it is commonly assumed that the mean firing rate of a neuron is the primary variable relating to sensory experience, and there is a quantitative relationship between the average firing rate of single cortical neurons and psychophysical judgements made by animals trained to perform specific tasks. An animal's behaviour in a visual discrimination task can be predicted by counting spikes over a long interval (typical one second or more) in a single neuron in the visual cortex. The highly variable temporal structure of neural spike trains strengthens the view that any statistic other than the averaged response is too random to convey information. However, the fine structure of spike intervals can potentially convey much more information than a firing rate code, and the precise relative temporal relationship between the outputs of different neurons appears to be relevant in certain cases. There may be no universal coding strategy; different neural networks may use different codes, or a combination of several coding strategies. It is clear nonetheless that the temporal precision of spike trains is an important limiting factor. The brain is not a unitary organ; we must start by selecting neural networks whose function is known and amenable to further investigation, and identify major relevant and tractable problems.

The way forward may be to focus on networks that subserve a particular, defined biological function, and use computational/theoretical strategies to build models that reproduce the reality of those networks, and then use those models to derive predictions amenable to empirical testing. The reality of the model networks must include frailty, imprecision, stochasticity, redundancy, and robustness; the reality need not extend to all biological details, but must follow a structured strategy aimed at including those elements of the phenotype that are essential and dispensing with those that are not, to evolve models that are computationally concise. To understand how neuronal networks process information, we will have to combine modern theory of nonlinear and stochastic dynamics and modern control theory with modern theory of neuroscience, and develop realistic models of highly complex systems to aid the process of developing an intuitive understanding of network behaviour and a neural network theory. Models should be inspired by biological systems, be constrained by the known biological limitations, and ideally should also be constrained by inspiration from two key principles alluded to above: (1) real neural networks are not preprogrammed but develop through application of a discrete number of bootstrap rules, and (2) real neural networks have emerged through an evolutionary design process. Using both abstract and biophysical models, we would gain more insights into how neural networks processes information.

Although so far we have mainly confined ourselves to neuronal systems, one of the most successful areas in systems biology approaches, all issues raised about neuronal networks hold true for gene networks and protein networks. How gene networks reliably control the development of a cell or a tissue, how protein networks function properly in response to internal and external perturbations, and how neuronal networks store and process information remain the grand challenges of biology. On the other hand, it is the basic paradigm of physics that, on the basis of deep and fundamental mathematical structures, general and universal laws can be formulated and experiments can then be devised for unambiguously confirming or rejecting those laws. In contrast to this, in modern biology we are usually confronted with large data sets in which we need to discern some structure. These structures are not self-evident, nor can they be deduced from general principles. Biological structures are aggregate structures of physicochemical constituents, and their function emerges from the interaction of many elements. Therefore, we need some guidance about which structures to look for, and in understanding how the biological functions are achieved.

Networks provide such a paradigm. They constitute an organizational principle for biological data sets. We need to identify the basic elements in a given framework. Depending on the biological context, these may be certain types of molecules like specific proteins or RNAs, or whole cells, like neurons. We also need to describe the interactions between them, forming bonds, catalyzing reactions, or exchanging spikes through synapses. On the one hand, we need to find out which elements interact with which specific other ones. On the other hand, we need to understand the dynamics of those interactions. The first aspect leads us to the mathematical structure of a graph. The second aspect brings in the theory of dynamical systems. A graph is a static structure and therefore not appropriate for capturing the process-like nature of biological systems. Therefore, we need to connect this static structure with a dynamical perspective. A dynamical network thus is a graph whose elements dynamically interact according to the structural pattern of the graph. From these local interactions, a collective dynamical pattern emerges that describes the system as a whole. Networks thus link structural and dynamical aspects. In mathematics, often the richest and most interesting patterns emerge from new connections between different theories. The present link between graph theory and dynamical systems in our opinion confirms this and leads to challenging research questions. Its ultimate aim is to contribute to the understanding of the formation of structures in biological and other systems.

Networks as such, however, are too general for reaching profound theories. There are simply too many types of graphs and dynamical systems, and arbitrary combinations of them seem of little use. Therefore, the theory needs the concrete biological data to identify the structures and dynamical patterns of interest and substance. In contrast to many areas of applied mathematics that simply provide tool boxes for data analysis, like statistics packages or numerical analysis software, here the theory itself is strongly driven by concrete biological findings.

Thus, this book aims at the same time at displaying a mathematical framework for organizing and understanding biological data, at providing a survey of such biological data, and showing the fruitful synthesis, the application of mathematical principles to those data, and the mathematical research stimulated by those data. We hope that this book will be a useful introduction to the theoretical aspects, as a survey of important biological networks (see, for example, Chapter 9 by Prohaska, Mosig, and Stadler on regulatory networks in eukaryotic cells), and as a sample of more concrete case studies of molecular and neural networks.

Since biological systems are aggregate and composite ones, in each case, we need to decide which aspects are relevant and crucial, and which can be neglected and perhaps even should be neglected because they obstruct the identification of the principles. In biological systems, as already discussed above for the case of a neural system, often details at some lower level average out at some higher level. This then leads to the question of which details should be included in a model and which ones omitted. While network theory agrees that the models are based on interactions between discrete entities, one then has to decide whether only discrete states are possible or whether we should rather admit continuous state values. In the first case, naturally, also the update is carried out at discrete time steps only. In the simplest case, there then are only two possible state values, labeled by 0 (rest) and 1 (activity). We then have a so-called Boolean network. Often such a network, even though it obviously represents a very simplified model, can still capture crucial qualitative aspects. Also, in a situation of only sparse data, it is advantageous to have a very simple model with as few parameters as possible, to avoid having to estimate such parameters without sufficient data. Some of the contributions in this book show the advantages of this approach. In other situations, one may know more about the temporal dynamics and can build a corresponding model. A prime example are networks of spiking neurons that have a much better correspondence with neurobiological reality than simple spin systems like the Hopfield network. Also, continuous state dynamics even with discrete temporal updating, like the so-called coupled map lattices, show new features like synchronization that are not so meaningful in discrete state systems.

A beautiful example where the question of which details to include in a model can be analyzed is transmission delays in networks. In many formal models they are, and can be, neglected because they do not affect the emerging collective dynamical patterns. In many cases, however, they can also lead to new dynamical phenomena. Atay (Chapter 2) presents three pertinent case studies, motivated by models in biology and neurobiology. In one of them, it is found that, surprisingly, transmission delays can facilitate synchronization of the individual dynamics in a global network of coupled oscillators. This provides a direct link to synchronization as a proposed mechanism for feature binding in neural systems. The idea, as originally proposed by von der Malsburg, is that a complex percept can be neurally represented by the coincident firing of the neurons responding to its specific features. Since there does not seem to exist a global temporal coordinator in neural systems, such a synchronization must emerge from local interactions between neurons, and the synchronization studies for dynamical systems provide novel insights. Nevertheless, synchronization is only the simplest type of collective dynamics, and nonlinear dynamical systems can also find richer global patterns. This is explored from an experimentally driven perspective in Chapter 7 by Eckhorn and his collaborators, who identify global dynamical patterns corresponding to specific sensory inputs. The same issue is approached from a theoretical perspective in the contribution of Ritter and his collaborators (Chapter 8), who construct neural networks that develop global dynamical patterns in response to their input from an interaction between neurons and a competition between neuron layers. This work is then connected with the principles of gestalt formation in cognitive psychology. This research reflects an important approach for elucidating the relationship between neural processes and cognitive phenomena.

It is a basic insight from the theory of networks, that such a dynamical pattern formation also depends on underlying structural aspects, and Chapter 3 by Jost lays some foundations for the analysis of the interplay between structure and dynamics. A basic set of structural parameters of a graph is provided by its Laplacian spectrum, which in turn characterizes the synchronizability of so-called coupled map lattice dynamics. This is an important example of the fact that one needs to be careful in isolating the correct parameters that determine the global dynamical features of the network supported by a graph. Since there far too many different graphs even of moderate size to describe all them exhaustively, we need to identify some graph classes that display fundamental qualitative features as observed in real data. At the same time, we should also be attentive to the specific features that distinguish graphs in some concrete biological domain from others. Huan Yu et al (Chapter 11) investigate protein–protein interaction networks from that perspective and they found that a hierarchical model captures the qualitative features of some species, but not of others. Also in this regard, Li et al (Chapter 10) demonstrate that real biological networks (their examples are the cell-cycle and life-cycle networks of protein–protein and protein–DNA interactions in budding yeast) can have structural and dynamical properties that are profoundly different from the ones of some general graph paradigms like random graphs studied in the literature. In particular, their examples show a dynamical attractor with a large basin of attraction that is reliably attained on a specific pathway. This secures the robust function of the network under perturbations, a crucial aspect for any biological system. While in the work of Li et al structural and dynamical stability of the important cycles is shown, Zhang and Qian (Chapter 13) study the same system from a stochastic point of view, and they show that the system is also stable against random perturbations. Finally, Huan Yu et al (Chapter 11) examine still another concept of stability, namely the one against experimental errors, and they reach the fortunate conclusion that protein–protein interaction networks are quite stable in that sense as well.

One specific class of graphs that has found important biological applications is trees. They are used for reconstructing and displaying evolutionary histories on the basis of observed similarities between present species. They are also used in other fields, such as linguistics for the reconstruction of historical divergences from common ancestral languages. Often the available data do not allow for the unambiguous reconstruction of such an evolutionary tree, however. Dress (Chapter 1) then argues that instead of forcibly suppressing those unresolvable ambiguities in a data set X, we should rather work with a class of graphs that is larger than the ones of X-trees. The endpoints of such a tree are the representatives of the data set to be grouped, and the internal nodes are putative common ancestors. Thus, the edges of a tree reflect unambiguous historical descendence relationships, and the lengths encode the degrees of dissimilarity between adjacent nodes. Dress introduces the category of X-nets, where the edges of a tree can get replaced by parallelograms that reflect alternative descendence relationships, and the lengths of the sides then describe again degrees of dissimilarity as contained in the data. More precisely, an important feature of an X-net is that for any neighbouring nodes u, v in the network and any other node x, one shortest connection from x to one of them (which one depends, of course, on x) has to pass through the other one. This theory constitutes a beautiful example of profound new mathematics developed in response to concrete challenges from biological and other data sets.

From a somewhat different perspective, certain ingredients in real dynamical systems are relegated to the role of noise or random perturbations, and we should then argue for the stability of the system against such perturbations. It is an important insight, however, that noise can also play a constructive role by providing energy that facilitates the distinction between weak signals. This aspect is taken up further in the contributions of Feng and his collaborations (Chapters 4 to 6) for networks of spiking neurons. They analyze the trade-off between speed and accuracy in neural decision making. Input correlations and inhibitory input improve the speed but increase the variability of neuronal output and thereby affect reproducibility of output patterns and thus accuracy of the decision. They develop a mathematical framework for the analysis of higher order statistics in neural networks. Tuckwell and Feng (Chapter 6) then investigate under which conditions such networks of spiking neurons can sustain spontaneous background activity and determine the neuronal firing rates. As demonstrated in the contribution of Zhang and Qian (Chapter 13), stochastic fluctuations may also play useful roles in intracellular and other networks, by allowing for transitions between different metastable states and optimizing the expected time spent in the deepest basin of attraction and following the most robust pathways for transitions between different states.

Sometimes, in molecular biology, network data are not directly available, and one only has sequence data instead. One then needs reliable methods for reconstructing the network from those sequence data. Qian and her collaborators (Chapter 11) provide Bayesian schemes that can solve that task successfully for protein–protein interaction networks. The basic ingredient to their approach are motifs of polypeptide subsequences of four amino acids. In another contribution from the group of Qian (Chapter 12), a new method for the identification of transcription factor binding sites in DNA sequences is introduced. Also, the complementary question for protein sequence motifs that can bind at regulatory sites at DNA level can be addressed with their methods.

As already mentioned, Prohaska, Mosig, and Stadler (Chapter 9) provide a comprehensive state-of-the-art survey of regulatory modes, modules, and motifs in eukaryotic cells, including in particular recent findings about regulation at RNA level as well as indications about deep phylogenies that conserved regulatory sequences can yield. Luo and Jia (Chapter 14) discover some surprising statistical correlations between RNA copy number or structure and protein structure. For example, the helices and strands in proteins are preferably coded by messenger RNA (mRNA) stem regions, but the protein coils by messenger RNS (mRNA) loops. One should notice here that RNA secondary structure can be reliably predicted by available algorithms, while protein folding is still a computationally very difficult and theoretically largely unsolved problem. This fact might provide a useful perspective for the results of Luo and Jia. They also find that RNA structure data available from different species show a significant preference for low folding energies when compared to randomized sequences. Moreover, they detect a dichotomy between intraspecific homogeneity and interspecific inhomogeneity of mRNA folding energies. Presumably, such results are just the tip of the iceberg, and many structural relationships in biology await discovery. We hope that the theory of networks will provide a useful framework within which to formulate hypotheses that can guide the conception of experiments and to organize and understand diverse biological data.

All the chapters in this book reflect our efforts to aim to have a concise mathematical theory of networks yet diverse applications in systems biology. Whereas in this introduction we have emphasized the cross-links for arranging the material, it seems natural to divide the book into three parts. The first part covers the mathematical theory. Although, as argued, much of the theory is motivated and inspired by biological or neurobiological questions, from the logical perspective it provides the foundation for the subsequent parts. Therefore, we naturally start with that. Next come applications in the neurosciences. As discussed above, neural systems represent a coherent class of intensively studied biological systems. We therefore choose them as our first class of systems and address them in the second part of this book. The third part is devoted to biological systems at the cell and molecular level and their investigation with the tools of bioinformatics. Bioinformatics is the field responding to the challenge of the huge data sets generated by modern techniques in molecular biology. While it developed out of computer science and utilizes predominantly mathematical techniques from the field of combinatorics and discrete mathematics, it obviously leads to many formal questions about networks and in turn benefits from the theoretical insights in that field. In fact, inside a cell there are various networks, in particular gene regulation and protein interaction networks. Here, we cover both of them.

The conference from which this book results was first conceived in October 2003 when one of us (J.J.) gave a public lecture, "Mathematical Perspectives in Biology and Neurobiology," and got in contact with the other two organizers at the Sino-German Centre in Beijing, a cooperative research institution of the DFG (German Research Foundation) and the NSFC (National Science Foundation of China). We are grateful to Robert Paul Königs and his colleagues from the Sino-German Centre for arranging that initial lecture, for supporting our scientific scheme, and for hosting our conference at the Sino-German Centre. All the lecturers and participants of the conference enjoyed the excellent facilities and the hospitality of the Center's staff. The conference stimulated both concrete cooperations between scientists from China and Germany and long-ranging scientific perspectives on which we plan to build in the future.

Jianfeng Feng
Jürgen Jost
Minping Qian
Warwick/Leipzig/Beijing
February 2006

Part I

Theory

1

The Category of X-Nets

Andreas Dress

Summary. The concept of X-*nets* is introduced as a convenient tool for dealing with taxonomic problems in terms of *phylogenetic networks*; in the same formalized quantitative fashion the concept of X-*trees* is used as a tool for dealing with taxonomic analysis in terms of *phylogenetic trees*. According to the definition proposed here, a net is considered to be a finite *metric space* (rather than a graph) that is a "bipartite L_1-space," that is, a metric space with point set M and metric $D : M \times M \to \mathrm{R} : (u, v) \mapsto uv$ that satisfies the following two rather technical conditions: $u, v, w, a, b, c \in M, uv + vb = ua + ab = ub, ab + bv = au + uv = av, vw + wc = vb + bc = vc$, and $bc + cw = bv + vw = bw$ always implies $ua + ac = uw + wc = uc$ and $au + uw = ac + cw = aw$, and $u, v, w \in M, uw < uv + vw$, and $vw < vu + uw$ implies $\{a \in M - \{u, v\} : ua + aw \neq \emptyset\}$; and an X-net is a metric space as above whose point set M contains X as well as "sufficiently many" shortest *geodesic paths* connecting the points in M with points in X. Though rather technical, these conditions allow us to develop a theory of X-nets that mimics all the results obtained in the theory of X-trees as well as to define "the category of X-nets" so that a canonical one-to-one correspondence between (1) the (isomorphism classes of) injective objects in that category and (2) "weighted systems of X-splits" can be derived generalizing the fundamental one-to-one correspondence between (1) the (isomorphism classes of) X-trees and (2) "weighted systems of pairwise compatible X-splits."

1.1 Introduction

Although in current discussions, networks are most often described in terms of (more or less ornamented) graphs, in this chapter on recent work done at the Center for Combinatorics at Nankai University, we prefer to describe networks in terms of *metric spaces*. The reason for this is that the concepts and results to be presented here using metric spaces as the basic notion are of some use in the context of phylogenetic analysis where the *length* of edges customarily used in pictorial representations of results not only are highly informative, as they indicate the presumed time span of evolutionary phases under investigation, but also are already quite essential for *deriving* such results algorithmically by searching for results that provide edge lengths that are "optimally" adapted to the given data.

More specifically, the *category of X-nets* described here in terms of metric spaces can be used as a natural framework for taxonomic analysis in terms of *phylogenetic networks* (cf. [2, 3, 4, 11, 14]) in analogy to the framework offered by the *theory of X-trees* supporting taxonomic analysis in terms of (the much more familiar) *phylogenetic trees* (cf. [5, 15]):

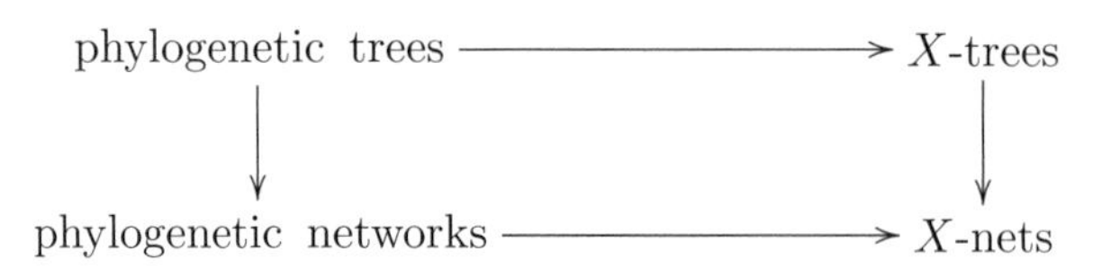

In the next section, we will present some basic terminology. In section 1.3, we collect some basic results (whose proofs need still to be written in publishable form, jointly with members of the Center for Combinatorics at Nankai University). Section 1.4 discusses how X-nets are "classified" by R-valued split systems. In the last section, some relevant examples are presented, extending from phylogenetic trees and networks to nets related to (1) subjectively perceived similarity of colours, (2) geographic data from a road atlas, and (3) the structural relatedness of world languages.

1.2 Basic Terminology

We consider finite metric spaces $\mathbf{M} = (M, D)$, that is, pairs consisting of a finite set M and a (proper) metric[1] $D : M \times M \rightarrow \mathbb{R} : (u, v) \mapsto uv$ defined on M.

[1] A metric or, more precisely, a *proper* metric defined on a set M is a bivariate map $D : M \times M \rightarrow \mathbb{R} : (u, v) \mapsto uv$ such that $uv = 0 \iff u = v$ and $uv + vw \geq wu$—and, therefore, also $uv = vu \geq 0$—holds for all $u, v, w \in M$. According to J. Isbell (cf. [13]), (1) the most appropriate way of defining the *Category of Metric Spaces* denoted by MET is to define, for any two metric spaces $\mathbf{M} = (M, D)$ and $\mathbf{M}' = (M', D')$, the set of morphisms from $\mathbf{M}$ into $\mathbf{M}'$ to consist of the set of all *non-expansive* maps from M into M', that is, all maps $\varphi : M \rightarrow M'$ with $D'(\varphi(u), \varphi(v)) \leq D(u, v)$ for all $u, v \in M$—with composition of morphisms defined in the obvious way; (2) there is a canonical class of monomorphisms in that category—including all isomorphisms—are the "isometric embeddings," that is, the maps $\varphi : M \rightarrow M'$ from a metric space $\mathbf{M} = (M, D)$ into a metric space $\mathbf{M}' = (M', D')$ for which $D'(\varphi(u), \varphi(v)) = D(u, v)$ holds for all $u, v \in M$; and that (3) two metric spaces are called *isometric* if they are isomorphic objects in that category.

Given any two points u, v in M, we define the interval $[u, v]$ spanned by u and v by

$$[u, v] = [u, v]_D := \{w \in M : uv = uw + wv\},$$

we write $u \leq_w v$ for some $w \in M$ whenever $u \in [w, v]$ holds and note that the binary relation "$\leq_u$" is a partial order of M for every $u \in M$, and we define a binary relation "$\|$" on M^2 by putting

$$uu' \| vv' \iff_{\text{def}} u', v \in [u, v'] \text{ and } u, v' \in [u', v]$$

for all pairs $(u, u'), (v, v')$ in M^2.

Next, the L_1-*product* of any two metric spaces $\mathbf{M} = (M, D)$ and $\mathbf{M}' = (M', D')$, denoted by $\mathbf{M} \times \mathbf{M}'$, is defined to be the metric space

$$\mathbf{M} \times \mathbf{M}' := (M \times M', D \oplus D')$$

whose point set is the Cartesian product $M \times M'$ of the point sets M and M' of the two given spaces, and whose metric $D \oplus D'$ is defined by putting

$$(D \oplus D')\big((u, u'), (v, v')\big) := D(u, v) + D'(u', v')$$

for all $(u, u'), (v, v') \in M \times M'$—note that the k-dimensional *standard* L_1-space $\mathbf{L_1(k)} := (\mathrm{R}^k, L_1^{(k)})$ whose metric $L_1^{(k)} : \mathrm{R}^k \times \mathrm{R}^k \rightarrow \mathrm{R}$ is given by

$$L_1^{(k)}(\mathbf{x}, \mathbf{y}) := \|\mathbf{x}, \mathbf{y}\|_1 := \sum_{i=1}^{k} |x_i - y_i|$$

for all $\mathbf{x} := (x_1, \ldots, x_k), \mathbf{y} := (y_1, \ldots, y_k) \in \mathrm{R}^k$ is nothing but the L_1-product of k copies of the space $\mathbf{L_1(1)}$, that is, the real line (endowed with its standard metric).

Further, given any metric space $\mathbf{M} = (M, D : V \times V \rightarrow \mathrm{R} : (u, v) \mapsto uv)$,

(1) we define $\mathbf{M}$ to be an (abstract) L_1-*space* if the relation "$\|$" is an equivalence relation on M^2—implying that the L_1-product of any two abstract L_1-spaces and, therefore, also (as good sense would require) the standard L_1-spaces $\mathbf{L_1(k)}$ are abstract L_1-spaces (as the real line is easily seen to be one; indeed, one has $uu' \| vv'$ for any two pairs $(u, u'), (v, v')$ of real numbers relative to $L_1^{(1)}$ if and only if either $\#\{u, u'\} = \#\{v, v'\} = 1$ or and $(u, v) = (u', v')$ holds);

(2) we define two elements $u, v \in M$ to be forming a *primitive pair* (in $\mathbf{M}$) if and only if $\#[u, v] = 2$—or, equivalently, $u \neq v$ and $\{u, v\} = [u, v]$—holds, and we denote the set of all primitive pairs in $\mathbf{M}$ by $\mathrm{Prim}(\mathbf{M})$, that is, we put

$$\mathrm{Prim}(\mathbf{M}) := \{\{u, v\} \subseteq M : \#[u, v] = 2\};$$

(3) we define a sequence $a_0, a_1, a_2, \ldots, a_k$ of points in M to be

(i) a *geodesic* sequence (in $\mathbf{M}$) if the identity $a_0a_k = \sum_{i=1}^{k} a_{i-1}a_i$ holds (in which case—even stronger—$a_ia_j = \sum_{\ell=i+1}^{j} a_{\ell-1}a_\ell$ holds for all $i, j \in \{0, 1, 2, \ldots, k\}$ with $i < j$),

(ii) a *path* (in $\mathbf{M}$) if all pairs $\{a_{i-1}, a_i\}$ $(i = 1, 2, \ldots, k)$ are primitive pairs in $\mathbf{M}$,

(iii) and, of course, a *geodesic path* (in $\mathbf{M}$) if it is a shortest geodesic sequence that is, simultaneously, a path in $\mathbf{M}$;

(4) we put

$$\mathbf{M}(u < v) := \{w \in M : u \leq_w v\} \quad (= \{w \in M : u \in [w, v]\})$$

for all $u, v \in M$, we define two elements $u, v \in M$ to be forming a *bipartitioning pair* (in $\mathbf{M}$) if $M = \mathbf{M}(u < v) \cup \mathbf{M}(v < u)$ holds—implying that a subset $\{u, v\} \subseteq V$ is a bipartitioning pair if and only if $u \neq v$ holds and $\{u, v\}$ is a *gated* subset of $\mathbf{M}$ (i.e., if and only if $\#\{u, v\} = 2$ holds and there exists, for every $x \in M$, some point $y \in \{u, v\}$—the *gate* of x in $\{u, v\}$—with $xw = xy + yw$ for each element $w \in \{u, v\}$), and that every bipartitioning must also be a primitive pair;

(5) we define $\mathbf{M}$ to be *bipartite* if, conversely, every primitive pair in $\mathbf{M}$ is also a bipartitioning pair;

(6) and we recall that $\mathbf{M}$ is said to be a *median* space if and only if one has $\#([u, v] \cap [v, w] \cap [w, u]) = 1$ for all $u, v, w \in M$, in which case the single element in $[u, v] \cap [v, w] \cap [w, u]$ is denoted by $\mathrm{med}(u, v, w) = \mathrm{med}_D(u, v, w)$ and dubbed the median of u, v, and w—note that the L_1-product of any two median spaces and, therefore, also the standard L_1-spaces $\mathbf{L_1(k)}$ are median spaces (as the real line $\mathbf{L_1(1)}$ is a median space, the median $\mathrm{med}(x, y, z)$ of any three real numbers x, y, z in $\mathbf{L_1(1)}$ being given by $\mathrm{med}(x, y, z) = x + y + z - \max(x, y, z) - \min(x, y, z)$).

Remarks: (R1) Note that the primitive pairs in a finite metric space correspond to the edges in a connected finite graph. More precisely, given any finite metric space $\mathbf{M} = (M, D)$, we can associate to $\mathbf{M}$ the necessarily finite and connected simple graph

$$G_{\mathbf{M}} := (M, \mathrm{Prim}(\mathbf{M}))$$

with vertex set M and edge set $\mathrm{Prim}(\mathbf{M}) \subseteq \binom{M}{2}$ and, conversely, to any finite and connected simple graph $G = (V, E)$ with vertex set V and edge set $E \subseteq \binom{V}{2}$, the finite metric space $\mathbf{M}_G := (V, D_G)$ with point set V whose metric D_G is the standard graph metric on V, that is, the (well-defined and unique) largest metric D defined on V for which $D(u, v) \leq 1$ holds for every edge $\{u, v\}$ in E.

This yields in particular a canonical one-to-one correspondence between

(1) the isometry classes of finite metric spaces $\mathbf{M}$ for which $D(u, v) = 1$ holds for every primitive pair $\{u, v\}$ in $\mathrm{Prim}(\mathbf{M})$ and

(2) the isomorphism classes of finite and connected simple graphs.

A subgraph $G' = (V', E')$ of a finite and connected simple graph $G = (V, E)$ is called an *isometric* subgraph of G if it is connected and $D_G(u', v') = D'_{G'}(u', v')$ holds for all $u', v' \in V'$.

(R2) More generally, given a *weighted* finite and connected simple graph, that is, a triple $\mathbf{G} = (V, E; L)$ consisting of a finite and connected simple graph $G = (V, E)$ together with an edge weighting $L : E \to \mathbb{R}_{>0} := \{\rho \in \mathbb{R} : \rho > 0\}$, we may associate to L the unique largest metric $D = D_L$ defined on V for which $D(u, v) \leq L(\{u, v\})$ holds for every edge $\{u, v\}$ in E, thus setting up a canonical one-to-one correspondence between the isometry classes of finite metric spaces $\mathbf{M} = (M, D)$ and the isomorphism classes of weighted finite and connected simple graphs $\mathbf{G} = (V, E; L)$ for which

$$L(\{u, v\}) < \sum_{i=1}^{k} L(\{u_{i-1}, u_i\})$$

holds for every edge $\{u, v\}$ in E and all sequences $u_0, u, \ldots, u_k$ of vertices in V of length $k > 1$ with $u_0 = u, u_k = v$, and $\{u_{i-1}, u_i\} \in E$ for all $i = 1, \ldots, k$ (or, equivalently, for which $D_L(u, v) < D_L(u, w) + D_L(w, v)$ holds for every edge $\{u, v\}$ in E and every $w \in V - \{u, v\}$). Indeed, if $\mathbf{M} = (M, D)$ is a finite metric space, the edge weighting $L = L_{\mathbf{M}}$ defined on the set $\mathrm{Prim}(\mathbf{M})$ of edges of the associated graph $G_{\mathbf{M}}$ by $L_{\mathbf{M}} : \mathrm{Prim}(\mathbf{M}) \to \mathbb{R} : \{u, v\} \mapsto D(u, v)$ satisfies the above condition and one has $D = D_{L_{\mathbf{M}}}$, while conversely, if an edge weighting L defined on the set E of edges of a finite and connected simple graph $G = (V, E)$ satisfies this condition, one has $G = G_{(\mathbf{M}_L)}$ for the associated finite metric space $\mathbf{M}_L = \mathbf{M}_{(G,L)} := (V, D_L)$ while L coincides with the corresponding edge weighting $L_{(\mathbf{M}_L)}$.

(R3) Note that, given a finite and connected simple graph $G = (V, E)$, the above condition holds for *every* edge weighting $L : E \to \mathbb{R}_{>0}$ if and only if G is a tree. The resulting metrics D_L will be called T-metrics, and the resulting metric spaces $\mathbf{M}_L$ will be called T-spaces. Consequently, a finite metric space $\mathbf{M} = (M, D)$ is a T-space if and only if the associated graph

$G_{\mathbf{M}} = (M, \text{Prim}(\mathbf{M}))$ is a tree in which case the metric D_L induced by the edge weighting

$$L := L_{\mathbf{M}} : \text{Prim}(\mathbf{M}) \to \mathbb{R} : \{u, v\} \mapsto D(u, v)$$

coincides with D.

(R4) Note also that a connected finite simple graph G is a bipartite or a median graph if and only if the associated metric space $\mathbf{M}_G$ is a bipartite or a median metric space, respectively, and that every T-space is a bipartite as well as a median metric space.

Finally, we'll need the following definitions:

(7) A finite metric space $\mathbf{H}$ is called a *hypercuboid* if it is isometric to the L_1-product $\mathbf{M}_1, \mathbf{M}_2, \ldots, \mathbf{M}_k$ of a finite number of metric spaces all of whose point sets have cardinality 2—implying that every hypercuboid is a median L_1-space (as any metric space of cardinality 2 is such a space), that any hypercuboid derived from k factors can be embedded isometrically into the k-dimensional standard L_1-space $\mathbf{L_1}(\mathbf{k})$, and that the graph $G_{\mathbf{H}}$ associated to a hypercuboid $\mathbf{H}$ is a hypercube, that is, it is isomorphic to a graph of the form $(\{0,1\}^k, E^{(k)})$ with $E^{(k)}$ defined by

$$E^{(k)} := \{\{(x_1, \ldots, x_k), (y_1, \ldots, y_k)\} \subseteq \{0,1\}^k : \sum_{i=1}^{k} |x_i - y_i| = 1\},$$

that is, the graph associated to the subspace of the standard L_1-space $\mathbf{L_1}(\mathbf{k})$ whose point set is $\{0,1\}^k$.

(8) Another, yet equivalent way to describe hypercuboids is to associate, to any weighted set $\mathbf{E}$, that is, to a pair $\mathbf{E} := (E, L)$ consisting of a finite set E and a map $L : E \to \mathbb{R}_{>0}$, the metric space $\mathbf{M}^{\mathbf{E}} := (\mathcal{P}(E), D^{\mathbf{E}})$ whose point set is the power set $\mathcal{P}(E)$ of E while its metric $D^{\mathbf{E}}$ is given by the map

$$D^{\mathbf{E}} : \mathcal{P}(E) \times \mathcal{P}(E) \to \mathbb{R} : (F, F') \mapsto L_+(F \triangle F')$$

(where, as usual, $F \triangle F'$ denotes the *symmetric difference* $(F - F') \cup (F' - F)$ of the two subsets F and F' of E, and $L_+(F)$ denotes, for a weighted set $\mathbf{E} = (E, L)$ and a subset F of E, the sum $L_+(F) := \sum_{e \in F} L(e)$), and then to define a finite metric space $\mathbf{H}$ to be a hypercuboid if it is isometric to a metric space of that form, that is, if a weighted set $\mathbf{E}$ as above exists so that $\mathbf{H}$ is isometric to $\mathbf{M}^{\mathbf{E}}$ (as $\mathbf{M}^{\mathbf{E}}$ is apparently isometric to the L_1-product $\Pi_{e \in E}(\{0,1\}, D_e)$ where $D_e : \{0,1\} \times \{0,1\} \to \mathbb{R}$ is, of course, defined by $D_e(0,1) = D_e(1,0) := L(e)$ and $D_e(0,0) = D_e(1,1) := 0$). So, $\mathbf{M}^{\mathbf{E}} = (\mathcal{P}(E), D^{\mathbf{E}})$ must, in particular, be a median space—and it is indeed also easily verified directly that the median of any three subsets F, F', F'' of E, considered as points in the point set $\mathcal{P}(E)$ of $\mathbf{M}^{\mathbf{E}}$, always exists, and always coincides with the subset $(F \cap F') \cup (F' \cap F'') \cup (F'' \cap F)$, independently of the choice of L.

(9) A *net* $\mathbf{N}$ is a metric space $\mathbf{N} = (N, D)$ with point set N and metric D that can be embedded into a hypercuboid $\mathbf{M}$ so that any two points in $\mathbf{N}$ can be connected by a geodesic path $a_0, a_1, a_2, \ldots, a_k$ in $\mathbf{M}$ all of whose points are points in $\mathbf{N}$.

(10) We define *the category* `NET` *of nets* to be the category whose objects are the nets while the morphisms from one net $\mathbf{N} = (N, D)$ into another net $\mathbf{N}' = (N', D')$ are defined to be exactly those morphisms from $\mathbf{N}$ into another net $\mathbf{N}'$ in the category `MET` (i.e., those non-expansive maps φ from N into N') that are *additive*, that is, one has

$$D'(\varphi(u), \varphi(v)) + D'(\varphi(v), \varphi(w)) = D'(\varphi(u), \varphi(w))$$

for all $u, v, w \in V$ with $uv + vw = uw$ (or, equivalently, $\varphi(w) \in [\varphi(u), \varphi(v)]$ holds for all $u, v, w \in N$ with $w \in [u, v]$), and for which

$$\{\varphi(u), \varphi(v)\} \in \mathrm{Prim}(V', D')$$

holds for every primitive pair $\{u, v\}$ in $\mathrm{Prim}(V, D)$ with $\varphi(u) \neq \varphi(v)$. And any such morphism φ is called an *isometric embedding* if φ, considered as a morphism in `MET`, is an isometric embedding, that is, if $D'(\varphi(u), \varphi(v)) = D(u, v)$ holds for all $u, v \in N$.

(11) Given a finite set X, we define an X-net $\mathcal{N}$ to be a pair $\mathcal{N} := (\mathbf{N}, \psi)$ consisting of a net $\mathbf{N} = (N, D)$ together with a map $\psi : X \to N$ such that

$$\psi(X) \cap \mathbf{N}(u < v) \cap \mathbf{N}(u' < v') \neq \emptyset$$

holds for all primitive pairs $\{u, v\}, \{u', v'\}$ in $\mathbf{N}$ for which the intersection $\mathbf{N}(u < v) \cap \mathbf{N}(u' < v')$ is non-empty.

(12) We define *the category* `X–NET` *of X-nets* to be the category whose objects are the X-nets while the morphisms from one X-net $\mathcal{N} = (\mathbf{N}, \psi)$ into another X-net $\mathcal{N}' = (\mathbf{N}', \psi')$ are those morphisms φ from $\mathbf{N}$ into $\mathbf{N}'$ in `NET` for which $\psi' = \varphi \circ \psi$ holds. As above, any such morphism will also be called an *isometric embedding* if it is one, considered as a morphism in `NET` (or, equivalently, in `MET`).

1.3 Some Basic Results

Clearly, the definition of a net given above is rather a "descriptive" or "constructive" than a structural or "intrinsic" definition. However, the other definitions collected above allow us to present the following seven characterizations of nets, two of which are "intrinsic":

Theorem 1. *Given a finite metric space* $\mathbf{M} = (M, D)$*, the following assertions all are equivalent:*

(i) $\mathbf{M}$ *is a net;*

(ii) $\mathbf{M}$ *is a bipartite* L_1*-space;*

(iii) $\mathbf{M}$ *is bipartite and the relation "$\|$" defined—by abuse of notation—on* $\mathrm{Prim}(\mathbf{M})$ *by putting*

$$\{u, u'\} \| \{v, v'\} \iff_{\mathrm{def}} uu' \| vv' \text{ or } uu' \| v'v$$

for all $\{u, u'\}, \{v, v'\} \in \mathrm{Prim}(\mathbf{M})$ *is an equivalence relation on* $\mathrm{Prim}(\mathbf{M})$*;*

(iv) *the graph* $G_{\mathbf{M}} = (M, \mathrm{Prim}(\mathbf{M}))$ *is an isometric subgraph of a hypercube, and* $uu' = vv'$ *holds for all* u, u', v, v' *in* M *for which* $\{u, u'\}, \{v, v'\}$ *are parallel edges in that hypercube;*

(v) *there exists a pair* $(\mathbf{E}, \Delta)$ *consisting of a weighted finite set* $\mathbf{E} = (E, L)$ *and a map* $\Delta : M^2 \to \mathcal{P}(E)$ *with*

$$\mathrm{Prim}(\mathbf{M}) = \{\{u, v\} \subseteq M : \#\Delta(u, v) = 1\}$$

such that

$$\Delta(u, v) = \Delta(u, w) \triangle \Delta(w, v)$$

and

$$D(u, v) = L_+(\Delta(u, v))$$

holds for all $u, v, w \in M$ *in which case*

(a) *the map*

$$\psi_v : M \to \mathcal{P}(E) : u \mapsto \Delta(u, v)$$

is an isometry from $\mathbf{M}$ *into the metric space* $\mathbf{M}^{\mathbf{E}}$ *for given any point* $v \in M$*,*

(b) $\Delta(u, v) = \Delta(v, u)$ *and "$\Delta(u, v) = \emptyset \iff u = v$" holds for all* u, v *in* M*,*

(c) *and* $uu' \| vv'$ *holds for some* u, u', v, v' *in* M *if and only if*

$$\Delta(u,u') = \Delta(v,v') \text{ and } \Delta(u,u') \cap \Delta(u',v') = \emptyset$$

and, hence, also

$$\Delta(u,v) = \Delta(u,v') \triangle \Delta(v',v) = \Delta(v',u) \triangle \Delta(u,u') = \Delta(u',v')$$

as well as

$$\Delta(u,v') = \Delta(u,u') \cup \Delta(u',v') = \Delta(u',u) \cup \Delta(u,v) = \Delta(u',v)$$

holds;

(vi) *there exists some* $k \in \mathrm{N}$ *and an isometric embedding* φ *of* $\mathbf{M}$ *into the standard* k*-dimensional* L_1*-space* $\mathbf{L_1(k)}$ *such that*

$$\mathrm{med}_{\mathbf{L_1(k)}}(\varphi(u), \varphi(v), \varphi(w)) \in \{\varphi(u), \varphi(v)\}$$

holds for all $u, v, w \in M$ *with* $\{u, v\} \in \mathrm{Prim}(\mathbf{M})$*;*

(vii) *there exists an isometric embedding* φ *of* $\mathbf{M}$ *into some median* L_1*-space* $\mathbf{M}'$ *with such that* $\mathrm{med}_{\mathbf{M}'}(\varphi(u), \varphi(v), \varphi(w)) \in \{\varphi(u), \varphi(v)\}$ *holds for all* $u, v, w \in M$ *with* $\{u, v\}$ *in* $\mathrm{Prim}(\mathbf{M})$.

To establish Theorem 1, the following more detailed results are required:

Theorem 2. *A path* $a_0, a_1, \ldots, a_k$ *in a finite bipartite metric space* $\mathbf{M}$ *is a geodesic path if and only if* $\{a_{i-1}, a_i\} \| \{a_{j-1}, a_j\}$ *implies* $i = j$ *for all* $i, j = 1, \ldots, k$*. Furthermore, if* $a_0, a_1, \ldots, a_k$ *is a geodesic path in* $\mathbf{M}$*, one has* $k' \geq k$ *for any other path* $a'_0, a'_1, \ldots, a'_{k'}$ *of points in* $\mathbf{M}$ *with* $a_0 = a'_0$ *and* $a_k = a'_{k'}$ *while equality* $k = k'$ *holds if and only if the path* $a'_0, a'_1, \ldots, a'_{k'}$ *is also a geodesic path in which case there exists a permutation* π *of the index set* $\{1, \ldots, k\}$ *such that* $a_{i-1}a_i = a'_{\pi(i)-1}a'_{\pi(i)}$ *holds for all* $i = 1, \ldots, k$*.*

Theorem 3. *If* $\mathbf{M} = (M, D)$ *is a finite bipartite metric space for which the binary relation "*$\|$*" defined above on* $\mathrm{Prim}(\mathbf{M})$ *is an equivalence relation on* $\mathrm{Prim}(\mathbf{M})$*, one has* $\{u, v\} \| \{u', v'\}$ *for two primitive pairs* $\{u, v\}, \{u', v'\}$ *in* $\mathrm{Prim}(\mathbf{M})$ *if and only if the two bipartitions* $\{\mathbf{M}(u < v), \mathbf{M}(v < u)\}$ *and* $\{\mathbf{M}(u' < v'), \mathbf{M}(v' < u')\}$ *of the point set* M *of* $\mathbf{M}$ *associated with* $\{u, v\}$ *and* $\{u', v'\}$ *coincide.*

Thus, denoting the set of "$\|$"-equivalence classes in $\mathrm{Prim}(\mathbf{M})$ by $\mathrm{E}(\mathbf{M})$ and associating to any path $a_0, a_1, \ldots, a_k$ in $\mathbf{M}$ the set

$$\Delta(a_0, a_1, \ldots, a_k) := \{e \in \mathrm{E}(\mathbf{M}) : \#\{i \in \{1, \ldots, k\} : \{a_{i-1}, a_i\} \in e\} \equiv 1 \mod 2\}$$

of "$\|$"-equivalence classes e in $\mathrm{E}(\mathbf{M})$ represented by an odd number of pairs of the form $\{a_{i-1}, a_i\}$ $(i = 1, \ldots, k)$, the following can be established:

Theorem 4. *If* $\mathbf{M}$ *is a finite bipartite metric space for which the binary relation "$\|$" defined above on* $\mathrm{Prim}(\mathbf{M})$ *is an equivalence relation on* $\mathrm{Prim}(\mathbf{M})$, *a path* $a_0, a_1, \ldots, a_k$ *in* $\mathbf{M}$ *is a geodesic path in* $\mathbf{M}$ *if and only if the cardinality of set* $\Delta(a_0, a_1, \ldots, a_k)$ *coincides with* k*, in which case one has* $\Delta(a_0, a_1, \ldots, a_k) = \Delta(a'_0, a'_1, \ldots, a'_{k'})$ *for any other path* $a'_0, a'_1, \ldots, a'_{k'}$ *in* $\mathbf{M}$ *with* $a_0 = a'_0$ *if and only if* $a_k = a'_{k'}$ *holds—allowing us to (well-)define the map* $\Delta : V^2 \to \mathcal{P}(\mathrm{E}(\mathbf{M}))$ *as described in* Theorem 1 (v) *by letting* $\Delta(u,v)$ *denote the set* $\Delta(a_0, a_1, \ldots, a_k)$ *for one—or, as well, for all—paths* $a_0 := u, a_1, \ldots, a_k; = v$ *from* u *to* v *in* $\mathbf{M}$.

Together, these results can be used to establish

Theorem 5. (i) *For any two* X*-nets* $\mathcal{N}$ *and* $\mathcal{N}'$*, there exists at most one morphism in* `X–NET` *from* $\mathcal{N}$ *into* $\mathcal{N}'$.
(ii) *Whenever a (necessarily unique) morphism* φ *from an* X*-net* $\mathcal{N} = (\mathbf{N}, \psi)$ *into an* X*-net* $\mathcal{N}' = (\mathbf{N}', \psi')$ *exists, this morphism is an isometric embedding if and only if* ψ *and* ψ' *induce the same metric on* X*, that is, if and only if*

$$D(\psi(x), \psi(y)) = D'(\psi'(x), \psi'(y))$$

holds for all $x, y \in X$.
(iii) *For every* X*-net* $\mathcal{N}$*, there exists an* X*-net* $\mathcal{N}^*$*, also called the* injective hull *of* $\mathcal{N}$*, together with an isometric embedding* $\varphi_{\mathcal{N}}$ *from* $\mathcal{N}$ *into* $\mathcal{N}^*$ *such that, for every isometric embedding* φ *of* $\mathcal{N}$ *into another* X*-net* $\mathcal{N}'$*, there exists a (necessarily unique) isometric embedding* φ' *from* $\mathcal{N}'$ *into* $\mathcal{N}^*$ *with* $\varphi_{\mathcal{N}} = \varphi' \circ \varphi$ *(implying, as usual, that both,* $\mathcal{N}^*$ *and* $\varphi_{\mathcal{N}}$ *are uniquely determined up to canonical isomorphism by* $\mathcal{N}$*, and that* $\mathcal{N}^*$ *is also the injective hull of every* X*-net* $\mathcal{N}'$ *for which an isometric embedding from* $\mathcal{N}$ *into* $\mathcal{N}'$ *exists). Moreover, any morphism* φ *in* `X–NET` *from an* X*-net* $\mathcal{N}_1$ *into an* X*-net* $\mathcal{N}_2$ *induces a morphism* φ^* *in* `X–NET` *from the injective hull* $\mathcal{N}_1^*$ *of* $\mathcal{N}_1$ *into the injective hull* $\mathcal{N}_2^*$ *of* $\mathcal{N}_2$.
(iv) *And, given an* X*-net* $\mathcal{N} = (\mathbf{N}, \psi)$*, the underlying metric space* N *of* $\mathcal{N}$ *is a median metric space if and only if every isometric embedding from* $\mathcal{N}$ *into any other* X*-net* $\mathcal{N}'$ *is an isomorphism if and only if the morphism* $\varphi_{\mathcal{N}} : \mathcal{N} \to \mathcal{N}^*$ *is an isomorphism, that is, if and only if* $\mathcal{N}$ *is its own injective hull (implying that the underlying metric space of the injective hull* $\mathcal{N}^*$ *of any* X*-net* $\mathcal{N}$ *is a median metric space).*

Corollary 1. *Every* T*-space is a net while a pair* $(\mathbf{N}, \psi)$ *consisting of* T*-space* $\mathbf{N} = (N, D)$ *and a map* $\psi : X \to N$ *from* X *into the point set* N *of* $\mathbf{N}$ *is an* X*-net if and only if the tree* $G_{\mathbf{N}}$ *together with the map* ψ *from* X *into its vertex set* N *is an* X*-tree, that is, if and only if every vertex in the graph*

$$G_{\mathbf{N}} = (N, \mathrm{Prim}(\mathbf{N}))$$

of degree less than 3 is contained in the image of ψ.

1.4 X-Nets and Split Systems Over X

Now, recall that given any finite set X, one denotes

- by $\mathcal{S}(X)$ the collection

$$\mathcal{S}(X) := \{\{A,B\} : A,B \subseteq X, A \cup B = X, A \cap B = \emptyset\}$$

  of all X-*splits*,

- by $S(x)$, for every X-split $S = \{A,B\} \in \mathcal{S}(X)$ and every element $x \in X$, that subset, A or B, in S that contains x,

- by $\mathcal{S}^*(X)$ the collection

$$\mathcal{S}^*(X) := \{\{A,B\} : A,B \subseteq X, A \cup B = X, A \cap B = \emptyset \neq A,B\}$$

  of all *bipartitions* of X, or *proper* X-*splits*,

- and by $\mathcal{S}^*(X|\,\mathrm{R})$ the R-vector space consisting of all maps μ from $\mathcal{S}(X)$ into R with $\mu(\{X,\emptyset\} = 0$, that is, all maps μ from $\mathcal{S}(X)$ into R whose support

$$\mathrm{supp}(\mu) := \{S \in \mathcal{S}(X) : \mu(S) \neq 0\}$$

  is contained in $\mathcal{S}^*(X)$.

Any such map μ will also be called an (R-weighted) split system over X, and it will be called an $\mathrm{R}_{\geq 0}$-weighted split system over X if $\mu(S) \geq 0$ holds for all $S \in \mathcal{S}(X)$.

There is a close connection between X-nets and $\mathrm{R}_{\geq 0}$-weighted split systems over X. To explicate this, note first that, given a finite set X and an X-net $\mathcal{N} = (\mathbf{N}, \psi)$, one can associate, to any primitive pair $\{u,v\} \in \mathrm{Prim}(\mathbf{N})$, the corresponding X-split

$$S_{u,v} = S^{\mathcal{N}}_{u,v} := \{X(u < v), X(v < u)\}$$

whose two parts $X(u < v)$ and $X(v < u)$ are the pre-images (relative to ψ) of the two parts of the split $\{\mathbf{N}(u < v), \mathbf{N}(v < u)\}$ associated to the pair $\{u,v\}$ in $\mathrm{Prim}(\mathbf{N})$, i.e., the two subsets

$$X(u < v) = X^{\mathcal{N}}(u < v) := \{x \in X : \psi(x) \in \mathbf{N}(u < v)\}$$

and

$$X(v < u) = X^{\mathcal{N}}(v < u) := \{x \in X : \psi(x) \in \mathbf{N}(v < u)\}$$

of X. The following is a simple corollary of the definitions and results collected above:

Corollary 2. *Given a finite set X, an X-net $\mathcal{N} = (\mathbf{N}, \psi)$, and two primitive pairs $\{u, v\}\{u', v'\} \in \mathrm{Prim}(\mathbf{N})$, one has $\{u, v\} \| \{u', v'\}$ if and only if $S_{u,v} = S_{u',v'}$ holds. In particular, one has $uv = u'v'$ for any two primitive pairs $\{u, v\}, \{u', v'\}$ in $\mathrm{Prim}(\mathbf{N})$ with $S_{u,v} = S_{u',v'}$.*

In consequence, one can associate, to any X-net $\mathcal{N} = (\mathbf{N}, \psi)$, a corresponding $\mathrm{R}_{\geq 0}$-weighted split system $\mu = \mu_{\mathcal{N}}$ over X that maps any split $S \in \mathcal{S}(X)$ of the form $S = S_{u,v}$ for some primitive pair $\{u, v\} \in \mathrm{Prim}(\mathbf{N})$, onto the positive real number uv, and all other splits $S \in \mathcal{S}(X)$ (including the split $\{X, \emptyset\}$) onto 0.

Conversely, given any $\mathrm{R}_{\geq 0}$-weighted split system μ over X, one can associate to μ the X-net $\mathcal{N} = \mathcal{N}_\mu = (\mathbf{N}_\mu, \psi_\mu) = ((N_\mu, D_\mu), \psi_\mu)$ for which $\mu = \mu_{\mathcal{N}}$ holds that is defined as follows: One defines N_μ to consist of all maps $v : X \to \mathcal{P}(\mathrm{supp}(\mu))$ with

$$\binom{\mathrm{supp}(\mu)}{2} = \bigcup_{x \in X} \binom{\mathrm{supp}(\mu) - v(x)}{2} \tag{1.1}$$

for which

$$v(x) \triangle v(y) = \Delta_\mu(x, y) := \{S \in \mathrm{supp}(\mu) : S(x) \neq S(y)\} \tag{1.2}$$

or, equivalently,

$$v(x) = v(y) \triangle \Delta_\mu(x, y)$$

holds for all $x, y \in X$—condition (1.1) just requiring that there exists, for any two splits $S_1, S_2 \in \mathrm{supp}(\mu)$, some $x \in X$ such that $S_1, S_2 \notin v(x)$ holds.

Next, one defines

$$\Delta_\mu(u, v) := \bigcup_{x \in X} u(x) \triangle v(x)$$

and

$$D_\mu(u, v) := \mu_+(\Delta_\mu(u, v))$$

for all maps $u, v : X \to \mathcal{P}(\mathrm{supp}(\mu))$ and, noting that

$$u(x) \triangle v(x) = (\Delta_\mu(x, y) \triangle u(y)) \triangle (\Delta_\mu(x, y) \triangle v(y)) = u(y) \triangle v(y)$$

holds for all $x, y \in X$ and $u, v \in N_\mu$, one sees that

$$\Delta_\mu(u, v) := u(x) \triangle v(x)$$

and

$$D_\mu(u, v) := \mu_+(u(x) \triangle v(x)) = \sum_{S \in u(x) \triangle v(x)} \mu(S)$$

holds for every $x \in X$ and any two maps $v, u \in N_\mu$. And one defines $\psi_\mu : X \to N_\mu$ by associating, to any $x \in X$, the map

$$\psi_\mu(x)(S) : X \rightarrow \mathrm{supp}(\mu) : y \mapsto \Delta_\mu(x, y).$$

Using these constructions, it can be shown that the $\mathbb{R}_{\geq 0}$-weighted split systems over X "classify" the *injective objects* in X–NET, that is, the X-nets $\mathcal{N}$ that "coincide" with their injective hull. More precisely, the following holds for any finite set X:

Theorem 6. (i) *Given any two $\mathbb{R}_{\geq 0}$-weighted split systems μ and μ' over X, there exists a morphism φ from $\mathcal{N}_\mu$ into $\mathcal{N}_{\mu'}$ in* X–NET *if and only if $\mu' \leq \mu$ holds (i.e., if and only if $\mu'(S) \leq \mu(S)$ holds for every X-split S in $\mathcal{S}^*(X)$); in particular, two X-nets of the form $\mathcal{N}_\mu$ and $\mathcal{N}_{\mu'}$ are isomorphic if and only if $\mu = \mu'$ holds.*

(ii) *Given any X-net $\mathcal{N}$, its injective hull $\mathcal{N}^*$ is canonically isomorphic to the X-net $\mathcal{N}_\mu$ for the $\mathbb{R}_{\geq 0}$-weighted split systems $\mu := \mu_\mathcal{N}$; in particular, $\mathcal{N}$ is an injective object in* X–NET *if and only if it is isomorphic to an X-net of the form $\mathcal{N}_\mu$ for some $\mathbb{R}_{\geq 0}$-weighted split system μ over X in which case there is only one such $\mathbb{R}_{\geq 0}$-weighted split system μ,* viz., *the $\mathbb{R}_{\geq 0}$-weighted split system $\mu_\mathcal{N}$; in particular,*

(1) *given any X-net $\mathcal{N}$, its injective hull $\mathcal{N}^*$ is canonically isomorphic to the X-net $\mathcal{N}_{\mu_\mathcal{N}}$,*

(2) *two X-nets $\mathcal{N}$ and $\mathcal{N}'$ have isomorphic injective hulls if and only if $\mu_\mathcal{N} = \mu_{\mathcal{N}'}$ holds,*

(3) *there exists a morphism from an X-net $\mathcal{N}$ into the injective hull of an X-net $\mathcal{N}'$ if and only if $\mu_\mathcal{N} \leq \mu_{\mathcal{N}'}$ holds.*

(iii) *Given any X-net $\mathcal{N}$, its injective hull $\mathcal{N}^*$ is canonically isomorphic to the X-net $\mathcal{N}_\mu$ for the $\mathbb{R}_{\geq 0}$-weighted split systems $\mu := \mu_\mathcal{N}$; in particular, $\mathcal{N}$ is an injective object in* X–NET *if and only if it is isomorphic to an X-net of the form $\mathcal{N}_\mu$ for some $\mathbb{R}_{\geq 0}$-weighted split system μ over X in which case there is only one such $\mathbb{R}_{\geq 0}$-weighted split system μ,* viz., *the $\mathbb{R}_{\geq 0}$-weighted split system $\mu := \mu_\mathcal{N}$; in particular,*

(1) *given any X-net $\mathcal{N}$, its injective hull $\mathcal{N}^*$ is canonically isomorphic to the X-net $\mathcal{N}_{\mu_\mathcal{N}}$,*

(2) *two X-nets $\mathcal{N}$ and $\mathcal{N}'$ have isomorphic injective hulls if and only if $\mu_\mathcal{N} = \mu_{\mathcal{N}'}$ holds,*

(3) *there exists a morphism from an X-net $\mathcal{N}$ into the injective hull of an X-net $\mathcal{N}'$ if and only if $\mu_\mathcal{N} \leq \mu_{\mathcal{N}'}$ holds.*

1.5 Examples

Here are some real-life examples of X-nets. The first two belong to the group of altogether more than 10 phylogenetic trees that were ever published. They were carefully drawn by Ernst Haeckel who published his work *Generelle Morphologie der Organismen* in 1866 (just 7 years after Charles Darwin published *The Origin of Species*), a book that Thomas Henry Huxley described in 1869 as "an attempt to put the Doctrine of Evolution, so far as it applies to the living world, into a logical form; and to work out its practical applications to their final results" (see also Preface and Table of Contents to Volume II, *Darwiniana*, of Huxley's *Collected Essays*).

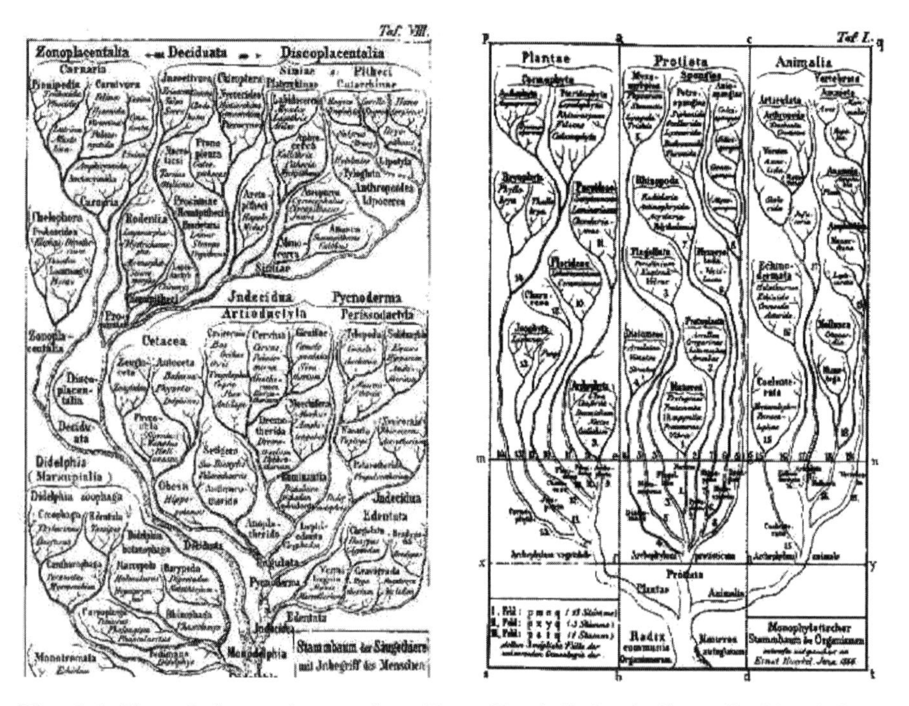

Fig. 1.1. Two phylogenetic trees from Ernst Haeckel's book *Generelle Morphologie der Organismen*, 2 vols., Berlin, 1866.

Figures 1.2 and 1.3 present *proper* networks, constructed using the program *SplitsTrees* based on data provided by Helms in his thesis on the perception of colour similarity, Hamburg, 1980.

Also Figures 1.4 to 1.5 present networks that do not refer to a biological context (even though manuscript copying has a number of interesting analogies with sequence evolution).

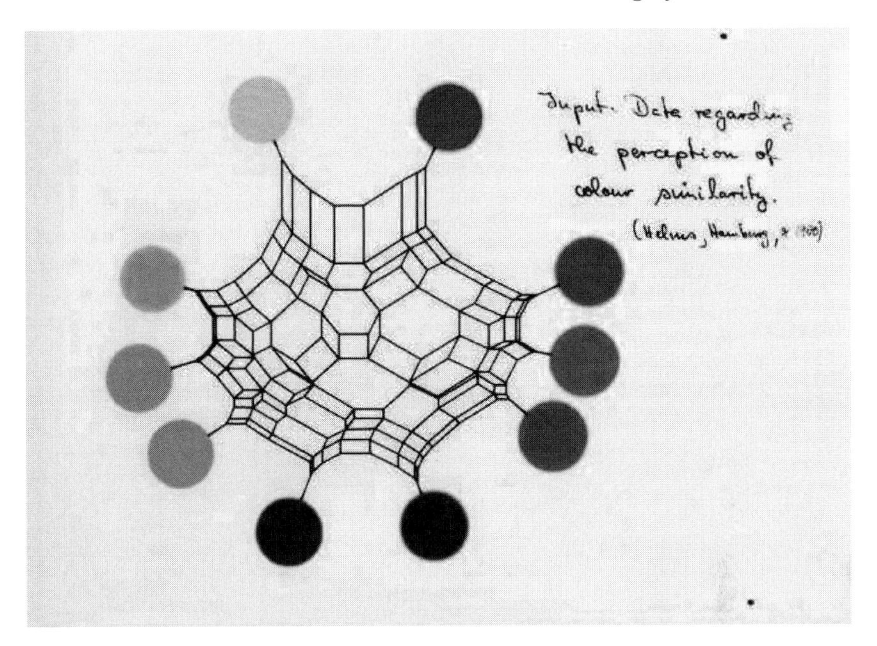

Fig. 1.2. An X-net constructed by applying the program *SplitsTrees* to data regarding the perception of colour similarity (Helms, Hamburg, 1980). The set X consists of 10 distinct colours.

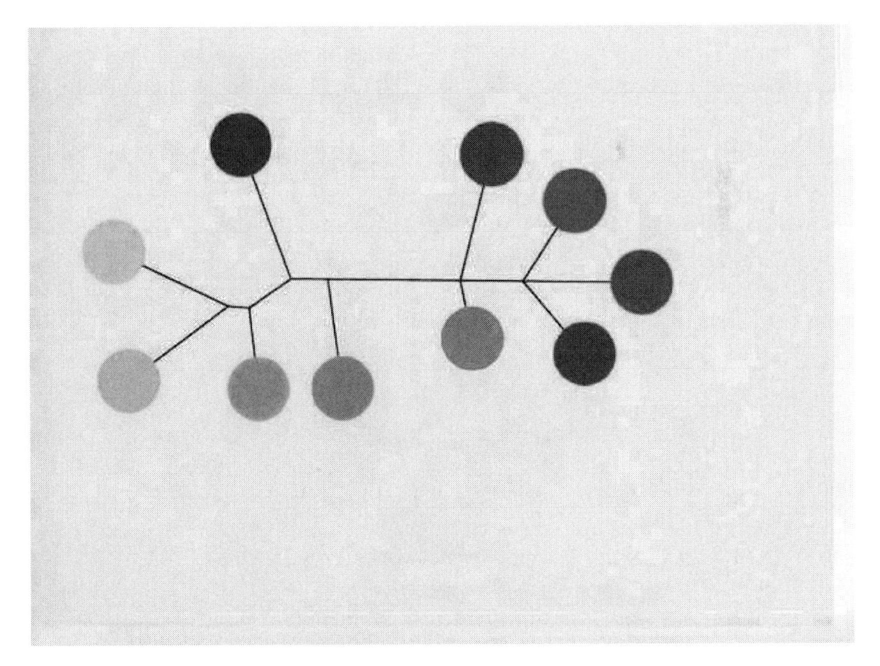

Fig. 1.3. Using data from the same source, now regarding a colour-blind subject's perception of colour similarity (also from data published by Helms, Hamburg, 1980): the X-net constructed by *SplitsTrees* now degenerates into an X-tree.

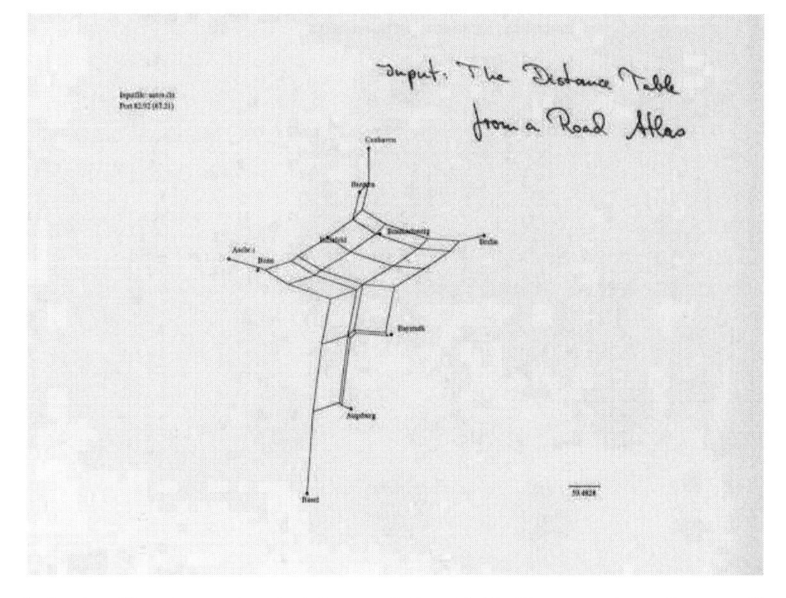

Fig. 1.4. An X-net constructed by applying *SplitsTrees* to data from a German Road Atlas. The set X consists of 10 distinct German cities.

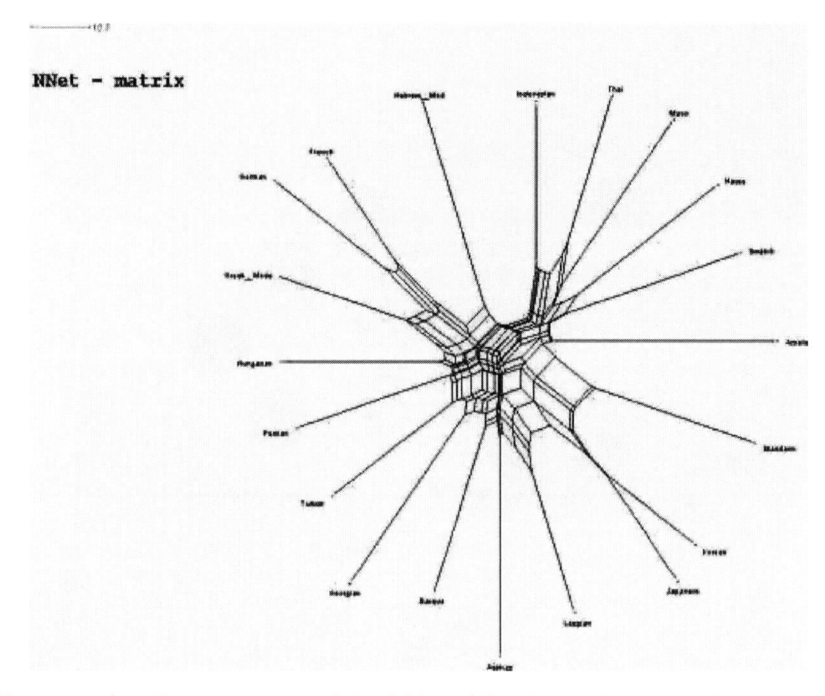

Fig. 1.5. An X-net constructed by Mihai Albu by applying *Neighbour Net* to data from *The World Atlas of Language Structure* depicting the overall structural (dis)similarity of 20 world languages (cf. Haspelmath, Martin & Dryer, Matthew & Gil, David & Comrie, Bernard (eds.) 2005. The World Atlas of Language Structures. Oxford University Press).

An X-net constructed by Peter Robinson et al by applying *SplitsTrees* to data that he derived by comparing distinct handwritten copies of the Prologue of Chaucer's *The Wife of Bath* (cf. *Nature*, August 1998). The set X consists of more than 40 such manuscripts. And the resulting X-net is, not unexpectedly, rather "tree-ish." Due to unresolved copyright questions, please look up the figure in the original *Nature* publication.

Figures 1.6 to 1.9 deal with proper biological data. Figure 1.6 deals with 16S rRNA sequences from all three *Kingdoms of Life*, the Eucariots, the Procariots, and the Archeae. Figures 1.7 and 1.8 deal with data regarding various variants of the AIDS virus, including HIV1, HIV2, and HIV sequences discovered in other primates. The same data have been analysed in two distinct ways, the first figure being based on (dis)similarity, taking into account the first position, only, from each coding triple; the second one taking into account all positions, yet registering only the difference between purins and purimidins, neglecting transitions between nucleotides. The resulting, rather tree-ish structures are surprisingly similar, corroborating each other, and demonstrat-

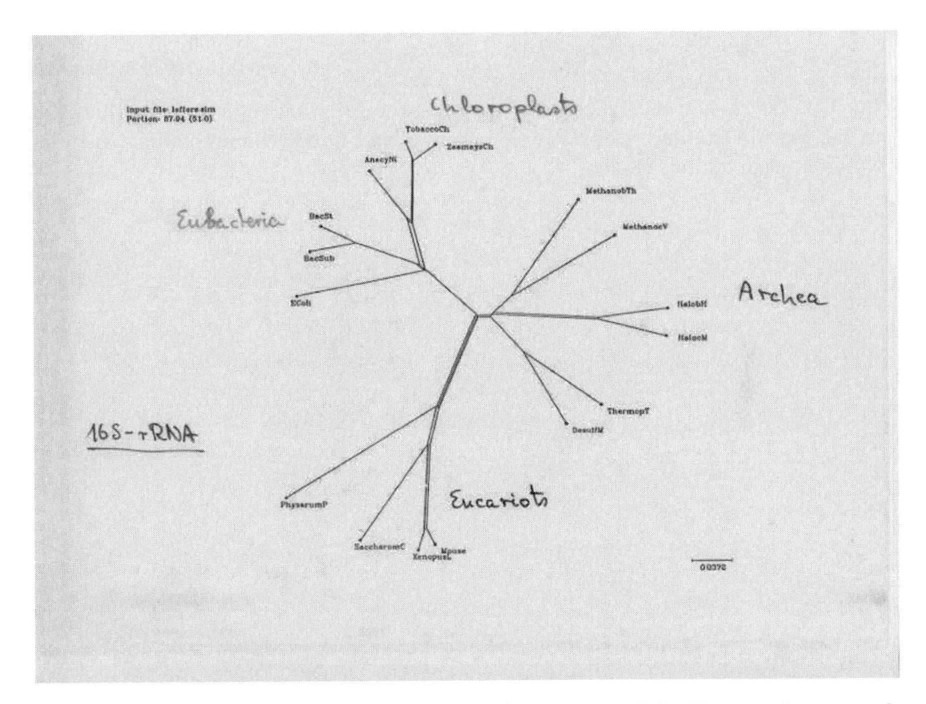

Fig. 1.6. An X-net constructed by applying the program *SplitsTrees* to data regarding the (dis)similarity of 16 distinct 16S rRNA sequences from Eucariots, Procariots, and Archeae. A proper net, though almost a tree as the length of edges not fitting in the "obvious" underlying tree structure is negligibly small. The underlying tree structure clearly supports Karl Woese's thesis claiming the existence of one common ancestor for all currently existing forms of Archeae.

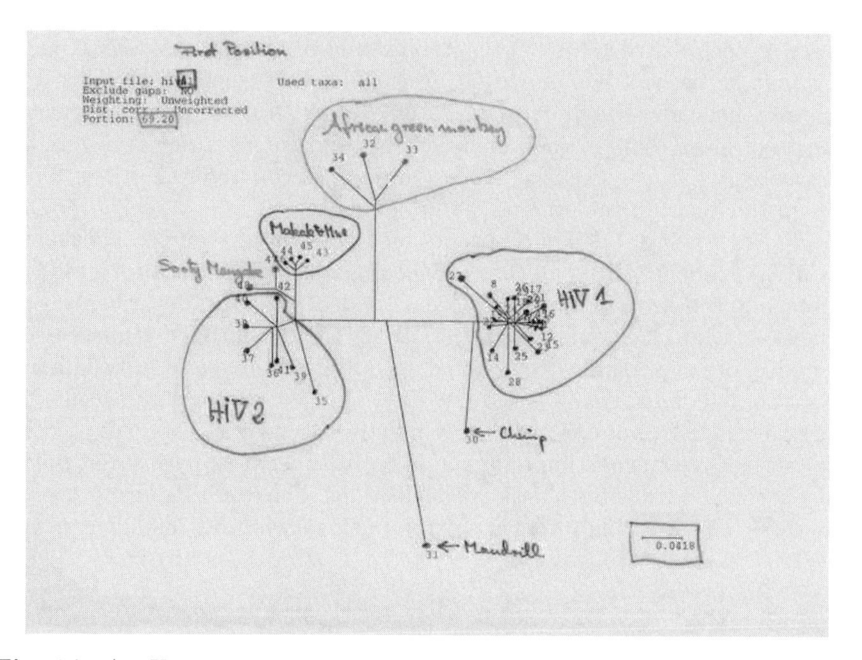

Fig. 1.7. An X-net constructed by applying the program *SplitsTrees* to data regarding the (dis)similarity of HIV sequences, taking into account the first position, only, from each coding triple.

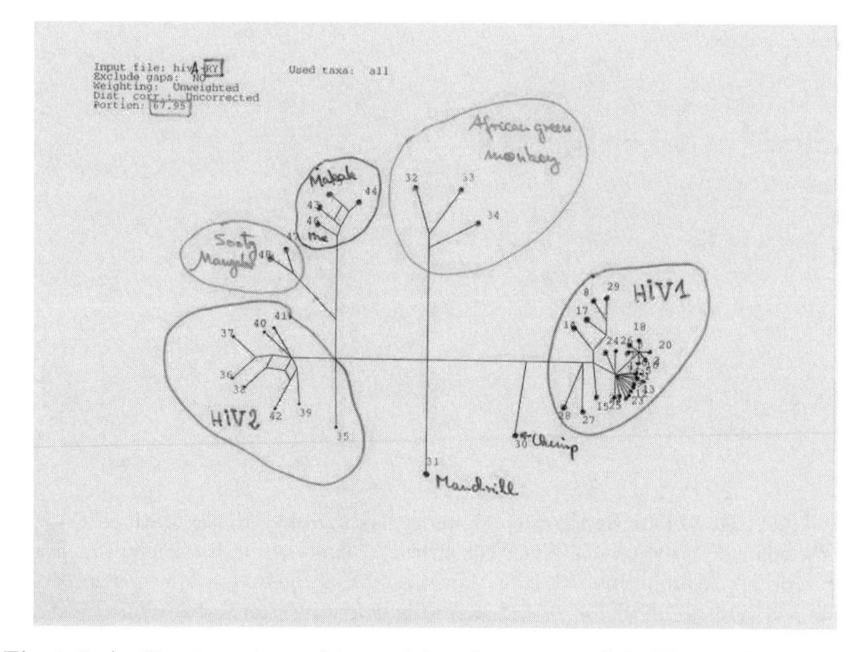

Fig. 1.8. An X-net constructed by applying the program *SplitsTrees* to data regarding the (dis)similarity of HIV sequences, taking into account the difference between purins and purimidins, only, neglecting all *transitions* between nucleotides.

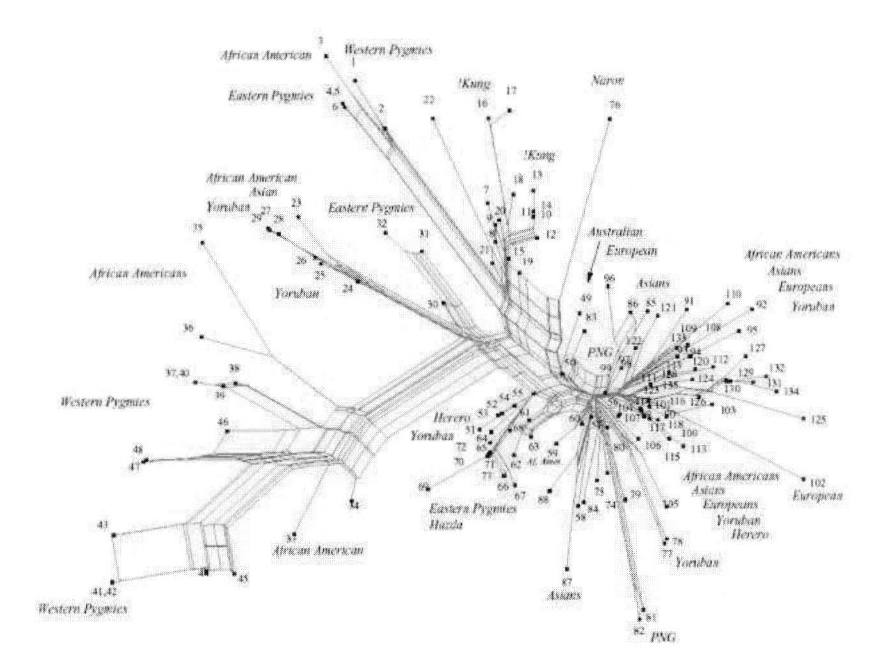

Fig. 1.9. An X-net constructed by applying the program *Neighbour-Net* (cf. [12]) to data from [16] regarding the (dis)similarity of Human mitochondrial DNA sequences. The resulting X-net clearly supports Allan Wilson's "Out of Africa" hypothesis as African sequences can be found all over in that net while all other groups are clearly localized.

ing that it is very unlikely that they both are just artefacts of the respective methods of quantifying (dis)similarity applied for deriving these two X-nets. Figure 1.9 deals with human mitochondrial DNA data, and the resulting X-net clearly supports Allan Wilson's "Out of Africa" hypothesis (as African sequences can be found all over in that net while all other groups are clearly localized).

References

1. Bandelt HJ (1990) Recognition of tree metrics. SIAM J Disc Math 3:1–6.
2. Bandelt HJ, Dress A (1989) Weak hierarchies associated with similarity measures—an additive clustering technique. Bul Math Biol 51:133–166.
3. Bandelt HJ, Dress A (1992) A canonical split decomposition theory for metrics on a finite set. Adv Math 92:47–105.
4. Bandelt HJ, Dress A (1992) Split decomposition: a new and useful approach to phylogenetic analysis of distance data. Mol Phylogenet Evol 1 (3):242–252.

5. Barthélemy JP, Guénoche A (1991) Trees and proximity representations. Wiley, New York.
6. Buneman P (1971) The recovery of trees from measures of dissimilarity. In: Hodson F, et al (ed) Mathematics in the Archeological and Historical Sciences. Edinburgh University Press, Edinburgh, pp. 387–395.
7. Dress A (1984) Trees, tight extensions of metric spaces, and the cohomological dimension of certain groups: a note on combinatorial properties of metric spaces. Adv Math 53:321–402.
8. Dress A, Huber KT, Moulton V (1997) Some variations on a theme by Buneman. Ann Combin 1:339–352.
9. Dress A, Huber KT, Moulton V (2002) An explicit computation of the injective hull of certain finite metric spaces in terms of their associated Buneman complex. Adv Math 168:1–28.
10. Dress A, Moulton V, Terhalle W (1996), T-theory: an overview. Eur J Combinatorics 17:161–175.
11. Huson D (1998) SplitsTree: a program for analyzing and visualizing evolutionary data. Bioinformatics 14:68–73; see also http://www-ab.informatik.uni-tuebingen.de/software/splits/welcome_en.html.
12. Huson DH, Bryant D (2006) Applications of phylogenetic networks in evolutionary studies. Mol Biol Evol 23:254–267.
13. Isbell J (1964) Six theorems about metric spaces. Comment Math Helv 39:65–74.
14. Lockhart PJ, Meyer AE, Penny D (1995) Testing the phylogeny of Swordtail fishes using split decomposition and spectral analysis. J Mol Evol 41:666–674.
15. Semple C, Steel M (2003) Phylogenetics. Oxford: Oxford University Press.
16. Vigilant L, Stoneking M, Harpending H, Hawkes K, Wilson A (1991) African populations and the evolution of human mitochondrial DNA. Science 253:1503–1507.

2

Networks with Delays

Fatihcan M. Atay

Summary. Information transmission over any spatial separation involves a time delay. For dynamical systems extended over a network, the network topology and temporal delays interact in a complicated way to shape the overall dynamical behaviour. We discuss the effects of delays on several prototypical systems, showing that the introduction of delays into the models can lead to the emergence of completely new behaviour that is not possible in the absence of delays.

2.1 Introduction

In any collection of interacting dynamical systems, the information flow between the individual units is constrained by the laws of nature. The information may be transmitted by chemical agents in biological systems, by the motion of electrons in electronic devices, and by light in optical equipment. In all cases, it is never conveyed instantaneously, but only after some time delay, across space. Indeed, it is a fundamental law of nature that the speed of light, about 299,792,458 m/s in vacuum, is an upper bound for the speed of mass and energy (and thus of information) flow in any system. Consequently, time delays are a fundamental reality for physical systems.

The description of physical reality, however, cannot and does not (and perhaps need not) include all the details. Mathematical models inevitably contain simplifications in order to be amenable to solution methods. The treatment of time delays is no exception, and it is usual that models of dynamical networks ignore the delays resulting from finite transmission speeds. This does not necessarily mean that such models are not useful. The situation can be compared to the role of Newtonian mechanics, which has been used to build most of our existing technology, although the theory of relativity tells us it is not the correct description of the universe. The important thing is to know when relativistic effects need to be included in our formulas, at the expense of making them more complicated. In a similar fashion, it is important to be aware of the possible effects of time delays so that a useful mathematical model can be chosen for the description of dynamical behaviour in a particular application.

While the speed of light is an upper limit for information transmission speed, many systems encountered in practice are considerably slower. In a spectrum of applications ranging from biology to optics, the range of time delays differ by orders of magnitude. The question then is, when to take delays into consideration and when to ignore them. One may be tempted to assume that delays might be more important for the slower biological systems, while they can be safely ignored in faster systems such as lasers or electro-optical devices. Upon more careful study it becomes apparent that the answer also depends on the particular dynamics being studied. Indeed, one sees that delays can only be neglected when they are of a smaller order of magnitude than the characteristic time scales of the system.

Several examples should be convincing that information transmission even near the speed of light may not be fast enough to ignore the effects of delays. For instance, signals from the earth station reaches a satellite in low earth orbit after about 0.2 s, which implies a round-trip feedback delay of 0.4 s [1]. This round-trip delay is about 3 s for spacecraft near the moon. It may be argued that these are extreme examples because of the astronomical distances involved. However, the effects can be significant also at the distance scales of biological entities. Consider, for instance, a pair of coupled lasers placed 1 m apart. At the speed of light, the information from one laser reaches the other in about 10^{-8} s. Is this a small delay? Considering that the period of a typical laser operating in the nanometer range is on the order of 10^{-14} s, the transmission delay between the lasers is about a million times larger, and certainly significant! (See [2] for some effects of delays in coupled lasers.) In biological systems delays are generally larger: The delays involved in neuromuscular action are typically on the order of 0.1 s, which can have important consequences for the dynamics, for example in the pupil light reflex [3, 4]. Similarly, the relatively slow speed of neural signal propagation along axons (0.1–10 m/s) gives rise to distance-dependent delays in the neural field, some of whose consequences will be briefly mentioned in Section 2.4.

What is it, then, that might be missed by neglecting delays in network models? Certainly there is a *quantitative* loss that manifests itself in decreased numerical accuracy of the results, which may or may not be significant for the particular application. More importantly, there is also the possibility of a *qualitative* loss, in the sense that the delays may cause a completely different type of behaviour that is absent in the undelayed model. The importance of qualitative discrepancies between the data and the model can hardly be overemphasized, as they bring into question the validity of experimental techniques, data analysis methods, and the assumptions underlying the model. It is thus important to know when the model can be reconciled with data by taking the time delays into account.

In the following sections we give an overview of the qualitatively different dynamics that may emerge in networks as a result of time delays. We focus on three prototypical examples of dynamical networks, each with a different character, that are often used to model physical and biological systems. We will see how delays enhance synchrony in coupled map lattices, suppress oscillations in coupled oscillators, and lead to traveling waves in coupled neural field models.

2.2 Coupled Map Lattices

A map, or a function, f on a suitable space (here $\mathbb{R}$) induces a discrete-time dynamical system by the iteration rule

$$x(t+1) = f(x(t))$$

where the iteration step $t \in \mathbb{Z}$ plays the role of discrete time. A coupled map lattice [5] is a collection of such systems that interact with each other in some specified way. A model of such a network where the interaction has a diffusive form is given by

$$x_i(t+1) = f(x_i(t)) + \kappa \frac{1}{d_i} \sum_{\substack{j \\ j \sim i}} (f(x_j(t-\tau)) - f(x_i(t))) \tag{2.1}$$

Here, $i = 1, \ldots, N$ indexes the units, $\kappa \in \mathbb{R}$ is the coupling strength, and $\tau \in \mathbb{Z}^+$ is the delay in information transmission. Thus the ith unit x_i interacts with its d_i neighbors, and the symbol $\sim$ denotes the neighborhood relation, that is, $i \sim j$ means that the ith and jth units are neighbors of each other. The neighborhood relation casts the system as one defined on a graph, and the diffusive nature of interaction is reflected in the *Laplacian* operator

$$\mathcal{L} = I - D^{-1}A,$$

where $A = [a_{ij}]$ is the adjacency matrix

$$a_{ij} = \begin{cases} 1 & \text{if } i \sim j \\ 0 & \text{otherwise} \end{cases}$$

and $D = \mathrm{diag}\{d_1, \ldots, d_N\}$ is the diagonal matrix of vertex degrees, that is, the number of neighbours of the units. The operator $\mathcal{L}$ encapsulates the information about the connection structure of the units.

A natural question is how the dynamics of (2.1) are related to the underlying graph structure. A particularly interesting type of dynamics is *synchronization*, whereby all units tend to behave identically regardless of initial

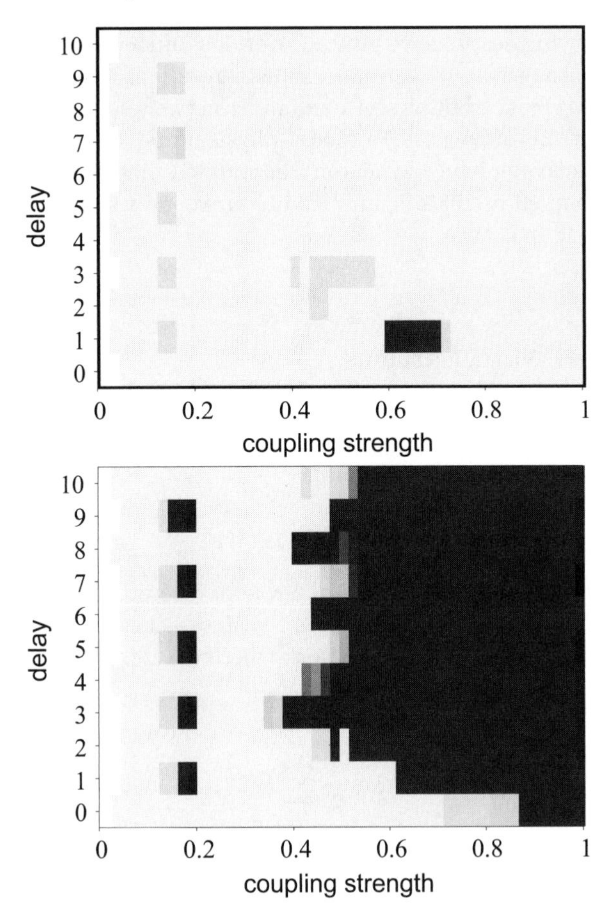

Fig. 2.1. Synchronization of coupled maps. The gray scale codes the degree of synchrony, black corresponding to complete synchronization. The upper graph corresponds to a small-world network obtained from a regular one by randomly reconnecting a few edges, and the lower graph is a random network, both having the same number of nodes and links.

conditions and external disturbances. Even chaotic units, which show sensitive dependence on initial conditions, can exhibit synchronous behaviour under appropriate conditions [6]. In the absence of delays ($\tau = 0$), synchronization is related to the network structure through the eigenvalues of the Laplacian operator, which encapsulates the connection structure of the network [7]. With nonzero delays, the relation between the two becomes more complicated. Figure 2.1 shows the parameter regions where chaotic logistic maps given by

$$f(x) = 4x(1 - x)$$

exhibit synchrony when coupled.

It is apparent that delays and the network topology interact in a nontrivial way to determine synchronization. Interestingly, it can be seen from the figure that the delays can enhance synchronization. For instance, in the random network, for the range of coupling strengths $0.6 < \kappa < 0.85$ the system synchronizes only when the delay τ is nonzero. The fact that the delayed system can manage to act in unison at all is a bit surprising in view of the fact that each unit is unaware of the present state of the others. Hence it is all the more surprising when synchrony occurs *only* with delays. Furthermore, it turns out that the synchronous behaviour itself can exhibit a wide range of dynamics in the presence of delays, including periodic and quasi-periodic oscilations, chaos with a different Lyapunov exponent than the original map, and hyperchaos (two or more positive Lyapunov exponents) [8, 9]. Even more possibilities arise if the delays in (2.1) are allowed to vary between pairs of units. When the delays are chosen from a random distribution with a sufficiently large variance, the coupled system can tend to a stable fixed point [10]. Thus, the delays can induce a very rich set of new behaviour into the network (2.1).

2.3 Coupled Oscillators

The coupled map model of the previous section assumes that the units update their states at the same instant, namely at integer values of time. In some cases it is more realistic to consider a continuous update mechanism. Such systems can be described by differential equations in continuous time, which may take the form

$$\dot{x}(t) = f(x(t); \varepsilon) \tag{2.2}$$

with $x \in \mathbb{R}^n$ and $\varepsilon \in \mathbb{R}$ is a parameter. A common example is the modeling of limit cycle oscillators that arise in biology, population dynamics, electronic circuits, etc. One way such stable oscillations can arise is through a supercritical Hopf bifurcation, where the system has a stable fixed point for $\varepsilon < 0$, which loses its stability when ε is increased as a pair of complex conjugate eigenvalues of the linearized system crosses to the right half complex plane at the bifurcation value $\varepsilon = 0$. It can then be shown that under quite general conditions an attracting periodic solution exists for $0 < \varepsilon \ll 1$. We assume this to be the case so that (2.2) models an oscillator. Analogous to (2.1), a coupled system of such oscillators may be represented by

$$\dot{x}_i(t) = f(x_i(t); \varepsilon) + \varepsilon\kappa \frac{1}{d_i} \sum_{\substack{j \\ j \sim i}} (x_j(t-\tau) - x_i(t)) \,. \tag{2.3}$$

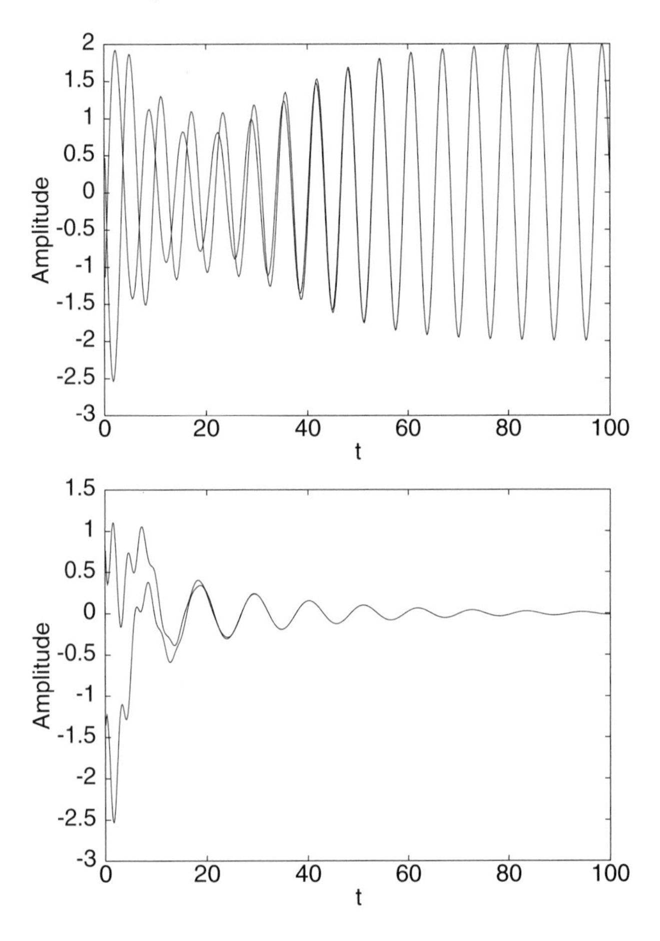

Fig. 2.2. Two coupled van der Pol oscillations. The upper graph shows the behaviour when there are no coupling delays ($\tau = 0$), where the system synchronizes. In the lower graph $\tau = 1$, and the oscillations are quenched.

As before, synchrony is again a possible behaviour for the coupled system, but we focus on another interesting behaviour that emerges in the presence of delays—the so-called *amplitude death* or *oscillator death.* The term refers to the quenching of oscillations as the system is attracted to the fixed point, which was unstable for the isolated system (2.2). We illustrate this behaviour for the familiar van der Pol oscillator, whose dynamics are described by

$$\ddot{y} + \varepsilon(y^2 - 1)\dot{y} + y = 0 \tag{2.4}$$

For (2.4), the zero solution is stable for negative values of ε, and undergoes a supercritical Hopf bifurcation at $\varepsilon = 0$ as a pair of complex conjugate eigenvalues of the linearized equation migrate to the right half of the complex plane. For small positive values of ε, there is a stable limit cycle with amplitude near 2, which attracts all initial conditions, except the unstable equilibrium at zero. We couple a pair of such oscillators through their velocities, obtaining the system

$$\begin{aligned} \ddot{y}_1 + \varepsilon(y_1^2 - 1)\dot{y}_1 + y_1 &= \varepsilon\kappa(\dot{y}_2(t-\tau) - \dot{y}_1(t)) \\ \ddot{y}_2 + \varepsilon(y_2^2 - 1)\dot{y}_2 + y_2 &= \varepsilon\kappa(\dot{y}_1(t-\tau) - \dot{y}_2(t)) \end{aligned} \tag{2.5}$$

Figure 2.2 illustrates delay-induced death for (2.5). When there are no delays in the coupling, the oscillators fall into step after starting from arbitrary initial conditions. For positive coupling delay and an appropriate range of coupling strengths, the oscilations are suppressed and replaced by an equilibrium solution.

Interestingly, in a network such as (2.3) where the oscillators are identical, amplitude death is possible only in the presence of delays [11, 12], and for non-identical oscillators, delays make amplitude death easier to occur [13]. This is clearly an important phenomenon since oscillations form a crucial function of many biological systems, such as cardiac cells, neurons, etc., which once again underscores the importance of delays. For systems near Hopf bifurcation, the role of the network topology can be expressed in terms of the largest eigenvalue of the Laplacian operator [14]. Hence, there is again an intimate relation between the delays and the network structure in shaping the dynamics of the system.

2.4 Neural Field Model

The coarse-grained activity of neural ensembles are often modeled by integro-differential equations of the form [15]

$$L(\partial/\partial t)V(x,t) = \kappa \int_{\Omega} K(|x-y|)S\left(V\left(y, t - \frac{|x-y|}{v}\right)\right) dy + E(x,t). \tag{2.6}$$

Here, $V(x,t)$ denotes the mean postsynaptic potential at time t and location x inside the spatial domain Ω (here $\mathbb{R}$), L is a first- or second-order temporal differential operator, $\kappa \geq 0$ is the coupling strength, S is a nonlinear transfer function, v is the axonal transmission speed, K is the connectivity kernel, and E is the external input. At first sight, (2.6) has little resemblance to the networks mentioned in the previous sections; but in fact it can be considered as a dynamical network over a continuum of nodes, labeled by the coordinate system in Ω. The propagation delay between two locations (or nodes) x and y is given by $|x-y|/v$. Thus, the delays in the model (2.6) are distributed over an interval, in contrast to the fixed delays of previous examples. Many studies

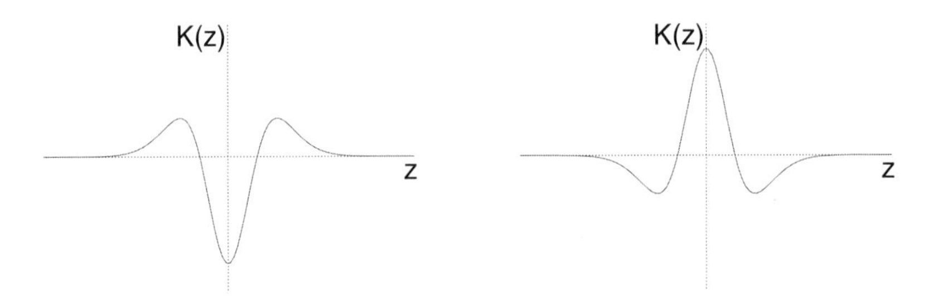

Fig. 2.3. Typical connectivity kernels K for the neural field model. The graph on the left corresponds to local inhibition-lateral excitation, and the one on the right gives local excitation-lateral inhibition.

ignore these delays by taking $v = \infty$. The function K describes the connection structure and plays a role similar to the Laplacian operator of (2.1) and (2.3). If K is a positive (respectively, negative) function, then the connections are excitatory (respectively, inhibitory). On the other hand, a kernel K with mixed signs allows the modeling of more general connectivity, such as local inhibition and lateral excitation or vice versa, as depicted in Figure 2.3.

A systematic analysis of (2.6) can proceed by finding spatially uniform equilibrium solutions for constant input level, and study their stability as parameters are varied. By following the behaviour near bifurcation values, one can discover such phenomena as spatial patterns (Turing patterns), oscillations, or traveling waves [16, 17]. The bifurcations can be classified as temporally constant or oscillatory, which can further be relegated as spatially constant or varying. Temporally constant bifurcations may lead to spatially uniform equilibria or spatial patterns, while oscillatory bifurcations may yield spatially uniform oscillations or traveling waves. Figure 2.4 summarizes the four basic possibilities for bifurcations. For the system (2.6) it turns out that oscillatory solutions can bifurcate from spatially homogeneous equilibria *only when* delays are present. We make this statement precise by the following theorem:

Theorem 1. *Suppose* $L(\frac{\partial}{\partial t}) = \eta\frac{\partial^2}{\partial t^2} + \gamma\frac{\partial}{\partial t} + \rho$, *where* $\eta = 0$ *or* 1 *and* $\gamma, \rho > 0$. *If*

$$\kappa \int_{\Omega} |zK(z)|\, dz < |\gamma| v \tag{2.7}$$

then there are no oscillatory bifurcations of spatially homogeneous equilibria.

Clearly, (2.7) is satisfied for instantaneous or very fast signal transmission, so in these cases only stationary solutions can bifurcate from equilibria. Stated

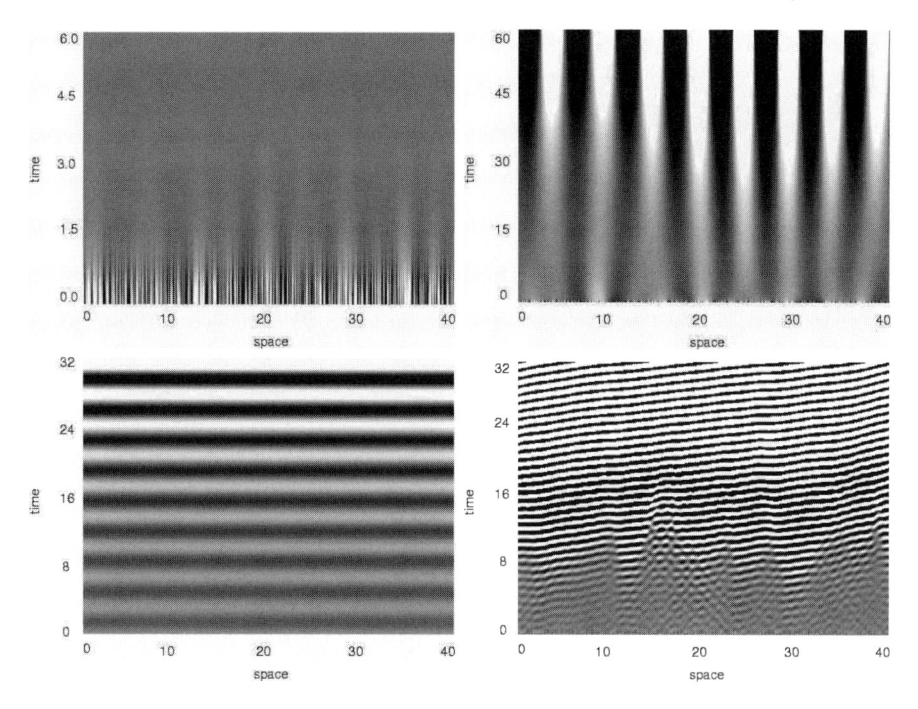

Fig. 2.4. Possible bifurcations from a spatially homogeneous equilibrium of the neural field. In the upper row are the stationary bifurcations leading to a new spatially constant equilibrium (left) or a Turing pattern (right). In the lower row are the oscillatory bifurcations yielding uniform oscillations (left) and traveling waves (right). In each graph, the vertical axis is the time and the horizontal axis is space.

differently, the two possibilities shown in the lower row of Figure 2.4 can only occur in the presence of (sufficiently large) delays. This once again shows the rich dynamics introduced by delayed interaction in the network.

2.5 Conclusion

We have looked at three examples of common networks arising from models of biological and physical systems. Each network is a dynamical system with a different character with respect to space and time, as summarized in Table 2.1. Each one can display quite a rich spectrum of dynamics, depending on the particular nonlinearities, system size, initial conditions, and the like. In each case we have seen that the introduction of transmission delays into the network enables the emergence of completely new behaviour, which is not possible in the absence of delays.

Table 2.1. Summary of example networks.

System	Space	Time
Coupled maps	Discrete	Discrete
Coupled differential equations	Discrete	Continuous
Field models	Continuous	Continuous

We have also noted that there is a nontrivial interaction between the delays and the network topology, or connectivity, which turns out to be an important factor in shaping the dynamical features of the system. In our examples of discrete networks, the connectivity is described by the Laplacian operator (which can be derived from the adjacency matrix), which supposes that the interaction between the units is of diffusive type. Similarly, in the continuous neural field (2.6), the connectivity given by the kernel K assumes a homogeneous and isotropic field, since it depends only on the absolute value of the distance betwen two locations. Of course, more general interaction types may arise in different application areas. On the other hand, in many applications, such as real neural networks, the precise form of the connectivity is not known. This underlines the fact that it is essential to have a theoretical understanding of the network dynamics for general connectivities, delays, and nonlinearities.

Realistic models of dynamical networks increasingly demand the inclusion of transmission delays for proper analysis, design, and control of their dynamics. As we have seen, novel dynamical behaviour can emerge as a result of delays, even though the undelayed network may itself be sufficiently complex. Note that we did not assume any particular form for the nonlinearities in the models considered. Hence, the observations are quite general and applicable to dynamical networks arising in diverse disciplines, showing that the effects of delays are not peculiar to a particular system. A mathematical theory of dynamical networks involving time delays is therefore important for understanding the general behaviour of a collection of interacting entities, be it cells, neurons, proteins, infectious agents, lasers, or stock markets.

References

1. Kolmanovskii V, Myshkis A (1992) Applied Theory of Functional Differential Equations. Kluwer Academic Publishers, Dordrecht.
2. Kuntsevich BF, Pisarchik AN (2001) Synchronization effects in a dual-wavelength class-B laser with modulated losses. Phys Rev E 64:046221.
3. Longtin A, Milton J (1988) Complex oscillations in the human pupil light reflex with "mixed" and delayed feedback. Math Biosci 90:183–199.
4. Atay FM, Mallet-Paret J (1998) Modeling reflex asymmetries with implicit delay differential equations. Bull Math Biol 60:999–1015.

5. Kaneko K (ed) (1993) Theory and applications of coupled map lattices. Wiley, New York.
6. Pecora LM, Carroll TL (1990) Synchronization in chaotic systems. Phys Rev Lett 64:821–824.
7. Jost J, Joy MP (2002) Spectral properties and synchronization in coupled map lattices. Phys Rev E 65:016201.
8. Atay FM, Jost J, Wende A (2004) Delays, connection topology, and synchronization of coupled chaotic maps. Phys Rev Lett 92:144101.
9. Atay FM, Jost J (2004) On the emergence of complex systems on the basis of the coordination of complex behaviors of their elements: synchronization and complexity. Complexity 10:17–22.
10. Masoller C, Marti AC (2005) Random delays and the synchronization of chaotic maps. Phys Rev Lett 94:134102.
11. Ramana Reddy DV, Sen A, Johnston GL (1998) Time delay induced death in coupled limit cycle oscillators. Phys Rev Lett 80:5109–5112.
12. Atay FM (2003) Total and partial amplitude death in networks of diffusively coupled oscillators. Physica D 183:1–18.
13. Atay FM (2003) Distributed delays facilitate amplitude death of coupled oscillators. Phys Rev Lett 91:094101.
14. Atay FM (2006) Oscillator death in coupled functional differential equations near Hopf bifurcation. J Differential Equations 221:190–209.
15. Wilson HR, Cowan JD (1972) Excitatory and inhibitory interactions in localized populations of model neurons. Biophys J 12:1–24.
16. Atay FM, Hutt A (2005) Stability and bifurcations in neural fields with finite propagation speed and general connectivity. SIAM J Appl Math 65:644–666.
17. Hutt A, Atay FM (2005) Analysis of nonlocal neural fields for both general and gamma-distributed connectivities. Physica D 203:30–54.

3

Dynamical Networks

Jürgen Jost

Summary. The theory of dynamical networks is concerned with systems of dynamical units coupled according to an underlying graph structure. It therefore investigates the interplay between dynamics and structure, between the temporal processes going on at the individual units and the static spatial structure linking them.
In order to analyse that spatial structure, formalized as a graph, we discuss an essentially complete system of graph invariants, the spectrum of the graph Laplacian, and how it relates to various qualitative properties of the graph. We also describe various stochastic construction schemes for graphs with certain qualitative features.
We then turn to dynamical aspects and discuss systems of oscillators with diffusive coupling according to the graph Laplacian and analyse their synchronizability.
The analytical tool here are local expansions in terms of eigenmodes of the graph Laplacian. This is viewed as a first step towards a general understanding of pattern formation in systems of coupled oscillators.

3.1 Introduction

The theory of dynamical networks is a combination of graph theory and nonlinear dynamics. It is concerned with elements or agents whose states are dynamical quantities, following some dynamical rule, and that dynamical rule includes interactions with neighbouring elements. These elements are considered as the nodes or vertices of a graph, and the edges connecting them with other vertices in the graph specify with which other elements they interact. Thus, from the point of view of dynamical systems, we have a coupled system of dynamical equations, and the emphasis is on the resulting global dynamics of the system emerging from the interactions between the local dynamics of the individual elements. Graph theory then analyses the coupling structure and its influence on those emerging global patterns. Here, one thinks about dynamical nodes, but with a fixed topology. In contrast to this, one may also consider how evolution rules shape the dynamics of a network, that is, admitting a dynamical topology. In that scenario, the underlying graph is not fixed, but changing, possibly in response to the dynamics supported by it.

Thus, the emphasis can be put either on the interplay between the local and the global dynamics or on the dynamically evolving topology. Most interesting is a combination of both. Typically, such a combination involves a coupling between two different time scales: a fast one on which the individual dynamic takes place, and a slow one on which the network responds to that dynamic and evolves. In neural networks, for example, one has a fast activity dynamic of the nodes, called neurons in this context, and a slow learning dynamic that changes the weights of the connections, called synapses, in response to the activity correlations between neurons.

Although, as discussed, the interplay between dynamic and structure is of ultimate interest, for the exposition we first need to describe the two aspects separately. We start with the theory of the underlying structure, that is, with graph theory, with a view toward dynamical patterns. In particular, we shall introduce and analyse the graph Laplacian and its spectrum. This will also provide a link with dynamical aspects. Namely, we shall consider dynamics at the nodes of the graph that are coupled via the graph Laplacian, or some generalizations thereof. We shall then conclude this survey by discussing the question of synchronization, that is, when the coupling is strong enough to make the dynamics at the nodes identical to each other.

3.2 Qualitative Properties of Graphs

As described in the introduction, we consider networks of interacting discrete agents or elements. Usually, these interactions follow not only some dynamical rule, but also some underlying structural pattern that encodes which elements interacts with which other one. This is typically specific, in the sense that the interaction partners of each element are specific, selected other elements. This structure may itself evolve in time, but if it does, then it will do so rather slowly in most situations.

This interaction pattern or structure is usually formalized as a graph. The nodes or vertices of the graph are the original elements themselves, whereas the edges or links encode the interaction pattern. In the simplest case, the graph is symmetric and without weights, and there is a link between the elements i and j when they interact. In a dynamical evolution, the influences need not be symmetric, however. That is, the state of i may be an input for the computation of the next state of j, but not conversely. In that case, we should represent this by a directed edge from i to j. Thus, we shall construct a directed graph. Also, the interaction or influence may occur with a certain strength w_{ji}, and we may also encode this by assigning the weight w_{ji} to the edge from i to j. We thus obtain a weighted graph; it is symmetric if always $w_{ji} = w_{ij}$.

Examples of unweighted, symmetric graphs are the ones that describe reciprocal social relationships like acquaintance, friendship, scientific collaboration, or the ones that describe infrastructures (roads, train or flight connections between cities, power lines, Internet connections between servers, etc). Directed graphs occur for such diverse patterns as Web links, gene regulatory networks, food webs, and flows of payments. Extreme cases are hierarchical or dependency, descendence structures and the like where links can only go in one direction. Weighted graphs occur, for example, as structures of neural networks.

A graph can be represented by its adjacency matrix. That matrix carries an entry 1 (or w_{ji} in the weighted case) at the intersection of the ith row and the jth column if there is a link from i to j. When there is no link, the entry will be 0.

We are interested in qualitative properties of the interaction graph and in quantities whose values can encode the significant such properties. Let us try to develop some of them. We consider the symmetric case without weights. We assume that the graph Γ is connected and finite. We say that two vertices are neighbours if they are connected by an edge. An extreme case is the complete graph of N vertices where each vertex is connected to every other one. Excluding self-connections, we thus have $\frac{N(N-1)}{2}$ links. That graph is said to be fully connected. In particular when N is large, this becomes unwieldy. Typically, large graphs are rather sparsely connected, that is, most entries in the adjacency matrix will be 0.

Another prototype that is a natural starting point is a regular graph. We consider some regular tessellation of Euclidean space, for example by unit cubes or simplices. We take all the corner points of this tessellation, for example all points in Euclidean space with integer coordinates in the case of a unit cube tessellation. We then select a regular connectivity pattern, that is, connect any such points with all the nearest other ones, and perhaps also with the second, third,..., nearest ones. To obtain a finite graph, we can identify points that are obtained from each other through a shift of some fixed size in any of the coordinate directions. Such a regular graph then possesses a transitive[1] symmetry group because the connectivity pattern of all the nodes is the same. While such a graph may describe the structure of certain crystals, in most other cases we do not encounter so much homogeneity, but rather some diversity and differences between the connectivity patterns of the various nodes. To provide a pattern that is opposite to the one of a regular graph, Erdös and Rényi introduced random graphs. Here, one starts with, say, N nodes and each pair of nodes gets a link with some fixed probability p. That is, we start with a fully connected graph of N nodes and delete any individual

[1] On the set of nodes.

edge with probability $1-p$, of course treating all edges independently of each other. Since this is a stochastic constructions, it then makes sense to consider the class of all such random graphs with fixed parameters N, p and derive typical properties, that is, properties that hold almost surely as N tends to ∞, with respect to some obvious natural measure on the space of all such random graphs.

For nodes i, j of a graph Γ, we let $d(i, j)$ denote their distance, that is, the smallest number of edges that establish a connection between them. In a regular graph, the distance between randomly drawn nodes grows like a power of the total number N of nodes. In most graphs occurring in applications, however, the average or maximal distance between nodes is much smaller. This has been called the small-world phenomenon, and Watts and Strogatz [31] proposed a simple algorithm for converting a regular graph into a small-world graph. Namely, given a regular graph and some p between 0 and 1, for any node i, select with probability p a random partner j somewhere in the graph and add a link from i to j. One variant of the construction then deletes one of the original links from i so as to keep the total number of edges constant. When p exceeds some critical threshold, the small-world phenomenon sets in, that is, the average distance between two nodes jumps to a much smaller value than for the original regular graph.

Another class of graphs that recently found much interest is the so-called scale-free graph. Here, the number of nodes of degree k, that is, those that possess k links, behaves like a power $k^{-\beta}$. That is, the number of nodes of degree k does not decay exponentially as a function of k, as many examples of graphs might lead one to expect, but rather only polynomially. Graphs modeling Web links, Internet connections between servers, airline connections between cities, and also biological networks like gene regulatory networks seem to exhibit this type of structure. Simon [29] and Barabasi and Albert [4] proposed a construction scheme for such scale-free graphs. One starts with some small graph and then adds new nodes successively. Any new node is allowed to make k links while any node already established has a probability proportional to the number of links it already possesses for becoming the recipient of one of those k new links. As the network grows it exhibits this type of scale-free behaviour. However, a scale-free structure can also be generated by completely different mechanisms, with different qualitative properties with regard to other aspects.

One such important network aspect is the clustering behaviour, both on the local and on the global scale. We start with the local aspect. For an Erdös-Rényi random graph, given a node i_0 and two of its neighbours i_1, i_2, the probability that they are connected, that is, that there exists an edge between i_1 and i_2, is given by the fixed number p underlying the construction and is not influenced at all by the fact that they possess a common neighbour, namely i_0. In many networks arising in applications, this is different, however. Namely, two nodes with a common neighbour have a higher probability of being directly connected than two arbitrary ones. Of course, it may also happen that that probability is lower. For example, for the regular graph formed from the points in Euclidean space with integer coordinates, connected when precisely one of these coordinates differs by 1 between them while the other ones coincide, that probability is 0 as the neighbours of a vertex are never directly connected. Likewise, two neighbours of a vertex are never connected for trees, that is, for graphs without cycles (loops), in other words, for simply connected graphs.

The average probability for two neighbours of some vertex to be directly connected is given by the clustering coefficient C of the graph Γ, defined as 3 times the ratio between the number of triangles in the graph (that is triples of vertices for which each is connected to the other two) and the number of connected triples (that is triples where one is connected to the other two). The factor 3 arises because each triangle contains three connected triples.

One can then also consider other graph motives that are larger than triangles, for example complete subgraphs of k vertices, that is, subgraphs in which any two vertices are connected by an edge, for some $k > 3$ (for $k = 3$, we have the triangles just discussed). Two such complete k-subgraphs are called adjacent when they share a complete $(k-1)$-subgraph. One can then search for clusters defined as maximal subgraphs consisting of uninterrupted chains of adjacent complete k-subgraph (see [12]). That means that each vertex is contained in a complete k-subgraph, and for any two vertices i, j inside such a cluster, we find a sequence $i_1 = i, i_2, ..., i_n = j$ of vertices inside the cluster such that i_ν and $i_{\nu+1}$ are always contained in adjacent complete k-subgraphs.

This was the local aspect of clustering. The global one is what is also called community structure. Here, one looks for groups of vertices with many connections within a group, but considerably fewer between groups. Perhaps this aspect is best understood by describing methods for breaking a network up into such subgroups, also called communities. We shall exhibit two methods that are kind of dual to each other. The first one is based on a quantity that is analogous to one introduced by Cheeger in Riemannian geometry. Letting n_i denote the degree of the node i (the number of its neighbours) and $|E|$ the number of edges contained in an edge set E, that quantity is

$$h(\Gamma) := \inf\left\{ \frac{|E_0|}{\min\left(\sum_{i \in V_1} n_i, \sum_{i \in V_2} n_i\right)} \right\} \tag{3.1}$$

where removing E_0 disconnects Γ into the components V_1, V_2. Thus, we try to break up the graph into two large components by removing only few edges. We may then repeat the process within those components to break them up further until we are no longer able to realize a small value of h. The other method was introduced by Girvan and Newman [14]. They define the "betweenness" of an edge as the number of shortest paths between vertices that contain it, that is, run through it. The underlying intuition is that severing an edge with a high value of that betweenness cuts many shortest paths between vertices on different sides and therefore is conducive to breaking up the graph. Their algorithm for breaking up a graph efficiently into communities then consists in the iterative removal of the (or an) edge with the highest betweenness where of course the value of that betweenness has to be recomputed after each removal. Rather obvious network parameters are the maximal and the average distance between nodes. As already mentioned, the values of those parameters are high for regular graphs and small for small-world ones. Random graphs also tend to have relatively small values.

So far, we have discussed qualitative features that can be simply described. In the next section, we shall turn to a more a technical set of invariants, the eigenvalues of the graph Laplacian. That will then enable us in subsequent sections to discuss the former qualitative aspects in a more profound manner.

3.3 The Graph Laplacian and Its Spectrum

So far, we have considered quantities that can be directly evaluated from an inspection of the graph. We now turn to constructions that depend on the choice of additional data. These data are functions on the set of vertices of the graph Γ. Of course, any such function can be extended to the edges by linear interpolation, and we shall occasionally assume that interpolation implicitly. For analysing such functions, it is in turn useful to have a basis of that function space. For that purpose, we introduce the L^2-product for functions on Γ:

$$(u,v) := \sum_{i\in\Gamma} n_i u(i)v(i) \tag{3.2}$$

where n_i is the degree of the vertex i. For purposes of normalization, one might wish to put an additional factor $\frac{1}{|\Gamma|}$ in front where $|\Gamma|$ is the number of elements of the graph, but we have decided to omit that factor in our conventions. We may then choose an orthonormal base of that space $L^2(\Gamma)$. To find such a basis that is also well adapted to dynamical aspects, we introduce the graph Laplacian

$$\Delta : L^2(\Gamma) \to L^2(\Gamma)$$

$$\Delta v(i) := \frac{1}{n_i} \sum_{j, j\sim i} v(j) - v(i) \tag{3.3}$$

where $j \sim i$ means that j is a neighbour of i.[2] The idea behind this operator is the comparison of the value of a function v at a vertex i with the average

[2] There are several different definitions of the graph Laplacian in the literature. Some of them are equivalent to ours inasmuch as the yield the same spectrum, but others are not. In the monograph [8], the operator $\mathcal{L}v(i) := v(i) - \sum_{j,j\sim i} \frac{1}{\sqrt{n_i}\sqrt{n_j}} v(j)$ is employed. Apart from the minus sign, it has the same eigenvalues as Δ: if $\Delta v(i) = \mu v(i)$, then $w(i) = \sqrt{n_i} v(i)$ satisfies $\mathcal{L}w(i) = -\mu w(i)$. While our operator Δ is symmetric w.r.t., the product $(u,v) = \sum_{i\in\Gamma} n_i u(i)v(i)$, $\mathcal{L}$ is symmetric w.r.t. $\langle u,v\rangle := \sum_{i\in\Gamma} u(i)v(i)$. The operator $Lv(i) := n_i v(i) -$

of the values at the neighbours of i. Thus, Δv detects local inhomogeneities in the values of the function v. In particular, since we assume that our graph is connected

- $$\Delta v \equiv 0 \text{ iff } v \equiv \text{const.} \tag{3.4}$$
  (If the graph is not connected, then $\Delta v \equiv 0$ when v is constant on each component.)

Other important properties of Δ are the following ones:

- Δ is selfadjoint w.r.t. $(.,.)$:
  $$(u, \Delta v) = (\Delta u, v) \tag{3.5}$$
  for all $u, v \in L^2(\Gamma)$.[3] This holds because the neighbourhood relation is symmetric.
- Δ is nonpositive:
  $$(\Delta u, u) \leq 0 \tag{3.6}$$
  for all u. This follows from the Cauchy-Schwarz inequality.

The preceding properties have consequences for the eigenvalues of Δ:

- By (3.5), the eigenvalues are real.
- By (3.6), they are nonpositive. We write them as $-\lambda_k$ so that the eigenvalue equation becomes
  $$\Delta u_k + \lambda_k u_k = 0. \tag{3.7}$$
- By (3.4), the smallest eigenvalue then is $\lambda_0 = 0$. Since we assume that Γ is connected, this eigenvalue is simple (see (3.4)), that is
  $$\lambda_k > 0 \tag{3.8}$$
  for $k > 0$ where we order the eigenvalues as
  $$\lambda_0 < \lambda_1 \leq ... \leq \lambda_K$$
  $(K = |\Gamma| - 1)$.

$\sum_{j,j\sim i} v(j)$ that is also often employed in the literature, however, has a spectrum different from Δ for general graphs.

[3] An operator $A = (A_{ij})$ is symmetric w.r.t., a product $\langle v, w \rangle := \sum_i b_i v(i) w(i)$, that is, $\langle Av, w \rangle = \langle v, Aw \rangle$ if $b_i A_{ij} = b_j A_{ji}$ for all indices i, j. The b_i are often called multipliers in the literature.

We next consider, for neighbours i, j

$$Du(i,j) := u(i) - u(j) \tag{3.9}$$

(Du can be considered as a function defined on the edges of Γ. Thus, D maps functions on the vertices to functions on the edges.) We also introduce the product

$$(Du, Dv) := \Big(\sum_{j \sim i} \big(u(i) - u(j)\big)\big(v(i) - v(j)\big)\Big) \tag{3.10}$$

where the sum is over edges[4] (note that here, in contrast to (3.2), we do not use any weights). We have

$$\begin{aligned}(Du, Dv) &= \frac{1}{2}\Big(\sum_i n_i u(i)v(i) + \sum_j n_j u(j)v(j) - 2\sum_{j \sim i} u(i)v(j)\Big) \\ &= -\sum_i u(i) \sum_{j \sim i} (v(j) - v(i)) \\ &= -(u, \Delta v).\end{aligned} \tag{3.11}$$

We may find an orthonormal basis of $L^2(\Gamma)$ consisting of eigenfunctions of Δ,

$$u_k, \ k = 1, ..., K.$$

This is achieved as follows. We iteratively define, with $H_0 := H := L^2(\Gamma)$ being the Hilbert space of all real-valued functions on Γ with the scalar product $(.,.)$,

$$H_k := \{v \in H : (v, u_i) = 0 \text{ for } i \leq k - 1\}, \tag{3.12}$$

starting with a constant function u_0 as the eigenfunction for the eigenvalue $\lambda_0 = 0$. Also

$$\lambda_k := \inf_{u \in H_k - \{0\}} \frac{(Du, Du)}{(u, u)}, \tag{3.13}$$

that is, we claim that the eigenvalues can be obtained as those infima. First, since $H_k \subset H_{k-1}$, we have

$$\lambda_k \geq \lambda_{k-1}. \tag{3.14}$$

Second, since the expression in (3.13) remains unchanged when a function u is multiplied by a nonzero constant, it suffices to consider those functions that satisfy the normalization:

$$(u, u) = 1 \tag{3.15}$$

whenever convenient. We may find a function u_k that realizes the infimum in (3.13), that is

$$\lambda_k = \frac{(Du_k, Du_k)}{(u_k, u_k)}. \tag{3.16}$$

---

[4] If we summed over pairs of vertices (i, j) with $i \sim j$, then each edge would be counted twice, and we would have to introduce a factor $\frac{1}{2}$ in front of the sum.

Since then for every $\varphi \in H_k, t \in \mathbb{R}$

$$\frac{(D(u_k + t\varphi), D(u_k + t\varphi))}{(u_k + t\varphi, u_k + t\varphi)} \geq \lambda_k, \tag{3.17}$$

the derivative of that expression w.r.t. t vanishes at $t = 0$, and we obtain, using (3.11)

$$0 = (Du_k, D\varphi) - \lambda_k(u_k, \varphi) = -(\Delta u_k, \varphi) - \lambda_k(u_k, \varphi) \tag{3.18}$$

for all $\varphi \in H_k$; in fact, this even holds for all $\varphi \in H$, and not only for those in the subspace H_k, since for $i \leq k - 1$

$$(u_k, u_i) = 0 \tag{3.19}$$

and

$$(Du_k, Du_i) = (Du_i, Du_k) = -(\Delta u_i, u_k) = \lambda_i(u_i, u_k) = 0 \tag{3.20}$$

since $u_k \in H_i$. Thus, if we also recall (3.11),

$$(\Delta u_k, \varphi) + \lambda_k(u_k, \varphi) = 0 \tag{3.21}$$

for all $\varphi \in H$ whence

$$\Delta u_k + \lambda_k u_k = 0. \tag{3.22}$$

Since, as noted in (3.15), we may require

$$(u_k, u_k) = 1 \tag{3.23}$$

for $k = 0, 1, ..., K$ and since the u_k are mutually orthogonal by construction, we have constructed an orthonormal basis of H consisting of eigenfunctions of Δ. Thus we may expand any function f on Γ as

$$f(i) = \sum_k (f, u_k) u_k(i). \tag{3.24}$$

We then also have

$$(f, f) = \sum_k (f, u_k)^2 \tag{3.25}$$

since the u_k satisfy

$$(u_j, u_k) = \delta_{jk}, \tag{3.26}$$

the condition for being an orthonormal basis. Finally, using (3.25) and (3.11), we obtain

$$(Df, Df) = \sum_k \lambda_k (f, u_k)^2. \tag{3.27}$$

We next derive **Courant's minimax principle**: Let P^k be the collection of all k-dimensional linear subspaces of H. We have

$$\lambda_k = \max_{L \in P^k} \min \left\{ \frac{(Du, Du)}{(u, u)} : u \neq 0, \ (u, v) = 0 \text{ for all } v \in L \right\} \tag{3.28}$$

and dually

$$\lambda_k = \min_{L \in P^{k+1}} \max \left\{ \frac{(Du, Du)}{(u, u)} : u \in L \backslash \{0\} \right\}. \tag{3.29}$$

To verify these relations, we recall (3.13)

$$\lambda_k = \min \left\{ \frac{(Du, Du)}{(u, u)} : \ u \neq 0, (u, u_j) = 0 \text{ for } j = 1, ..., k-1 \right\}. \tag{3.30}$$

Dually, we have

$$\lambda_k = \max \left\{ \frac{(Du, Du)}{(u, u)} : \ u \neq 0 \text{ linear combination of } u_j \text{ with } j \leq k \right\}. \tag{3.31}$$

The latter maximum is realized when u is a multiple of the kth eigenfunction, and so is the minimum in (3.30). If now L is any $k+1$-dimensional subspace, we may find some v in L that satisfies the k conditions

$$(v, u_j) = 0 \text{ for } j = 0, ..., k-1. \tag{3.32}$$

From (3.25) and (3.27), we then obtain

$$\frac{(Dv, Dv)}{(v, v)} = \frac{\sum_{j \geq k} \lambda_j (v, u_j)^2}{\sum_{j \geq k} (v, u_j)^2} \geq \lambda_k. \tag{3.33}$$

This implies

$$\max_{v \in L \backslash \{0\}} \frac{(Dv, Dv)}{(v, v)} \geq \lambda_k. \tag{3.34}$$

We then obtain (3.29). (3.28) follows in a dual manner.

For linking the function theory on Γ with the underlying topological structure of Γ, it is important to understand how the eigenvalues of Δ depend on the properties of Γ. Again, we start with some extreme cases that are easy to analyse. For a fully connected graph, we have

$$\lambda_1 = ... = \lambda_K = \frac{|\Gamma|}{|\Gamma| - 1} \tag{3.35}$$

since

$$\Delta v = -v \tag{3.36}$$

for any v that is orthogonal to the constants, that is

$$\frac{1}{|\Gamma|} \sum_{i \in \Gamma} n_i v(i) = 0. \tag{3.37}$$

We also recall that since Γ is connected, the trivial eigenvalue $\lambda_0 = 0$ is simple. If Γ had two components, then the next eigenvalue λ_1 would also become 0. A corresponding eigenfunction would be equal to a constant on each component, the two values chosen such that (3.37) is satisfied; in particular, one of the two would be positive, the other one negative. We therefore expect that for graphs with a pronounced community structure, that is, for ones that can be broken up into two large components by deleting only a few edges as discussed above, the eigenvalue λ_1 should be close to 0. Formally, this is easily seen from the variational characterization

$$\lambda_1 = \min\left\{ \frac{\sum_{j\sim i}\big(v(i) - v(j)\big)^2}{\sum_i n_i v(i)^2} : \sum_i n_i v(i) = 0 \right\}, \tag{3.38}$$

[see (3.13) and observe that $\sum_i n_i v(i) = 0$ is equivalent to $(v, u_0) = 0$ as the eigenfunction u_0 is constant]. Namely, if two large components of Γ are only connected by few edges, then one can make v constant on either side, with opposite signs so as to respect the normalization (3.37) with only a small contribution from the numerator.

The strategy for obtaining an eigenfunction for the first eigenvalue λ_1 is, according to (3.38), to do the same as one's neighbours. Because of the constraint $\sum_i n_i v(i) = 0$, this is not globally possible, however. The first eigenfunction thus exhibits oscillations with the lowest possible frequency.

By way of contrast, according to (3.29), the highest eigenvalue is given by

$$\lambda_K = \max_{u \neq 0} \frac{(Du, Du)}{(u, u)}. \tag{3.39}$$

Thus, the strategy for obtaining an eigenfunction for the highest eigenvalue is to do the opposite of what one's neighbours are doing, for example to assume the value 1 when the neighbours have the value -1. Thus, the corresponding eigenfunction will exhibit oscillations with the highest possible frequency. Here, the obstacle can be local. Namely, any triangle, that is, a triple of three mutually connected nodes, presents such an obstacle. More generally, any cycle of odd length makes an alternation of the values 1 and -1 impossible. The optimal situation here is represented by a bipartite graph, that is, a graph that consists of two sets Γ_+, Γ_- of nodes without any links between nodes in the same such subset. Thus, one can put $u_m = \pm 1$ on $\Gamma_\pm$. The highest eigenvalue λ_K becomes smallest on a fully connected graph, namely

$$\lambda_K = \frac{|\Gamma|}{|\Gamma| - 1} \tag{3.40}$$

according to (3.37). For graphs that are neither bipartite nor fully connected, this eigenvalue lies strictly between those two extremal possibilities.

Perhaps the following caricature can summarize the preceding: For minimizing λ_1—the minimal value being 0—one needs two subsets that can internally be arbitrarily connected, but that do not admit any connection between each other. For maximizing λ_K—the maximal value being 2—one needs two subsets without any internal connections, but allowing arbitrary connections between them. In either situation, the worst case—that is the one of a maximal value for λ_1 and a minimal value for λ_K—is represented by a fully connected graph. In fact, in that case, λ_1 and λ_K coincide.

We also have the following version of Courant's nodal domain theorem of Gladwell-Davies-Leydold-Stadler [15]:

Lemma 1. *Let u_k be an eigenfunction for the eigenvalue λ_k, with our above ordering, $0 = \lambda_0 < \lambda_1 \le \lambda_2 \le .... \le \lambda_K$. Delete from Γ all edges that connect points on which the values of u_k have opposite signs. This divides Γ into connected components $\Gamma_1, ..., \Gamma_l$. Then $l \le k+1$, where we need to order the eigenvalues appropriately when they are not simple.*

Systematic questions:

- What is the relationship between global properties or quantities (diameter, clustering coefficient, Cheeger constant, community structure,...), local quantities (pointwise clustering, upper and lower bounds for vertex degrees,...) and eigenvalues?
- How do operations on graphs affect eigenvalues, in particular (in the case of a connected graph) λ_1 as given in (3.38)? Let us consider some examples for the latter question:

  1. We add a new node, labeled j_0, with a single edge to some existing node i_0 in the graph Γ. This does not increase λ_1. This is seen as follows: Let $u = u_1$ be a first eigenfunction; thus, u realizes the infimum in (3.38) and satisfies the constraint $\sum_i n_i u(i) = 0$. However, on the new graph Γ' obtained by adding the node j_0, this constraint is no longer satisfied, even if we put $u(j_0) = 0$, because the degree of i_0 has been increased by 1 by the new edge from j_0 to i_0. We therefore construct a new function u' by putting $u'(i) = u(i) + \eta$ for all nodes $i \in \Gamma$ and $u'(j_0) = u'(i_0)$, for some constant η to be determined by the constraint, which now becomes $\sum_{j\in\Gamma'} n'_j u'(j) = 0$ where n'_j of course denotes the vertex degrees in Γ'. This becomes

$$\sum_{i\in\Gamma} n_i\big(u(i)+\eta\big) + 2\big(u(i_0)+\eta\big) \tag{3.41}$$

  where the last term arises from the increase in n_{i_0} by 1 and the presence of the additional node j_0. This yields $\eta = -\frac{2u(i_0)}{\sum n_i + 2}$. Since we are changing u by a global additive constant and since $u'(i_0) = u'(j_0)$, the numerator in (3.38) is the same for u' on Γ' as it was for u on Γ. We now compute the denominator as

$$\sum_{i\in\Gamma} n_i\big(u(i)+\eta\big)^2 + 2\big(u(i_0)+\eta\big)^2 > \sum_{i\in\Gamma} n_i u(i)^2$$

since $\sum_{i\in\Gamma} n_i\eta u(i) = 0$ by the constraint condition for u and the additional contributions are squares and therefore positive (unless $u(i_0) = 0$ in which case they vanish).

2. We add an edge between existing nodes. Here, however, the effect on λ_1 depends on the structure of Γ and where the edge is added. If Γ has two connected components, then $\lambda_1 = 0$ as noted above, and if we add an edge that connects those two components, then λ_1 becomes positive, and so, it increases. Also, we may find graphs that are almost disconnected and have small λ_1 and if we add more and more edges the graph eventually approaches a fully connected graph for which all eigenvalues except λ_0 are $=1$. Thus, again, during this process of adding edges, we expect λ_1 to increase. However, in certain situations, adding an edge may decrease λ_1 instead. Namely, if there exist nodes i_1, i_2 that are not linked, but for which the first eigenfunction satisfies $u_1(i_1) = u_1(i_2)$, then linking them decreases λ_1 by the same computation as in 1. The same then of course also holds when $u_1(i_1)$ and $u_1(i_2)$ are not quite equal, but their difference is sufficiently small.

3. We rewire the graph: We choose nodes i_1, i_2, j_1, j_2 with $i_1 \sim i_2$ and $j_1 \sim j_2$, but with no further neighbouring relation, that is neither i_1 nor i_2 is a neighbour of j_1 or j_2. We then delete the two edges, that is, the one between i_1 and i_2 and the one between j_1 and j_2 and insert new edges between i_1 and j_1 and between i_2 and j_2. Obviously, since this is a reversible process, in general this will not have a systematic effect on λ_1. However, we may try to find such pairs i_1, i_2 and j_1, j_2 to influence certain characteristic properties of the graph like its clustering through systematic such rewirings, and we may then study the influence on λ_1 as well.

We now derive elementary estimates for λ_1 from above and below in terms of the constant $h(\Gamma)$ introduced in (3.1). Our reference here is [8] (that monograph also contains many other spectral estimates for graphs, as well as the original references). We start with the estimate from above and use the variational characterization (3.38). Let the edge set E divide the graph into the two disjoint sets V_1, V_2 of nodes, and let V_1 be the one with the smaller vertex sum $\sum n_i$. We consider a function v that is $=1$ on all the nodes in V_1 and $= -\alpha$ for some positive α on V_2. α is chosen so that the normalization $\sum_\Gamma n_i v(i) = 0$ holds, that is, $\sum_{i\in V_1} n_i - \sum_{i\in V_2} n_i\alpha = 0$. Since V_2 is the subset with the larger $\sum n_i$, we have $\alpha \le 1$. Thus, for our choice of v, the quotient in

(3.38) becomes $\leq \frac{(1+\alpha)^2|E|}{\sum_{i\in V_1} n_i + \sum_{i\in V_2} n_i\alpha^2} = \frac{(\alpha+1)|E|}{\sum_{V_1} n_i} \leq 2\frac{|E|}{\sum_{V_1} n_i}$. Since this holds for all such splittings of our graph Γ, we obtain from (3.1) and (3.38)

$$\lambda_1 \leq 2h(\Gamma). \tag{3.42}$$

The estimate from below is slightly more subtle. We consider the first eigenfunction u_1. Like all functions on our graph, we consider it to be defined on the nodes. We then interpolate it linearly on the edges of Γ. Since u_1 is orthogonal to the constants (recall $\sum_i n_i u(i) = 0$), it has to change sign, and the zero set of our extension then divides Γ into two parts Γ' and Γ''. Without loss of generality (W.l.o.g.), Γ' is the part with fewer nodes. The points where (the extension of) $u_1 = 0$ are called boundary points. We now consider any function φ that is linear on the edges, 0 on the boundary, and positive elsewhere on the nodes and edges of Γ'. We also put $h'(\Gamma') := \inf\{\frac{|E|}{\sum_{i\in\Omega} n_i}\}$ where removing the edges in E cuts out a subset Ω that is disjoint from the boundary. We then have

$$\begin{aligned}
\sum_{i\sim j} |\varphi(i) - \varphi(j)| &= \int_\sigma \sharp_e(\varphi = \sigma) d\sigma \\
&= \int_\sigma \frac{\sharp_e(\varphi = \sigma)}{\sum_{i:\varphi(i)\geq\sigma} n_i} \sum_{i:\varphi(i)\geq\sigma} n_i \, d\sigma \\
&\geq \inf_\sigma \frac{\sharp_e(\varphi = \sigma)}{\sum_{i:\varphi(i)\geq\sigma} n_i} \int_s \sum_{i:\varphi(i)\geq s} n_i \, ds \\
&= \inf_\sigma \frac{\sharp_e(\varphi = \sigma)}{\sum_{i:\varphi(i)\geq\sigma} n_i} \sum_i n_i |\varphi(i)| \\
&\geq h'(\Gamma') \sum_i n_i |\varphi(i)|
\end{aligned}$$

when the sets $\varphi = \sigma$ and $\varphi \geq \sigma$ satisfy the conditions in the definition of $h'(\Gamma)$; that is, the infimum has to be taken over those $\sigma < \max \varphi$. Here, $\sharp_e(\varphi = \sigma)$ denotes the number of edges on which φ attains the value σ. Applying this to $\varphi = v^2$ for some function v on Γ' that vanishes on the boundary, we obtain

$$\begin{aligned}
h(\Gamma') \sum_i n_i |v(i)|^2 &\leq \sum_{i\sim j} |v(i)^2 - v(j)^2| \\
&\leq \sum_{i\sim j} (|v(i)| + |v(j)|)|v(i) - v(j)| \\
&\leq 2\Big(\sum_i n_i |v(i)|^2\Big)^{1/2} \Big(\sum_{i\sim j} |v(i) - v(j)|^2\Big)^{1/2}
\end{aligned}$$

from which

$$\frac{1}{4}h(\Gamma')^2 \sum_i n_i |v(i)|^2 \leq \sum_{i \sim j} |v(i) - v(j)|^2. \tag{3.43}$$

We now apply this to $v = u_1$, the first eigenfunction of our graph Γ. We have $h'(\Gamma') \geq h(\Gamma)$, since Γ' is the component with fewer nodes. We also have

$$\lambda_1 \sum_{i \in \Gamma'} n_i u_1(i)^2 = \frac{1}{2} \sum_{i \in \Gamma'} \sum_{j \sim i} \left(u_1(i) - u_1(j)\right)^2, ^5 \tag{3.44}$$

cf. (3.16) (this relation holds on both Γ' and Γ'' because u_1 vanishes on their common boundary).[6] (3.43) and (3.44) yield the desired estimate

$$\lambda_1 \geq \frac{1}{2} h(\Gamma)^2. \tag{3.45}$$

From (3.42) and (3.45), we also observe the inequality

$$h(\Gamma) \leq 4 \tag{3.46}$$

for any connected graph.

3.4 Other Graph Parameters

One set of useful parameters that encode important qualitative properties comes from the metric on the graph generated by assigning every edge the length 1, that is, letting neighbours in the graph have distance 1. The diameter of the graph (assumed to be connected, as always) then is the maximal distance between any two of its nodes. Most graphs of N nodes have a diameter of order $\log N$. More precisely, there exists a constant c with the property that the fraction of all graphs with N nodes having diameter exceeding $c \log N$ tends to 0 for $N \to \infty$. Of course, a fully connected graph has diameter 1. However, one can realize a small diameter already with much fewer edges; namely, one selects one central node to which every other node is connected. In that manner, one obtains a graph of N nodes with $N - 1$ edges and diameter 2. Of course, the central node then has a very large degree, namely $N - 1$. It is a big hub. Similarly, one can construct graphs with a few hubs, so that none of them has to be quite that big, efficiently distributed so that the

[5] We obtain the factor $\frac{1}{2}$ because we are now summing over vertices so that each edge gets counted twice.

[6] To see this, one adds nodes at the points where the edges have been cut, and extends functions by 0 on those nodes. These extended functions then satisfy the analogue of (3.11) on either part, as one sees by looking at the derivation of that relation and using the fact that the functions under consideration vanish at those new "boundary" nodes.

diameter is still rather small. Such graphs can be realized as so-called scale free graphs to be discussed below. Another useful quantity is the average distance between nodes in the graph. The property of having a small diameter or average distance has been called the small-world effect.

A rather different quantity that was already described in the introduction is the clustering coefficient that measures how many connections there exist between the neighbours of a node. Formally, it is defined as

$$C = \frac{3 \times \text{ number of triangles}}{\text{number of connected triples of nodes}}. \tag{3.47}$$

The normalization is that C becomes one for a fully connected graph. It vanishes for trees and other bipartite graphs.

A triangle is a cycle of length 3. One may then also count the number of cycles of length k, for integers > 3. A different generalization consists in considering complete subgraphs of order k. For example, for $k = 4$, we would have a subset of four nodes that are all mutually connected. One may then associate a simplicial complex to our graph by assigning a k-simplex to every such complete subgraph, with obvious incidence relations. This is the basis of topological combinatorics, enabling one to apply tools from simplicial topology to graph theory.

3.5 Generalized Random Graphs

A generalized random graph is characterized by its number N of nodes or vertices and real numbers $0 \leq p_{ij} \leq 1$ (with the symmetry $p_{ij} = p_{ji}$) that assign to each pair i, j of nodes the probability for finding an edge between them. Self-connections of the node i are excluded when $p_{ii} = 0$. The expected degree of i then is

$$\nu_i = \sum_j p_{ij}. \tag{3.48}$$

This construction generalizes the random graphs introduced by Erdös and Rényi [13]; their important idea was not to specify a graph explicitly, but rather only its generic type by selecting edges between nodes randomly. In their construction, for any pair of nodes, there was a uniform probability p for an edge between them. If the network has N nodes, then, if we do not allow self-links, each node has $N - 1$ possible recipients for an edge, while if self-links are permitted, there are N of them. Thus, the average degree of a node is

$$z := (N-1)p \text{ or } Np, \tag{3.49}$$

and this difference of course becomes insignificant for large N. Moreover, the probability that a given node has degree k in an Erdös-Rényi graph is

$$p_k = \binom{N-1}{k} p^k (1-p)^{N-1-k} \tag{3.50}$$

because the degree happens to be k when precisely p out of the $N-1$ possible edges from the given node are chosen, and each of them is chosen with probability p and not chosen with probability $1-p$. Thus, the degree distribution is binomial, and for $N \gg kz$, this is approximated by the Poisson distribution

$$p_k = \frac{z^k e^{-z}}{k!} \tag{3.51}$$

(and so $z = \langle k \rangle = \sum_k kp_k$).

For an Erdös-Rényi graph, one can also compute the distribution of the number of second neighbours of a given node, that is, the number of neighbours of its neighbours, discarding of course the original node itself as well as all its direct neighbours that also happen to be connected with another neighbour. However, since there is no tendency to clustering in the construction, the probability that a second neighbour is also a first neighbour behaves like $1/N$ and so becomes negligible for large N. Now, however, the degree distribution of first-order neighbours of some node is different from the degree distribution of all the nodes in the random graph, because the probability that an edge leads to a particular node is proportional to that node's degree so that a node of degree k has a k-fold increased chance of receiving an edge. Therefore, the probability distribution of our first neighbours is proportional to kp_k, that is, given by $\frac{kp_k}{\sum_l lp_l}$, instead of p_k, the one for all the nodes in the graph. Such a first neighbour of degree k has $k-1$ edges leading away from the original node. Therefore, when we shift the index by 1, the distribution for having k second neighbours via one particular neighbour is then given by

$$q_k = \frac{(k+1)p_{k+1}}{\sum_l lp_l}. \tag{3.52}$$

Thus, to obtain the number of second neighbours, we need to sum over the first neighbours, since, as argued, we can neglect clustering in this model. So, the mean number of second neighbours is obtained by multiplying the expected number of second neighbours via a particular first neighbour, that is, $\sum kq_k$, by the expected number of first neighbours, $z = \sum kp_k$. So, we obtain for that number

$$\sum_l lp_l \sum_k kq_k = \sum_{k=0}^{\infty} k(k+1)p_{k+1} = \sum_{k=0}^{\infty} (k-1)kp_k = \langle k^2 \rangle - \langle k \rangle. \tag{3.53}$$

Following the exposition in [24], such probability distributions can be encoded in probability generating functions. If we have a probability distribution p_k as above on the nonnegative integers, we have the generating function defined as

$$G_p(x) = \sum_{k=0}^{\infty} p_k x^k. \tag{3.54}$$

Likewise, the above distribution for the number of second neighbours then is encoded by

$$G_q(x) = \sum_{k=0}^{\infty} q_k x^k = \frac{\sum_k (k+1) p_{k+1}}{\sum_l l p_l} = \frac{G'_p(x)}{z}. \tag{3.55}$$

When we insert the Poisson distribution (3.51), we obtain

$$G_p(x) = e^{-z} \sum_{k=0}^{\infty} \frac{z^k}{k!} x^k = e^{z(x-1)} \tag{3.56}$$

and from (3.55) then also

$$G_q(x) = e^{z(x-1)} \tag{3.57}$$

Thus, for an Erdös-Rényi graph, the two generating functions agree. This is quite useful for deriving analytical results.

To also include scale-free graphs, [5] introduced the following generalization of this procedure. One starts with an N-tuple $\nu = (\nu_1, ..., \nu_N)$ of positive numbers satisfying

$$\max_i \nu_i^2 \leq \sum_j \nu_j; \tag{3.58}$$

when the ν_i are positive integers, this is necessary and sufficient for the existence of a graph with nodes i of degree ν_i, $i = 1, ..., N$. When putting $\gamma := \frac{1}{\sum_i \nu_i}$ and $p_{ij} := \gamma \nu_i \nu_j$, then $0 \leq p_{ij} \leq 1$ for all i, j. We then insert an edge between the nodes i and j with probability p_{ij} to construct the (generalized) random graph Γ. By (3.48), the expected degree of node i in such a graph is ν_i. When all the ν_i are equal, we obtain an Erdös-Rényi graph. For other types, the number of nodes i with $\nu_i = k$ will decay as a function of k, at least for large k, for example exponentially. When that number behaves like a power $k^{-\beta}$ instead, we obtain a so-called scale free graph.

Scale-free graphs were apparently first considered by H. Simon [29]; they have been popularized more recently by Barabasi and Albert [4]. Their construction can be simply described: Prescribe some positive integer $m \geq 2$ and start with some (small) connected graph. Add each iteration step, add a new node that can make connections to m nodes already existing in the network. The probability for each of those nodes to be a recipient of a connection from the new node is proportional to the number of connections it already has. By this scheme, those nodes that already possess many connections are favoured to become recipients of further connections, and in this manner, many hubs are generated in the network, and in the final network the number of nodes with degree k decays like a power of k, instead of exponentially as is the case for random graphs. It should be pointed out that the construction just described is not the only one that can generate graphs with such a power law behaviour. For example, Kleinberg et al [19] introduced a copying model that in contrast to the preceding, but like the subsequently mentioned models needs only local information for deciding the connections of new nodes. In [30] and [26], a local exploration model is introduced. Jost and Joy [18] constructed graphs by the make-friends-with-the-friends-of-your-friends principle. According to this principle, new nodes added to the network that are again allowed to make a specified number m of connections make their first connection at random and then preferentially make connections with neighbours of those nodes they are already connected with. Since the probability of a node i being a neighbour of another node the new node is already connected to is proportional to the degree of i, this principle again favours highly connected nodes and leads to a power law degree sequence. With respect to important graph parameters, however, the graphs so constructed are rather different from the ones produced by the Barabasi-Albert scheme. For example, their diameter tends to be much larger, and their first eigenvalue significantly smaller than the corresponding quantities for the latter. For a more systematic analysis of the spectral properties, see [1]. See also the discussion below. More generally, [18] discuss attachment rules where the preference for a node to become a recipient of a new connection depends on the distance to the node forming those connections.

One can also consider networks not only where new nodes are added, but also where rewiring takes place. For example, Klemm and Eguíluz [20, 21] consider a growing network model based on the scale-free paradigm, with the distinctive feature that older nodes become inactive at the same rate that new ones are introduced. This is interpreted as a finite memory effect, in the sense that older contributions tend to be forgotten when they are not frequently enough employed. This results in networks that are even more highly clustered than regular ones. Davidsen et al [11] consider a network that rewires itself through triangle formation. Nodes together with all their links are randomly removed and replaced by new ones with one random link. The resulting network again is highly clustered, has small average distance, and can be tuned toward a scale-free behaviour.

In the construction of [10], the probability to find a connection between i and j depends only on a property intrinsic to i and j, namely their expected degrees. In the more general construction we would like to propose here, that probability rather encodes a relationship between i and j that may well be special to them. This seems to capture the essential aspect underlying most constructions of graphs in specific applications where they are supposed to represent a particular structure of relations between individual elements. Each element can form relations with selective other elements, and whether another element is chosen need not only depend on a property intrinsic to the latter, but also on some affinity, similarity, or dissimilarity to the former. In this regard, see also the Cameo principle of Blanchard and Krüger [6].

The appropriate approach to analysing properties of random graphs of course is not to consider an individual such graph, but rather to derive properties that hold for all or almost all such graphs for a given collection p_{ij} or class of such collections with specified properties, at least perhaps asymptotically for $N \to \infty$. This applies in particular to the eigenvalues of the graph Laplacian. Such properties, in general, will not only depend on the N-tuple ν above. Two graphs with the same ν can have rather different geometric properties, as for example encoded by the clustering coefficient or the average distance between nodes, and also their spectrum can be rather different. For instance, we can apply the rewiring rule 3 from Section 3.3 to change some of those quantities in a systematic manner without affecting the degrees of nodes. In particular, since we can affect the constant $h(\Gamma)$ in that manner, for instance making it arbitrarily small, we cannot expect nontrivial estimates for the first eigenvalue λ_1 that hold uniformly for all graphs with given ν. This is systematically investigated in [1].

3.6 Interactions

Above, we have considered the graph Laplacian Δ. It can be considered as the prototype of an interaction operator. We consider a function f representing the dynamics of the individual elements. This means that an element i in isolation obeys the dynamical rule

$$u(i, n+1) = f(u(i,n)); \tag{3.59}$$

here, $n \in \mathbb{N}$ stands for the discrete time. $u(i,n)$ is the state of element i at time n. Prototypes of such reaction functions f are

$$f_1(x) = \rho x(1-x) \tag{3.60}$$

with $0 < \rho \leq 4$, which, for sufficiently large ρ generates chaotic dynamics, see for example [16] for details (in our investigations, we have mostly considered the value $\rho = 4$), and

$$f_2(x) = \frac{1}{1+e^{-\kappa(x-\theta)}}, \text{ with } \kappa > 0, \tag{3.61}$$

the so-called sigmoid function in neural networks, which in contrast to f_1 is monotonically increasing and leads to a regular, nonchaotic behaviour, and which has one or three fixed points (depending on κ, θ).

When considering networks of interacting elements, we need to add an interaction term to (3.59):

$$u(i, n+1) = f\big(u(i,n)\big) + \epsilon\mu_i\Big(\sum_j c_{ji} g\big(u(j,n)\big) - \sum_k c_{ik} g\big(u(i,n)\big)\Big) \tag{3.62}$$

for a network with nodes i, with multipliers μ_i, coupling strengths c_{jk} from j to k, and global interaction strength ϵ, for some function g. For simplicity of presentation, we restrict ourselves here to the case $g = f$, even though the general case can be treated by the same type of analysis. Thus, we shall consider

$$u(i,n+1) = f\big(u(i,n)\big) + \epsilon\mu_i\Big(\sum_j c_{ji} f\big(u(j,n)\big) - \sum_k c_{ik} f\big(u(i,n)\big)\Big). \tag{3.63}$$

A question that we shall discuss is under which conditions the solution u synchronizes, that is when $\lim_{n\to\infty} |u(i,n) - u(j,n)| = 0$ for all vertices i, j (in computer simulations, when synchronization occurs, it can be seen already at finite values of n, that is, $u(i,n) = u(j,n)$ for all n larger than some $N \in \mathbb{N}$). The fundamental reference on this topic that we shall also use in the sequel is [27].

When we put $\mu_i = \frac{1}{n_i}$ and $c_{jk} = 1$ when j and k are neighbours and 0 otherwise, the interaction operator becomes our graph Laplacian Δ studied above. In fact, that operator serves as a prototype for the interaction, and we now derive conditions that allows us to generalize the key features of Δ to some larger class of interaction operators. There are three types of conditions that we shall now discuss in turn:

1. A balancing condition

$$\sum_j c_{ji} = \sum_k c_{ik} \quad \text{for all } i. \tag{3.64}$$

This condition states that what comes in at a node is balanced by what is flowing out of that node.

For simplicity, we also assume

$$c_{ii} = 0 \quad \text{for all } i. \tag{3.65}$$

This condition is not essential for the formal analysis, but notationally convenient in the sequel. It excludes a self-interaction term from the interaction operator, and this is easily justified by declaring that any self-interaction is already contained in the reaction term, the first term on the right-hand side of (3.62).

The balancing condition can be rewritten as a zero-row-sum condition when we define an operator $L = (l_{xy})$ by putting

$$l_{ik} := \mu_i c_{ki} \quad \text{for } i \neq k \tag{3.66}$$

(note the reversal of the indices), and

$$l_{ii} := -\mu_i \sum_j c_{ji} = -\mu_i \sum_k c_{ik}. \tag{3.67}$$

The balancing condition (3.64) then is equivalent to

$$\sum_j l_{ij} = 0 \quad \text{for all } i. \tag{3.68}$$

In terms of the operator L, our dynamics (3.63) becomes

$$u(i,n+1) = f\big(u(i,n)\big) + \epsilon \sum_j l_{ij} f\big(u(j,n)\big) \tag{3.69}$$

or, more abstractly,

$$u(.,n+1) = f\big(u(.,n)\big) + \epsilon L f\big(u(.,n)\big). \tag{3.70}$$

The important point of the balancing condition (3.64) or the equivalent zero-row-sum condition (3.68) is that the synchronized solutions, that is,

$$u(i,n) = u(j,n) \quad \text{for all } i,j,n, \tag{3.71}$$

satisfying the individual equation (3.59) are also solutions of the collective equation (3.63) (or, equivalently, (3.69), (3.70)).

2. Another useful condition is a nonnegativity condition: Let the coupling matrix $C = (c_{xy})$ have **nonnegative entries**, that is,

$$c_{ij} \geq 0 \quad \text{for all } i,j. \tag{3.72}$$

We shall **not** assume this condition in most of the sequel, but we discuss it here because it allows us to derive some kind of asymptotic stability condition for the globally synchronized solution. We consider the adjoint operator L^* with coefficients

$$l^*_{ik} = \frac{\mu_i}{\mu_k} l_{ki} \ (= \mu_i c_{ik} \ \text{ for } i \neq k), \tag{3.73}$$

assuming, of course, that the multipliers $\mu_i \neq 0$. We then have, from (3.68),

$$\sum_k l^*_{ik} = 0. \tag{3.74}$$

The crucial quantity that will be seen as a Lyapunov function then is

$$\begin{aligned} & -\sum_i \frac{1}{\mu_i} L^* u(i,n+1) u(i,n+1) \\ = & -\sum_i \frac{1}{\mu_i} L^* \big((1+\epsilon L) f\big(u(i,n)\big)\big)(1+\epsilon L) f\big(u(i,n)\big). \end{aligned} \tag{3.75}$$

We have

$$-\sum_i \Big(\sum_k \frac{1}{\mu_i} l^*_{ik}\big((1+\epsilon L)f\big(u(k,n)\big)\big)\Big)(1+\epsilon L)f\big(u(i,n)\big)$$
$$= \frac{1}{2}\sum_{i \neq k} \frac{1}{\mu_i} l^*_{ik}(1+\epsilon L)\big(f\big(u(i,n)\big) - f\big(u(k,n)\big)\big)$$
$$(1+\epsilon L)\big(f\big(u(i,n)\big) - f\big(u(k,n)\big)\big)$$
$$\leq -\sup|f'|^2\|1+\epsilon L\|^2 \sum_i \frac{1}{\mu_i} L^* u(i,n)u(i,n) \tag{3.76}$$

by (3.74) and since the nondiagonal entries of L are nonnegative. Here, the norm of the operator $1+\epsilon L$ is evaluated on those functions that are orthogonal to the constants as the difference in the preceding formula vanishes on the constants. Therefore, when this norm is smaller than $|f'|^{-1}$, that is, when we have the global stability condition,

$$\sup|f'|^2\|1+\epsilon L\|^2 < 1, \tag{3.77}$$

$$-\sum_i \frac{1}{\mu_i} L^* u(i,n)u(i,n) \tag{3.78}$$

is exponentially decreasing as a function of n under our dynamical iteration (3.63). Thus, for $n \to \infty$, it goes to 0. Thus, the limit $u(i) := \lim_{n\to\infty} u(i,n)$ (or, more precisely, the limit of any subsequence) satisfies

$$\sum_i \Big(\sum_j c_{ji}u(j) - \sum_j c_{ik}u(i)\Big)u(i) = 0, \tag{3.79}$$

which, by (3.72), (3.64), and the Schwarz inequality, implies that $u(i)$ is constant, that is, independent of i. This means that under dynamical iteration (3.63) synchronizes as $n \to \infty$.
In the symmetric case to be discussed below, such an estimate had been derived in [17]. In the general case, similar conditions have been obtained by Wu [32] and Lu and Chen [23]. For recent work in this direction, see [22].

3. A symmetry condition on operator L will guarantee that its spectrum is real, and then, when also the nonnegativity condition (3.72) will be assumed, that it is nonnegative. To obtain such a condition, we need a scalar product on the space of functions on Γ, to make it into an L^2-space. Such a product is given by

$$(u,v) := \sum_{i,j} a_{ij}u(i)v(j). \tag{3.80}$$

Here, $A := (a_{ij})$ should be positive definite and symmetric:

$$a_{ij} = a_{ji} \quad \text{for all } i,j. \tag{3.81}$$

L is symmetric with respect to this product if

$$(Lu, v) = (u, Lv) \quad \text{for all } u, v, \tag{3.82}$$

that is, if

$$\sum_{i,j,k} a_{ij} l_{ik} u(k) v(j) = \sum_{i,j,k} a_{ij} u(i) l_{jk} v(k). \tag{3.83}$$

This requires

$$a_{ij} l_{ik} = a_{ik} l_{ij}, \tag{3.84}$$

using the symmetry (3.81). (3.84) is the condition yielding the symmetry of L, that is, we can infer that L has a real spectrum when we can find a symmetric matrix $A := (a_{ij})$ satisfying (3.84).

If we want to have a scalar product that is more intimately tied to the graph structure, we should require that it admits an orthonormal basis of functions of the form u_i with $u_i(j) = 0$ for $j \neq i$. In that case, the scalar product has to be of the form already given in (3.2)

$$(u, v) = \sum_i n_i u(i) v(i) \tag{3.85}$$

where, however, the n_i can be arbitrary positive numbers, not necessarily the vertex degrees. In that situation, the symmetry condition for L becomes

$$n_i l_{ik} = n_k l_{ki} \quad \text{for all } i, k. \tag{3.86}$$

When we have

$$n_i = \frac{1}{\mu_i} \tag{3.87}$$

for the above multipliers μ_i, this simply becomes the symmetry of the interaction matrix,

$$c_{ik} = c_{ki} \quad \text{for all } i, k, \tag{3.88}$$

see (3.66). In that case, that is, when L is self-adjoint with respect to the product (3.85) with (3.87), it coincides with the operator L^* defined in (3.73), as the latter is the adjoint of L with respect to this product.

3.7 Local Stability of the Synchronized Solution

In the preceding section, we have derived a global stability condition, (3.77). We now discuss the local stability of a synchronized solution with the help of the eigenvalue expansion obtained in Section 3.3, following the presentation in [17]. A similar analysis has been carried out and related to the Gershgorin disk theorem in [7, 28]. We shall only consider the case of the graph Laplacian as our interaction operator, although the subsequent analysis readily extends to other symmetric operators that are negative definite on the subspace orthogonal to the constant functions.

We consider a solution $\bar{u}(n)$ of the uncoupled equation (3.59),

$$\bar{u}(n+1) = f\big(\bar{u}(n)\big). \tag{3.89}$$

$u(i,n) = \bar{u}(n)$ then is a synchronized solution of the coupled equation

$$u(i,n+1) = f\big(u(i,n)\big) + \frac{\epsilon}{n_i} \sum_{j,j\sim i} \big(f\big(u(j,n)\big) - f\big(u(i,n)\big)\big). \tag{3.90}$$

The local stability question then is whether a perturbation

$$u(i,n) = \bar{u}(n) + \delta\alpha_k(n)u_k(i) \tag{3.91}$$

by an eigenmode u_k for some $k \geq 1$, and small enough δ, $\alpha_k(n)$ goes to 0 for $n \to \infty$, if $u(i,n)$ solves (3.90). We now perform a linear stability analysis. We insert (3.3) into (3.2) and expand about $\delta = 0$. This yields

$$\alpha_k(n+1) = (1-\epsilon\lambda_k)f'\big(\bar{u}(n)\big)\alpha_k(n). \tag{3.92}$$

The sufficient local stability condition[7]

$$\lim_{N\to\infty} \frac{1}{N} \log \frac{\alpha_k(N)}{\alpha_k(0)} = \lim_{N\to\infty} \frac{1}{N} \log \prod_{n=0}^{N-1} \frac{\alpha_k(n+1)}{\alpha_k(n)} < 0 \tag{3.93}$$

therefore becomes

$$\log|1-\epsilon\lambda_k| + \lim_{N\to\infty} \frac{1}{N} \sum_{n=0}^{N-1} \log\big|f'\big(\bar{u}(n)\big)\big| < 0. \tag{3.94}$$

Here,

$$\mu_0 = \lim_{N\to\infty} \frac{1}{N} \sum_{n=0}^{N-1} \log\big|f'\big(\bar{u}(n)\big)\big|$$

is the Lyapunov exponent of f. Therefore, the stability condition (3.94) is

$$|e^{\mu_0}(1-\epsilon\lambda_K)| < 1. \tag{3.95}$$

The fundamental observation is that we may find synchronization, that is, may have (3.95) for all $k \geq 1$, even in the presence of temporal instability, that is,

$$\mu_0 > 0. \tag{3.96}$$

In that case, the individual, and therefore also the synchronized global dynamics, may exhibit chaotic behaviour. Synchronization then means that the

[7] Stability is understood here in the sense of Milnor. See [27] for a detailed analysis.

chaotic dynamics of the individual nodes are identical. We shall now investigate this issue further. By our ordering convention for the eigenvalues, (3.95) holds for all $k \geq 1$ if

$$\frac{1-e^{-\mu_0}}{\lambda_1} < \epsilon < \frac{1+e^{-\mu_0}}{\lambda_K}. \tag{3.97}$$

For that condition, we need

$$\frac{\lambda_K}{\lambda_1} < \frac{e^{\mu_0}+1}{e^{\mu_0}-1}. \tag{3.98}$$

When (3.98) holds, from the local stability analysis, we expect the following behaviour of the coupled system as ϵ increases. For very small values of ϵ, the solution is not synchronized because the positive Lyapunov exponent of the individual dynamics is not compensated by a sufficiently strong diffusive interaction. As ϵ increases, first the highest modes become stabilized until at some critical value of ϵ, all spatially inhomogeneous modes become stable directions, and global synchronization sets in. When ϵ is further increased, first the highest mode becomes unstable, and then others follow, and the solution gets desynchronized again. The important point here is that a unidirectional variation of the coupling strength parameter leads from a desynchronized to a synchronized state and then back to a—different—desynchronized state.

We now describe this process in some more detail, following again [17]. For very small values of $\epsilon > 0$, as we assume (3.97)

$$e^{\mu_0}(1-\epsilon\lambda_k) > 1,$$

and so, all spatial modes $u_k, k \geq 1$, are unstable, and no synchronization occurs. Unless we have a global all-to-all coupling (see Section 3.3) λ_1 is smaller than λ_K. Let ϵ_k be the solution of

$$e^{\mu_0}(1-\epsilon_k\lambda_k) = 1.$$

The smallest among these values is ϵ_K, the largest ϵ_1. If then, for $k_1 < k_2$,

$$\epsilon_{k_2} < \epsilon < \epsilon_{k_1}$$

the modes $u_{k_2}, u_{k_2+1}, ..., u_K$ are stable, but the modes $u_1, u_2, ..., u_{k_1}$ are unstable. Recalling Lemma 1, we see that desynchronization can lead to at most k_2+1 subdomains on which the dynamics are either advanced or retarded.

In particular, if ϵ increases, first the highest modes, that is, the ones with the most spatial oscillations, become stabilized, and the mode u_1 becomes stabilized the last. So if $\epsilon_2 < \epsilon < \epsilon_1$, then any desynchronized state consists of two subdomains.

We then let $\bar{\epsilon_k}$ be the solution of

$$e^{\mu_0}(\bar{\epsilon_k}\lambda_k - 1) = 1.$$

Again,

$$\bar{\epsilon_k} \leq \bar{\epsilon}_{k-1}.$$

Because of (3.97),

$$\epsilon_1 < \bar{\epsilon}_K.$$

If

$$\epsilon_1 < \epsilon < \bar{\epsilon}_K,$$

then all modes $u_k, k = 1, 2, ..., K$, are stable, and the dynamics synchronizes.

If ϵ increases beyond $\bar{\epsilon}_K$, then the highest frequency mode u_K becomes unstable and we predict spatial oscillations of high frequency of a solution of the dynamics. If ϵ increases further, then more and more spatial modes become destabilized.

References

1. Atay F, Biyikoglu T, Jost J (2006) On the synchronization of networks with prescribed degree distributions. IEEE Transactions on Circuits and Systems I-Regular Papers Vol 53 (1):92–98.
2. Atay F, Jost J (2004) On the emergence of complex systems on the basis of the coordination of complex behaviors of their elements: synchronization and complexity. Complexity 10:17–22.
3. Atay F, Jost J, Wende A (2004) Delays, connection topology, and synchronization of coupled chaotic maps. Phys Rev Lett 92:144101–144104.
4. Barabasi A-L, Albert R (1999) Emergence of scaling in random networks. Science 286:509–512.
5. Bender E, Canfield E (1978) The asymptotic number of labeled graphs with given degree sequences. J Comb Th A 24:296–307.
6. Blanchard P, Krüger T (2004) The "Cameo" principle and the origin of scale-free graphs in social networks. J Stat Phys 114:1399–1416.
7. Chen YH, Rangarajan G, Ding MZ (2003) General stability analysis of synchronized dynamics in coupled systems. Phys Rev E 67:26209–26212.
8. Chung F (1997) Spectral graph theory. Regional Conference Series in Mathematics 92, Amer Math Soc, Providence.
9. Leader I (1991) Discrete isoperimetric inequalities. In: Bollobás B (ed) Probabilistic Combinatorics and Its Applications. AMS.
10. Chung F, Lu L, Vu V (2003) Spectra of random graphs with given expected degrees. PNAS 100 (11):6313–6318.
11. Davidsen J, Ebel H, Bornholdt S (2002) Emergence of a small world from local interaction: modeling acquaintance networks. Phys Rev Lett 88:128701.
12. Derényi I, Palla G, Vicsek T (2005) Clique percolation in random networks. Phys Rev Lett 94:160202.
13. Erdös P, Rényi A (1959) On random graphs I. Publ Math Debrecen 6:290–291.
14. Girvan M, Newman M (2002) Community structure in social and biological networks. Proc Nat Acad Sci 99:7021–7026.

15. Gladwell G, Davies EB, Leydold J, Stadler PF (2001) Discrete nodal domain theorems. Lin Alg Appl 336:51–60.
16. Jost J (2005) Dynamical Systems. Springer.
17. Jost J, Joy MP (2001) Spectral properties and synchronization in coupled map lattices. Phys Rev E 65: 16201–16209.
18. Jost J, Joy MP (2002) Evolving networks with distance preferences. Phys Rev E 66:36126–36132.
19. Kleinberg J (1999) The web as a graph: measurements, methods, and models. In: Lecture Notes in Computer Science. Springer, New York.
20. Klemm K, Eguíluz VM (2002) Growing scale-free networks with small-world behavior. Phys Rev E 65:057102.
21. Klemm K, Eguíluz VM (2002) Highly clustered scale-free networks. Phys Rev E 65:036123.
22. Lu WL (2005) Chaos synchronization in coupled map networks via time varying topology. Preprint MPIMIS.
23. Lu WL, Chen TP (2004) Synchronization analysis of linearly coupled map networks with discrete time systems. Physica D 198:148–168.
24. Newman M (2002) Random graphs as models of networks. In: Bornholdt S, Schuster HG (eds) Handbook of Graphs and Networks. Wiley-VCH, Berlin.
25. Newman M (2003) The structure and function of complex networks. SIAM Rev 45 (2):167–256.
26. Pastor-Satorras R, Vazquez A, Vespignani A (2001) Dynamical and correlation properties of the internet. Phys Rev Lett 87:258701-4.
27. Pikovsky A, Rosenblum M, Kurths J (2003) Synchronization: A Universal Concept in Nonlinear Sciences. Cambridge University Press, Cambridge.
28. Rangarajan G, Ding MZ (2002) Stability of synchronized chaos in coupled dynamical systems. Phys Lett A 296:204–212.
29. Simon H (1955) On a class of skew distribution functions. Biometrika 42:425–440.
30. Vázquez A (2002) Growing networks with local rules: preferential attachment, clustering hierarchy and degree correlations. Phys Rev E 67:056104.
31. Watts D, Strogatz S (1998) Collective dynamics of 'small-world' networks. Nature 393:440–442.
32. Wu CW (2005) Synchronization in networks of nonlinear dynamical systems coupled via a directed graph. Nonlinearity 18:1057–1064.

Part II

Applications in Neuroscience

4

Neuronal Computation Using High-Order Statistics

Jianfeng Feng

Summary. The neuron is a stochastic unit: it receives and sends out random spikes. Usually the second-order statistic is thought of as noise: useless and harmful. Here we first review some results in the literature and point out that in fact the second-order statistic plays an important role in computation. Both positive and negative correlations are observed in *in vivo* recorded data. Positive correlation (~synchronization) has been extensively discussed in the literature. We mainly focus on some functional roles of negative correlations. Based on these results, a novel way of neuronal computation of moment neuronal networks is proposed and some applications are included.

4.1 Introduction

During the past 20 years, we have witnessed the development of the *artificial neural network (ANN)* and its impact on neuroscience. In comparison with the development of ANN in the 1950s where a neuron takes two states only, we have used a continuous input-output relationship (sigmoidal function) to describe the activity of a neuron. Using a simple sigmoidal function to characterize the complex input-output activity of a neuron is oversimplified, and such an approach never really brings us anywhere near a biological reality.

It is clear from the past few years of research that neuron activity will be decided not only by its mean input activity (mean firing rate), but also by its higher order statistics of inputs. For example, in the scenario of balanced inhibitory and excitatory inputs, the mean input to a neuron is zero and in the ANN approach the neuron will be simply silent (higher order statistics are not taken into account). Can we develop a theory that will naturally include the mean (first-order statistics), the second-order statistics, and so on to be computed? In other words, can we develop a theory of a moment neuronal networks (MNNs)? The advantage of such an approach over the ANN is obvious. For example, we know that synchronized neuronal activity might play an important role in information processing in the nervous systems. In the MNN, we can naturally include the synchronized firing case since it corresponds to the fully correlated activity, that is, with a correlation coefficient being unity. There are other advantages of the MNN over the ANN. First, since our theory is developed based on spiking neurons, we can relate all of our parameters to biologically measurable quantities. We can always test our theory in biological experiments. Second, from a purely theoretical point of view, we know that in a stochastic system, any computation based on the mean is somewhat similar to the law of large numbers in probability theory. To have a complete description of a stochastic system, we need the central limit theorem (second-order statistics calculations). The MNN is equivalent to the development of the central limit theory in probability. Third, although our theory is developed in terms of spiking neuronal networks, the learning theory developed in ANN will be mostly applicable and hence MNN will serve as a bridge between ANN theory and neurobiology.

This chapter develops a theory of moment computation. But first we present a few examples to emphasize the importance of the second-order statistics. The examples include the input and output surfaces of a single neuron (the integrate-and-fire model and the Hodgkin-Huxley model), optimal control models, discrimination with correlated inputs, noise reduction and storage capacity with negative correlations.

Our general setup is as following. In section 4.2, the input-output surfaces of a single neuron are discussed. In section 4.3, we discuss an optimal control model based on the second-order statistics. In section 4.4, we address the issue of discrimination among mixed input signals. In section 4.5, the importance of negative correlations is presented. In section 4.6, we develop the moment neural networks. The exact relationship of input and output of an integrate-and-fire neuron is known. The renewal theorem is applied to the integrate-and-fire model, which enables us to have complete equations to describe the input and output relationship of spiking neuronal networks. The similar idea to investigate the input-output relationship of spiking neuronal networks has been widely investigated in the literature [2]; however, to the best of our knowledge, our approach is novel. In fact, as pointed out in [69] at page 435, final paragraph, all approaches before are based on two key assumptions: the output of the integrate-and-fire model is again a Poisson process, which is only approximately true in very limited parameter regions; the input is independent. Our general theory requires neither of the assumptions.

We confine ourselves here to neuronal computation with the first- and the second-order statistics. The reason is that to completely describe a random system the first- and the second-order statistics are enough in most cases. We can, of course, tailor our approach to general cases: including the third-, the fourth- etc. order statistics. Furthermore, we only consider feedforward networks, and it is easy to generalize our results to recurrent networks.

4.2 Single Neuron Activity: Variance

Let us consider a neuron activity. Traditionally, an $F-I$ curve is used to *fully* characterize its behaviour, where F means its output frequency and I is the input current. The $F-I$ curve usually takes the form of a sigmoidal function, which serves as the basis of classical neural computation theory.

The simplest model of a spiking neuronal network, the leaky integrate-and-fire model[69], is defined by

$$dV_t = -\frac{V_t}{\gamma} + I$$

where V_t is the membrane potential (in mV), γ is the decay time (in msec), and I is the external input. The firing time is then given by

$$T = \gamma \log \left(1 - \frac{V_{thre}}{I\gamma}\right)$$

where V_{thre} is the threshold. The output frequency F (Hz) is

$$F = \frac{1000}{T + T_{ref}}$$

which shows a typical sigmoidal shape, where T_{ref} is the refractory time.

If we agree that a neuron is a stochastic unit, then the output of a single neuron could be totally different from the above scenario. Now the membrane potential of a neuron takes the form

$$dV_t = -\frac{V_t}{\gamma} + dS_t$$

where S_t could be pulsed synaptic inputs, or its approximation taking the form $\mu dt + \sigma dB_t$ with μ, σ being constants and B_t the noise. In Figure 4.1, we plot the output firing rate and CV (standard deviation/mean) of interspike intervals for the integrate-and-fire neuron (upper panel) and the Hodgkin-Huxley neuron (bottom panel). It is clearly seen, not surprisingly, that the output firing rate depends on not only the input firing rate μ, but also the input variance σ. In fact, with different neuron models, the surfaces are quite different and we refer the reader to [23] for more details.

To further illustrate the importance of the second-order statistics, we turn our attention to a high-level model: a model on movement control.

4.3 Control Tasks

The experimental study of movements in humans and other mammals has shown that voluntary reaching movements obey two fundamental psychophys-

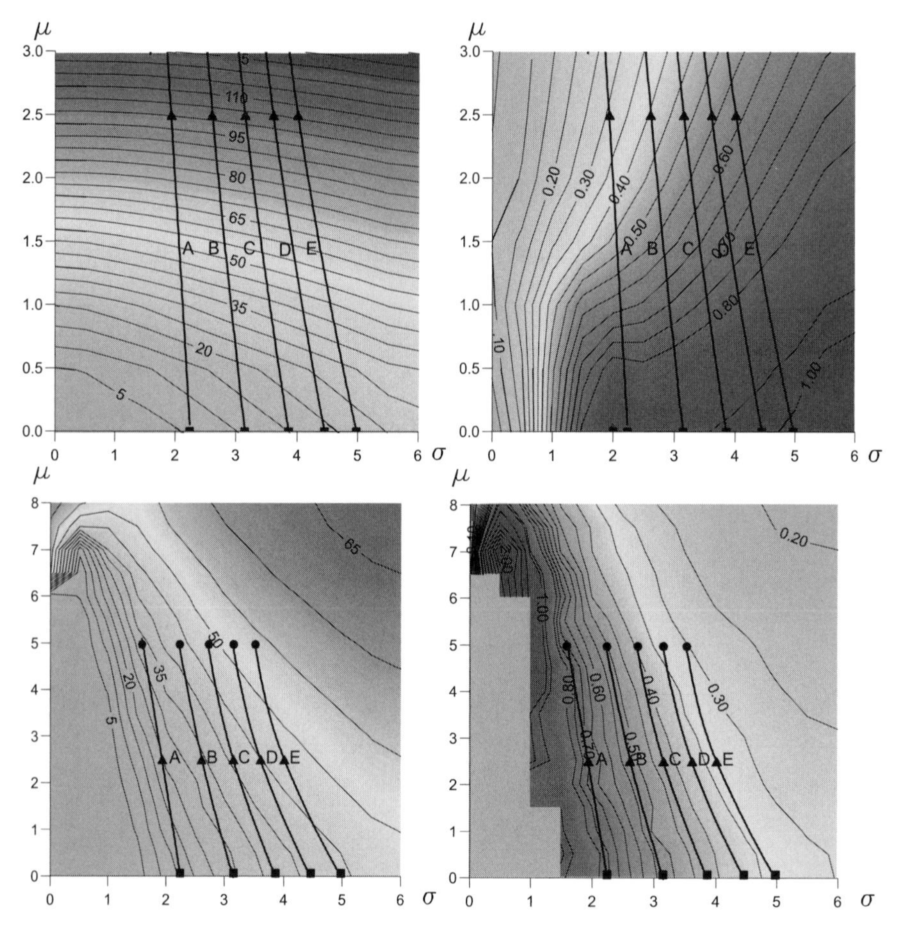

Fig. 4.1. Top panel: the response surface of the integrate-and-fire model, left, contour of the output firing rate; right, contour of the CV. Bottom panel: the response surface of the Hodgkin-Huxley model: left, contour of the output firing rate; right, contour of the CV. We refer the reader to [23] for details of simulations.

ical principles from the point of view of trajectory formation, and of the dependence of movement time on relative accuracy, as reviewed in [70].

1. **Trajectory formation**: It is found that the trajectory of arm (hand) movement from a starting point to an end point is almost a straight line and the velocity profile is bell-shaped.

2. **Fitts law [68]**: The longer the time taken for a reaching movement, the more accurate the hand is in arriving at the end point.

These principles are of paramount importance and of great interest both in theory and in applications. In theory, how our brain controls our daily movements remains elusive, despite a century of intensive research in neuroscience. In comparison with an open-loop approach, where we try to directly understand how a brain works, a close-loop approach to connect sensory inputs with motor outputs (for example, voluntary movements) might be the only way to make genuine progress in this area. In engineering applications, if we can unravel the mechanisms underlying these principles, we could apply them to the control of (humanoid) robots. For these reasons, there is considerable interest in investigating these principles (see [69] for a recent review, [71, 73]). However, as far as we are aware, no rigorous results are available in the literature and many issues remain elusive.

In this section we present an analytical and rigorous approach to tackling the problem. Within the framework of optimal stochastic control principles and in three-dimensional space [72], it is shown that the straight line trajectory is a natural consequence of optimal stochastic control principles, under the condition that we have a nondegenerate optimal control signal. Numerically we conclude that the bell-shaped velocity profile follows as well. Furthermore, we also numerically demonstrate that the optimal variance is proportional to the time interval in which the movements are performed, that is, Fitts's law is verified.

4.3.1 The Model

We consider a simple model of (arm) movements. Let $X(t) = (x(t), y(t), z(t))'$ be the position of the hand at time t. We then have

$$\ddot{X} = -\frac{1}{\tau_1\tau_2}X - \frac{\tau_1+\tau_2}{\tau_1\tau_2}\dot{X} + \frac{1}{\tau_1\tau_2}[\Lambda(t)dt + \Lambda(\alpha,t)d\mathbf{B}(t)] \tag{4.1}$$

where $\tau_1, \tau_2, \alpha > 0$ are parameters, $\Lambda(t) = (\lambda_x(t), \lambda_y(t), \lambda_z(t))'$ is the control signal, $\Lambda(\alpha, t)$ is a 3×3 diagonal matrix with diagonal elements as $\lambda_x^\alpha(t), \lambda_y^\alpha(t), \lambda_z^\alpha(t)$, respectively, and $\mathbf{B}_t = (B_x(t), B_y(t), B_z(t))'$ is the standard Brownian motion. In physics, we know that (4.1) is the well-known Kramers' equation. In neuroscience, it is observed in all *in vivo* experiments that the noise strength is proportional to the signal strength $\Lambda(t)$ and hence the signals received by muscle take the form of (4.1) (see for example, [57, 69, 67]).

For a point $D = (d_x, d_y, d_z)' \in I\!R^3$ and two positive numbers T, R, we intend to find a control signal $\Lambda^*(t)$ that satisfies

$$\langle X(t) \rangle = D, \text{ for } t \in [T, T+R] \tag{4.2}$$

and

$$
\begin{aligned}
I(\Lambda^*) = \min_{\Lambda \in \mathcal{L}^{2\alpha}[0,T+R]} I(\Lambda) &= \min_{\Lambda \in \mathcal{L}^{2\alpha}[0,T+R]} \int_T^{T+R} \text{var}(X(t))dt \\
&= \min_{\Lambda \in \mathcal{L}^{2\alpha}[0,T+R]} \int_T^{T+R} [\text{var}(x(t)) + \text{var}(y(t)) + \text{var}(z(t))]dt \quad (4.3)
\end{aligned}
$$

where $\Lambda \in \mathcal{L}^{2\alpha}[0, T+R]$ means that each component of it is in $\mathcal{L}^{2\alpha}[0, T+R]$. To stablize the hand, we further require that the hand will stay at D for a while, that is, in time interval $[T, T+R]$, which also naturally requires that the velocity should be zero at the end of movement. The physical meaning of the problem we considered here is clear: *at time T the hand will reach the position D ((4.2)), and as precisely as possible ((4.3)).* Without loss of generality we assume that $d_x > 0, d_y > 0$ and $d_z > 0$.

4.3.2 Optimal Control

The optimal control problem posed in the previous section is a highly non-linear problem, as we are going to show below.

Solving (4.1) we obtain

$$
\begin{aligned}
(x(t), \dot{x}(t))' &= \left(\int_0^t \frac{b_{12}(t-s)}{\tau_1 \tau_2} \lambda_x(s) ds, \int_0^t \frac{b_{22}(t-s)}{\tau_1 \tau_2} \lambda_x(s) ds \right)' \\
&+ \left(\int_0^t \frac{b_{12}(t-s)}{\tau_1 \tau_2} \lambda_x^\alpha(s) dB_x(s), \int_0^t \frac{b_{22}(t-s)}{\tau_1 \tau_2} \lambda_x^\alpha(s) dB_x(s) \right)' \quad (4.4)
\end{aligned}
$$

where

$$
\begin{aligned}
b_{12}(t) &= \frac{\tau_1 \tau_2}{\tau_2 - \tau_1} \left[\exp\left(-\frac{t}{\tau_2}\right) - \exp\left(-\frac{t}{\tau_1}\right) \right] \\
b_{22}(t) &= \frac{\tau_1 \tau_2}{\tau_2 - \tau_1} \left[\frac{1}{\tau_1} \exp\left(-\frac{t}{\tau_1}\right) - \frac{1}{\tau_2} \exp\left(-\frac{t}{\tau_2}\right) \right].
\end{aligned}
$$

A similar equation holds true for $y(t)$ and $z(t)$.

Note that

$$
\begin{aligned}
\int_T^{T+R} \text{var}(x(t))dt &= \left\langle \int_T^{T+R} \left[\int_0^t b_{12}(t-s) \lambda_x(s)^\alpha dB_x(s) \right]^2 dt \right\rangle \\
&= \int_T^{T+R} \left[\int_0^t b_{12}^2(t-s) |\lambda_x(s)|^{2\alpha} ds \right] dt \quad (4.5)
\end{aligned}
$$

The original control problem defined by (4.2) and (4.3) is reduced to the following optimization problem:

Find $\Lambda^*(s) \in \mathcal{L}^{2\alpha}[0, T+R]$ that minimizes

$$\int_T^{T+R} \left[\int_0^t b_{12}^2(t-s)[|\lambda_x(s)|^{2\alpha} + |\lambda_y(s)|^{2\alpha} + |\lambda_z(s)|^{2\alpha}]ds \right] dt \tag{4.6}$$

subject to the constraint

$$\int_0^t b_{12}(t-s)\Lambda(s)ds = \tau_1\tau_2 D, \ \text{ for } t \in [T, T+R]. \tag{4.7}$$

Since $R > 0$, by differentiating (4.7), we obtain

$$\begin{aligned} -\frac{1}{\tau_2} \exp\left(-\frac{t}{\tau_2}\right) \int_0^t \exp\left(\frac{s}{\tau_2}\right) \Lambda(s)ds \\ + \frac{1}{\tau_1} \exp\left(-\frac{t}{\tau_1}\right) \int_0^t \exp\left(\frac{s}{\tau_1}\right) \Lambda(s)ds = 0 \end{aligned} \tag{4.8}$$

for $t \in (T, T+R)$. Solving (4.7) and (4.8) we see that

$$\begin{cases} \int_0^t \exp\left(\frac{s}{\tau_2}\right) \Lambda(s)ds = D\tau_2 \exp\left(\frac{t}{\tau_2}\right) \\ \int_0^t \exp\left(\frac{s}{\tau_1}\right) \Lambda(s)ds = D\tau_1 \exp\left(\frac{t}{\tau_1}\right) \end{cases} \tag{4.9}$$

for $t \in [T, T+R]$, which implies that $\Lambda(t) = D$ and in particular

$$\begin{cases} \int_0^T \exp\left(\frac{s}{\tau_2}\right) \Lambda(s)ds = D\tau_2 \exp\left(\frac{T}{\tau_2}\right) \\ \int_0^T \exp\left(\frac{s}{\tau_1}\right) \Lambda(s)ds = D\tau_1 \exp\left(\frac{T}{\tau_1}\right) . \end{cases} \tag{4.10}$$

Now let us find the optimal signal $\Lambda^*(t)$ for $t \in [0, T]$.

It is easily seen that

$$I(\Lambda) = I_x(\lambda_x) + I_y(\lambda_y) + I_z(\lambda_z) \tag{4.11}$$

$$\begin{aligned} &= \int_0^T \left[\int_T^{T+R} b_{12}^2(t-s)dt \right] [|\lambda_x|^{2\alpha}(s) + |\lambda_y|^{2\alpha}(s) + |\lambda_z|^{2\alpha}(s)]ds \\ &+ \int_T^{T+R} \left[\int_T^s b_{12}^2(t-s)dt \right] [|\lambda_x|^{2\alpha}(s) + |\lambda_y|^{2\alpha}(s) + |\lambda_z|^{2\alpha}(s)]ds . \end{aligned} \tag{4.12}$$

Since to minimize each term in (4.12) implies minimizing $I(\Lambda)$, we apply the calculus of variations to the first term in (4.12), as the second term is a constant (for $t \in [T, T+R]$ $\Lambda(t) = D$, see (4.10).) To this end, let us define

$$\left\{\lambda_x, \int_0^T b_{12}(T-s)\lambda_x(s)ds = \tau_1\tau_2 d_x, \lambda_x(t) = d_x, t \in [T, T+R]\right\} = \mathcal{U}_D.$$

For a small τ, consider $\lambda_x + \tau\phi \in \mathcal{U}_D$, that is,

$$\begin{aligned}\phi \in \{\phi, \quad & \int_0^T \exp(s/\tau_1)\phi(s)ds = 0, \\ & \int_0^T \exp(s/\tau_2)\phi(s)ds = 0, \phi(t) = 0, \text{ for } t \in [T, T+R]\} = \mathcal{U}_D^0.\end{aligned}$$

The first two constraints in $\mathcal{U}_D^0$ are from (4.10). We then have

$$\frac{dI_x(\lambda_x + \tau\phi)}{d\tau}|_{\tau=0} = 0,$$

which gives

$$\int_0^T \left\{\left[\int_T^{T+R} b_{12}^2(t-s)dt\right] |\lambda_x(s)|^{2\alpha-1}\text{sgn}(\lambda_x(s))\phi(s)\right\} ds = 0. \qquad (4.13)$$

Comparing (4.13) with the first two constraints in $\mathcal{U}_D^0$, we conclude that

$$\left[\int_T^{T+R} b_{12}^2(t-s)dt\right] |\lambda_x(s)|^{2\alpha-1}\text{sgn}(\lambda_x(s)) = \xi_x \exp\left(\frac{s}{\tau_1}\right) + \eta_x \exp\left(\frac{s}{\tau_2}\right) \qquad (4.14)$$

almost surely for $s \in [0, T]$ with two parameters $\xi_x, \eta_x \in R$. Hence the solution of the original problem is

$$\lambda_x^*(s) = \frac{\left|\xi_x \exp\left(\frac{s}{\tau_1}\right) + \eta_x \exp\left(\frac{s}{\tau_2}\right)\right|^{1/(2\alpha-1)} \text{sgn}\left[\xi_x \exp\left(\frac{s}{\tau_1}\right) + \eta_x \exp\left(\frac{s}{\tau_2}\right)\right]}{\left(\int_T^{T+R} b_{12}^2(t-s)dt\right)^{1/(2\alpha-1)}} \qquad (4.15)$$

with ξ_x, η_x being given by

$$\begin{cases} d_x \tau_2 \exp\left(\frac{T}{\tau_2}\right) = \int_0^T \exp\left(\frac{s}{\tau_2}\right) \\ \cdot \frac{\left|\xi_x \exp\left(\frac{s}{\tau_1}\right) + \eta_x \exp\left(\frac{s}{\tau_2}\right)\right|^{1/(2\alpha-1)} \operatorname{sgn}\left[\xi_x \exp\left(\frac{s}{\tau_1}\right) + \eta_x \exp\left(\frac{s}{\tau_2}\right)\right]}{\left(\int_T^{T+R} b_{12}^2(t-s)dt\right)^{1/(2\alpha-1)}} ds \\ d_x \tau_1 \exp\left(\frac{T}{\tau_1}\right) = \int_0^T \exp\left(\frac{s}{\tau_1}\right) \\ \cdot \frac{\left|\xi_x \exp\left(\frac{s}{\tau_1}\right) + \eta_x \exp\left(\frac{s}{\tau_2}\right)\right|^{1/(2\alpha-1)} \operatorname{sgn}\left[\xi_x \exp\left(\frac{s}{\tau_1}\right) + \eta_x \exp\left(\frac{s}{\tau_2}\right)\right]}{\left(\int_T^{T+R} b_{12}^2(t-s)dt\right)^{1/(2\alpha-1)}} ds. \end{cases} \tag{4.16}$$

A similar equation is true for λ_y and λ_z.

From the results above, we see that the optimal problem is in general a highly nonlinear problem. However, we arrive at the following surprising conclusions.

Theorem 1 *Under the optimal control framework as we set up here and $\alpha > 1/2$, the optimal mean trajectory is a straight line. When $\alpha \leq 1/2$ the optimal control problem is degenerate, that is, the optimal control signal is a delta function; (4.12) with $\Lambda = \Lambda^*$ gives us an exact relationship between time T and variance.*

Proof We only need to show that if, say, $d_x = kd_y$, we then have $\langle x(t)\rangle = k\langle y(t)\rangle$ with a positive constant k. Let us assume that for $\alpha > 1/2$, (ξ_x, η_x) is the solution of (4.16). It is easily seen that $k^{2\alpha-1}(\xi_x, \eta_x)$ is the solution of (4.16) with d_x being replaced by kd_y. From (4.15), together with the fact above, we have $\lambda_y(s) = k\lambda_x(s)$. From (4.4), we finally conclude that $\langle x(t)\rangle = k\langle y(t)\rangle$ holds true. The proof of the second part of Theorem 1 is easy and is left to the reader.

For a deterministic problem similar to the optimal control problem defined by (4.2) and (4.3), that is, $\alpha \to 0$, the optimal signal is a delta function. The results in Theorem 1 tell us that when the noise is not strong enough, $\alpha \leq 1/2$, and the optimal control signal is still degenerate. Nevertheless, when $\alpha > 1/2$, that is, the noise is strong enough, the solution of the optimal control problem turns out to be nondegenerate. This is different from our usual intuition that any tiny noise could smooth an action, that is, any tiny noise ensures that the system is ergodic.

A straight-line trajectory indicates that the traveling distance between the starting point and the end point is the shortest. In other words, the shortest distance principle is the consequence of the optimal control problem defined by (4.2) and (4.3), provided that $\alpha > 1/2$.

In Figure 4.2 [1], we plot $\langle y(t)\rangle$, the optimal trajectory $\langle x(t)\rangle, \langle y(t)\rangle$, velocity profile, and variance for fixed $R = 0.2T$, $\tau_1 = 5, \tau_2 = 15$, and various D and T as shown in the figure for $\alpha = 0.75, 1, 1.5$ [2]. As we proved in Theorem 1, $(\langle x(t)\rangle, \langle y(t)\rangle)$ shows a straight line trajectory starting from (0,0) in Figure 4.2) and arriving at an end point (indicated by the arrow in Figure 4.2 upper right trace). It is worthwhile pointing out that $\langle x(t)\rangle$ and $\langle y(t)\rangle$ (Figure 4.2 upper left) are highly nonlinear functions of time t. Figure 4.2 middle left shows the velocity profiles vs. distance. The bell shape is obvious. In fact, this might not be surprising in comparison with the trajectory formation, which is a consequence of a highly nonlinear optimal control problem. The optimal control problem we considered here requires that the velocity be zero at the beginning and the end point of the movement, that is, $(\langle x(t)\rangle, \langle y(t)\rangle)$ is a constant for $t \in [T, T + R]$. We would then expect that a bell-shaped velocity profile will be the most natural outcome, provided that the velocity is continuous. In Figure 4.2 middle right trace, we also depict the optimal variance $I_x(\lambda_x^*) + I_y(\lambda_y^*)$ vs. distance. We see that the optimal variance is an increasing function of distance, as one would naturally expect. To compare our results with Fitts's law, we plot $I_x(\lambda_x^*) + I_y(\lambda_y^*)$ vs. time T in Figure 4.2, bottom right. It is easily seen that the longer the time is, the smaller the variance. In fact, from our results, we could also assess the exact relationship between the optimal variance and the time. Finally at the bottom left we show the optimal control signals vs. t for various times T.

It is clear now that the second-order statistics play an important role in neuronal computation, but we have not taken into account another important second-order statistic here: the correlation between neuronal activities. In the next few sections, we will concentrate on it.

4.4 Discrimination

In this section we present a study on the discrimination capacity of the simplest neuron model: the integrate-and-fire model. Suppose that a neuron receives two sets of signal. Both of them are contaminated by noise, as shown in Figure 4.3. After neuronal transformations, we want to know whether the signals become more mixed or more separated. This is a typical scenario in decision theory.

[1] Due to the difficulty of plotting three-dimensional figures, we will simply consider the two-dimensional case. However, it is readily seen that all conclusions below are true for movements in three-dimensional space.

[2] Note that we do not want to fit our results with experimental data (such a fitting would be straightforward), but intend to present a general theory. All quantities are unitless.

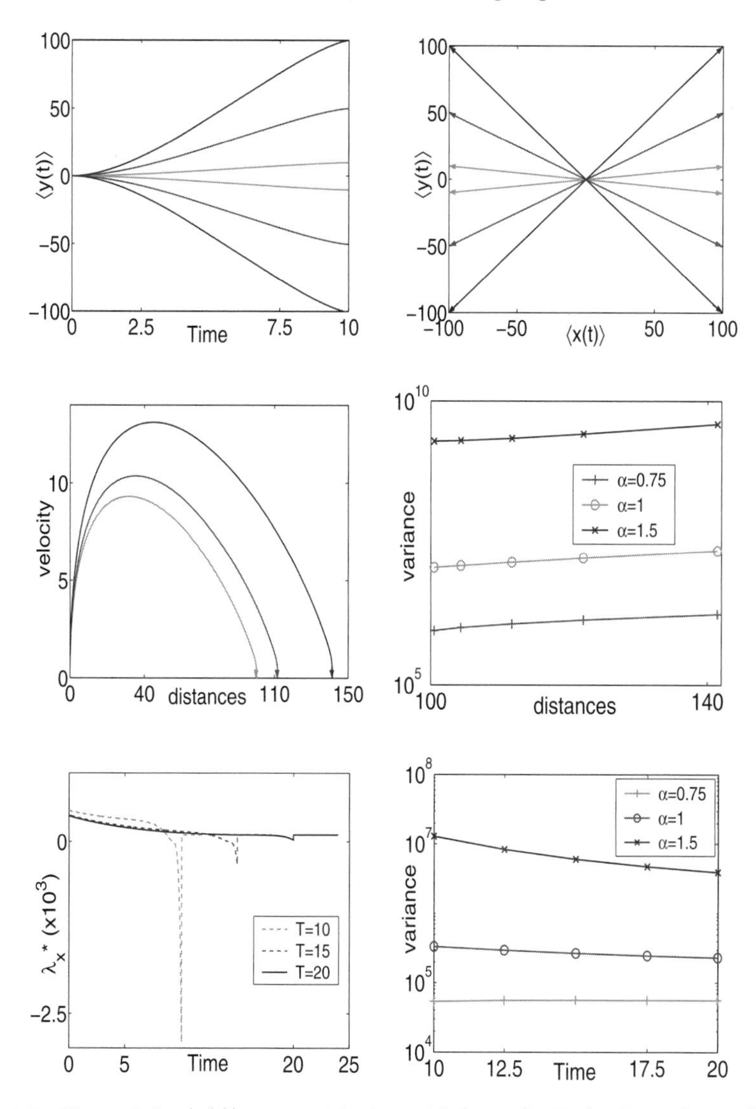

Fig. 4.2. Upper left, $\langle y(t)\rangle$ vs. t with $d_y = 10$ (green), 50 (red), and 100 (blue). Upper right, $(\langle y(t)\rangle, \langle x(t)\rangle)$ for $t \in [0, T]$ with $T = 10, \alpha = 0.75$. It is easily seen that the trajectory is a straight line. Middle left, velocity $(\langle \dot{x}(t)\rangle\langle x(t)\rangle + \langle \dot{y}(t)\rangle\langle y(t)\rangle)/\sqrt{(\langle x(t)\rangle)^2 + (\langle y(t)\rangle)^2}$ vs. distance $\sqrt{(\langle x(t)\rangle)^2 + (\langle y(t)\rangle)^2}$ with $\alpha = 0.75, T = 20$. Middle right the optimal variance $I_x(\lambda_x^*) + I_y(\lambda_y^*)$ vs. distance with $T = 20$. Bottom left, λ_x^* vs. time t for fixed $d_x = d_y = 100$ with $\alpha = 0.75$. Bottom right, the optimal variance $I_x(\lambda_x^*) + I_y(\lambda_y^*)$ vs. time T with $d_x = d_y = 10$.

We conclude that without correlation between signals, the output histograms are separable if and only if the input histograms are separable. With positively correlated input signals, the output histograms become more separable than input histograms. With negatively correlated input signals, the output histograms are more mixed than the input histograms. This is a clear-cut and interesting result. In fact, in recent years, many publications explored the functional role of correlated neuronal activity. For example, synchronization is a special case of correlated neuronal activity[30, 35, 75]. Researchers have extensively investigated the functional roles of correlations in neurons from information theoretic approaches [8, 47, 48], from experimental approaches [52, 53] and from modeling approaches [1, 15, 51, 54].

In the classical theory of neural networks, we only take into account the excitatory inputs. However, in recent years, we have found many intriguing functional roles of inhibitory inputs, ranging from linearizing input-output relationship of a neuron [38], to synchronizing a group of neurons [58] and to actually increasing neuron firing rates [19]. In particular, neuronal and neuronal network models with an exactly balanced inhibitory and excitatory input are intensively studied in the literature [55, 59]. For these two most interesting cases, an independent input case and an exactly balanced input case, we are able to find out the exact value of neuronal discrimination capacity. Roughly speaking, here the discrimination capacity is the minimal number of synapses carrying signals so that the output histograms of the neuron are separable, provided that input signals are different (see Section 4.3 for the definition). Interestingly, the obtained analytical discrimination capacity is *universal* for the model. It is independent of the decay rate, the threshold, the magnitude of the excitatory postsynaptic potential (EPSP) and the inhibitory postsynaptic potential (IPSP), and the input signal distributions.

This section is organized as follows. In subsection 4.4.1, the model is ex-

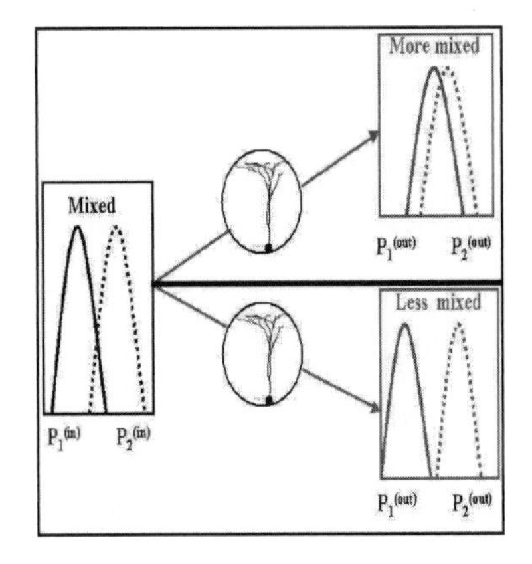

Fig. 4.3. For two possibly mixed input signals, after neuronal transformation, will they become more separated or mixed?

actly defined. In subsection 4.4.2, we consider the worst case, and the exact value of discrimination capacity is obtained. In subsection 4.4.3, we generalize the results from subsection 4.4.2. In subsection 4.4.4, some numerical results are included. We have presented numerical results for the integrate-and-fire model and the IF-FHN model [19] in a meeting report[21]. In subsection 4.4.6, we briefly discuss related issues. Detailed theoretical proofs are omitted and can be found in [14].

4.4.1 The Models

The neuron model we use here is the classical integrate-and-fire model [10, 18, 28, 57]. When the membrane potential $V_t^{(k)}$ is below the threshold V_{thre}, it is given by

$$dV_t^{(k)} = -L(V_t^{(k)} - V_{rest})dt + dI_{syn}^{(k)}(t) \tag{4.17}$$

where $L = 1/\gamma$ (section 4.2) is the decay coefficient and the synaptic input is

$$I_{syn}^{(k)}(t) = a\sum_{i=1}^{p} E_i^{(k)}(t) - b\sum_{j=1}^{q} I_j^{(k)}(t)$$

with $E_i^{(k)}(t), I_i^{(k)}(t)$ as the Poisson processes with rates $\lambda_{i,E}^{(k)}$ and $\lambda_{i,I}^{(k)}$, respectively, $a > 0, b > 0$ are the magnitudes of each EPSP and IPSP, p and q are the total number of active excitatory and inhibitory synapses, $k = 1, 2$ represent different input signals, and we aim at discriminating between them in terms of an observation of efferent firing rates. Once $V_t^{(k)}$ crosses V_{thre} from below, a spike is generated and $V_t^{(k)}$ is reset to V_{rest}, the resting potential. This model is termed the intergrate and fire (IF) model. The interspike interval of efferent spikes is

$$T^{(k)} = \inf\{t : V_t^{(k)} \geq V_{thre}\}.$$

For simplicity of notation we assume that $q = p$ and $\lambda_{i,I}^{(k)} = r\lambda_{i,E}^{(k)}$, where $0 \leq r \leq 1$ is the ratio between inhibitory and excitatory inputs.

Furthermore, we suppose that p_c out of p synapses carry the true signal and the rest $p - p_c$ synapses are noise (or distorted signals). Synapses that code true signals are correlated, but synapses that code noise are independent. For simplicity of notation, we assume that the correlation coefficient between the ith excitatory (inhibitory) synapse and the jth excitatory (inhibitory) synapse is a constant c, where $i, j = 1, \cdots, p_c$. The correlation considered here reflects the correlation of activity of different synapses, as discussed and explored in [19, 76]. More specifically, synaptic inputs take the following form ($p = q$)

$$I_{syn}^{(k)}(t) = a\sum_{i=1}^{p} E_i^{(k)}(t) - b\sum_{j=1}^{p} I_j^{(k)}(t)$$
$$= a\sum_{i=1}^{p_c} E_i^{(k)}(t) + a\sum_{i=p_c+1}^{p} E_i^{(k)}(t) - b\sum_{i=1}^{p_c} I_i^{(k)}(t) - b\sum_{i=p_c+1}^{p} I_i^{(k)}(t)$$

where $E_i^{(k)}(t), i = 1, \cdots, p_c$ are correlated Poisson processes with an identical rate $\lambda^{(k)}$ (signal), $E_i^{(k)}(t), i = p_c + 1, \cdots, p$ are Poisson processes with a firing rate λ_i of independently and identically distributed random variables from $[0, \lambda_{max}]$Hz (noise), and $I_i^{(k)}(t), i = 1, \cdots, p$ have the same properties as $E_i^{(k)}(t)$, but with a firing rate of $r\lambda^{(k)}$ or $r\lambda_i$ for $r \in [0, 1]$, representing the ratio between inhibitory and excitatory inputs. It was pointed out in [55] that the ratio of inhibitory and excitatory synapses is around 15/85. Of course, in general, inhibitory inputs are larger than excitatory inputs. All conclusions below can be easily extended to the case of $r > 1$. On the other hand, for the simplicity of notation, we have introduced a single parameter r to describe the relationship between inhibitory and excitatory inputs. From the proofs below, we can see that this assumption can be easily relaxed. Without loss of generality we simply assume that $\lambda^{(1)}, \lambda^{(2)} \in [0, \lambda_{max}]$Hz.

Hence the neuron model receives two set of inputs: one is

$$a\sum_{i=1}^{p_c} E_i^{(1)}(t) + a\sum_{i=p_c+1}^{p} E_i^{(1)}(t) - b\sum_{i=1}^{p_c} I_i^{(1)}(t) - b\sum_{i=p_c+1}^{p} I_i^{(1)}(t)$$

where the signal term

$$a\sum_{i=1}^{p_c} E_i^{(1)}(t) - b\sum_{i=1}^{p_c} I_i^{(1)}(t)$$

is masked by the noise term

$$a\sum_{i=p_c+1}^{p} E_i^{(1)}(t) - b\sum_{i=p_c+1}^{p} I_i^{(1)}(t);$$

the other is

$$a\sum_{i=1}^{p_c} E_i^{(2)}(t) + a\sum_{i=p_c+1}^{p} E_i^{(2)}(t) - b\sum_{i=1}^{p_c} I_i^{(2)}(t) - b\sum_{i=p_c+1}^{p} I_i^{(2)}(t)$$

where the signal term

$$a\sum_{i=1}^{p_c} E_i^{(2)}(t) - b\sum_{i=1}^{p_c} I_i^{(2)}(t)$$

is masked by the noise term

$$a \sum_{i=p_c+1}^{p} E_i^{(2)}(t) - b \sum_{i=p_c+1}^{p} I_i^{(2)}(t).$$

In the following, we further use diffusion approximations for synaptic inputs [57] and assume $a = b$:

$$\begin{aligned} i_{syn}^{(k)}(t) &= ap_c\lambda^{(k)}t + a \sum_{i=p_c+1}^{p} \lambda_i t - bp_c r\lambda^{(k)}t - b \sum_{i=p_c+1}^{p} r\lambda_i t \\ &\quad + \sqrt{(a^2+b^2r)\lambda^{(k)}p_c(1+c(p_c-1)) + (a^2+b^2r) \sum_{i=p_c+1}^{p} \lambda_i} \cdot B_t \\ &= a(1-r)t \left[p_c\lambda^{(k)} + \sum_{i=p_c+1}^{p} \lambda_i \right] \\ &\quad + a\sqrt{(1+r)[\lambda^{(k)}p_c(1+c(p_c-1)) + \sum_{i=p_c+1}^{p} \lambda_i]} \cdot B_t \end{aligned} \tag{4.18}$$

where B_t is the standard Brownian motion.

Therefore, the term

$$a(1-r)t \left[p_c\lambda^{(k)} + \sum_{i=p_c+1}^{p} \lambda_i \right], \qquad k = 1, 2 \tag{4.19}$$

in (4.18) is the mean input signal to the cell. Without loss of generality, we always assume that $\lambda^{(1)} < \lambda^{(2)}$. Denote $p_k^{(in)}(\lambda)$ as the distribution density of random variables [ignore the constants $a(1-r)t$ in (4.19)].

$$p_c\lambda^{(k)} + \sum_{i=p_c+1}^{p} \lambda_i. \tag{4.20}$$

In summary, we consider the case that a neuron receives p synaptic inputs, with p_c out of p carrying the signals and $p - p_c$ being noise (distorted signals). The setup here roughly corresponds to the experiments of Newsome and his colleagues. We will explore this aspect in further publications [26].

4.4.2 Discrimination Capacity: The Worst Case

For a fixed $\lambda^{(1)} < \lambda^{(2)}$ we have corresponding two histograms $p_1^{(out)}(\lambda)$ and $p_2^{(out)}(\lambda)$ of output firing rates as shown in Figure 4.4. Let

$$R_{\min}^{(out)}(\lambda^{(2)}) = \min\{\lambda : p_2^{(out)}(\lambda) > 0\}$$

and

$$R_{\max}^{(out)}(\lambda^{(1)}) = \max\{\lambda : p_1^{(out)}(\lambda) > 0\}$$

and denote

$$\alpha(\lambda^{(1)}, \lambda^{(2)}, c, r) = \{p_c : R_{\min}^{(out)}(\lambda^{(2)}) = R_{\max}^{(out)}(\lambda^{(1)})\}. \tag{4.21}$$

Hence for fixed $(\lambda^{(1)}, \lambda^{(2)}, c, r)$, $\alpha(\lambda^{(1)}, \lambda^{(2)}, c, r)$ gives us the critical value of p_c: when $p_c > \alpha(\lambda^{(1)}, \lambda^{(2)}, c, r)$ and the input patterns are perfectly separable in the sense that the output firing rate histograms are not mixed; when $p_c < \alpha(\lambda^{(1)}, \lambda^{(2)}, c, r)$, the input patterns might not be separable. For fixed $(\lambda^{(1)}, \lambda^{(2)}, c, r)$, α is termed the *(worst) discrimination capacity* of the neuron.

For input signals let us introduce more notation. Define

$$R_{\min}^{(in)}(\lambda^{(2)}) = \min\{\lambda : p_2^{(in)}(\lambda) > 0\}$$

and

$$R_{\max}^{(in)}(\lambda^{(1)}) = \max\{\lambda : p_1^{(in)}(\lambda) > 0\}.$$

Therefore as soon as $R_{\min}^{(in)}(\lambda^{(2)}) > R_{\max}^{(in)}(\lambda^{(1)})$ the two masked inputs are perfectly separable. Otherwise the two masked inputs are mixed. Hence the relationship between $R_{\min}^{(in)}(\lambda^{(2)}) - R_{\max}^{(in)}(\lambda^{(1)})$ and $R_{\min}^{(out)}(\lambda^{(2)}) - R_{\max}^{(out)}(\lambda^{(1)})$ characterizes the input-output relationship of signals.

Behaviour of $\alpha(\lambda^{(1)}, \lambda^{(2)}, c, r)$

First, we note that the output firing rate is given by [57]:

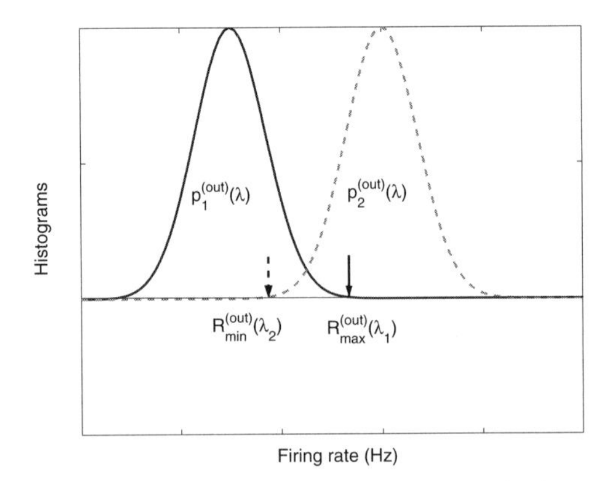

Fig. 4.4. A schematic plot of two output histograms, $R_{\min}^{(out)}(\lambda_2)$ and $R_{\max}^{(out)}(\lambda^{(1)})$. As soon as $R_{\min}^{(out)}(\lambda_2) - R_{\max}^{(out)}(\lambda^{(1)}) > 0$, the two histograms are separable.

$$\langle T^{(k)} \rangle = \frac{2}{L} \int_{A(V_{rest}, \sum_{j=p_c+1}^{p} \lambda_j)}^{A(V_{thre}, \sum_{j=p_c+1}^{p} \lambda_j)} g(x) dx, \tag{4.22}$$

where

$$A(x, y) = \frac{xL - a[p_c \lambda^{(k)} + y](1-r)}{a\sqrt{[\lambda^{(k)} p_c (1 + c(p_c - 1)) + y](1+r)}}$$

and

$$g(x) = \left[\exp(x^2) \int_{-\infty}^{x} \exp(-u^2) du \right].$$

Let us define

$$\tilde{T}^{(k)}(x) = \frac{2}{L} \int_{A(V_{rest}, x)}^{A(V_{thre}, x)} g(u) du. \tag{4.23}$$

We know that the output firing rate is calculated via

$$1000/(\langle T^{(k)} \rangle + T_{ref})$$

where T_{ref} is the refractory period. It is obvious to see that $\tilde{T}(x)$ is a monotonic function of input $x \geq 0$, that is the output firing rate of a neuron is an increasing function of input. We conclude that $\alpha(\lambda^{(1)}, \lambda^{(2)}, c, r)$ is the solution of the following equation about p_c.

$$\begin{aligned} &\int_0^{V_{thre}L} g\left(\frac{y - a[p_c\lambda^{(1)} + (p - p_c)\lambda_{\max}](1-r)}{a\sqrt{[\lambda^{(1)} p_c (1 + c(p_c - 1)) + (p - p_c)\lambda_{\max}](1+r)}} \right) dy \\ &= \frac{\sqrt{[\lambda^{(1)} p_c (1 + c(p_c - 1)) + (p - p_c)\lambda_{\max}]}}{\sqrt{[\lambda^{(2)} p_c (1 + c(p_c - 1))]}} \\ &\quad \cdot \int_0^{V_{thre}L} g\left(\frac{y - a(p_c\lambda^{(2)})(1-r)}{a\sqrt{[\lambda^{(2)} p_c (1 + c(p_c - 1))](1+r)}} \right) dy. \end{aligned} \tag{4.24}$$

The critical value $\alpha(\lambda^{(1)}, \lambda^{(2)}, c, r)$ can be found analytically in the two most interesting cases: $c = 0$ and $r = 1$. Define

$$0 \leq \Lambda = \frac{\lambda^{(2)} - \lambda^{(1)}}{\lambda_{max}} \leq 1 .$$

We then have the following conclusions:

Theorem 2 *We assume* $\Lambda > 0, 0 < r < 1$,

- *When $c > 0$ we have*

$$\alpha(\lambda^{(1)}, \lambda^{(2)}, c, 1) < \alpha(\lambda^{(1)}, \lambda^{(2)}, 0, r) \tag{4.25}$$

and furthermore

$$\alpha(\lambda^{(1)}, \lambda^{(2)}, c, 1) = \frac{\sqrt{[\Lambda(1-c)+1]^2 + 4pc\Lambda} - (1-c)\Lambda - 1}{2c\Lambda}. \tag{4.26}$$

- *When $c = 0$ we have*

$$\alpha(\lambda^{(1)}, \lambda^{(2)}, 0, r) = \frac{p}{1+\Lambda} \tag{4.27}$$

independent of r.

- *When $c < 0$ we have*

$$\alpha(\lambda^{(1)}, \lambda^{(2)}, c, 1) > \alpha(\lambda^{(1)}, \lambda^{(2)}, 0, r). \tag{4.28}$$

The proof of Theorem 2 is quite tricky and we refer the reader to [14]. In fact, from all our numerical results (see Figure 4.6, bottom panel), we have the following stronger conclusions than Theorem 2 [(4.25) and (4.28)].

- When $c > 0$ we have

$$\alpha(\lambda^{(1)}, \lambda^{(2)}, c, r_2) < \alpha(\lambda^{(1)}, \lambda^{(2)}, c, r_1) \tag{4.29}$$

where $1 \geq r_2 > r_1 > 0$.

- When $c < 0$ we have

$$\alpha(\lambda^{(1)}, \lambda^{(2)}, c, r_2) > \alpha(\lambda^{(1)}, \lambda^{(2)}, c, r_1) \tag{4.30}$$

where $1 \geq r_2 > r_1 > 0$.

However, we are not able to theoretically prove the stronger conclusions [(4.29) and (4.30)].

It is very interesting to note that (4.27) and (4.26) are independent of a, V_{thre}, and L, three essential parameters in the integrate-and-fire model. In other words, the results of (4.27) and (4.26) of the integrate-and-fire model are *universal*. In Figure 4.5 we plot α vs. Λ for $c = 0$ and $c = 0.1$ according to (4.27) and (4.26). For a given Λ and $c = 0.1$, the solid line in Figure 4.5 gives us the smallest number of coherently synaptic inputs for an integrate-and-fire model to discriminate between input signals if we assume that $r \in [0, 1]$. Hence the solid line in Figure 4.5 is the smallest discrimination capacity of an

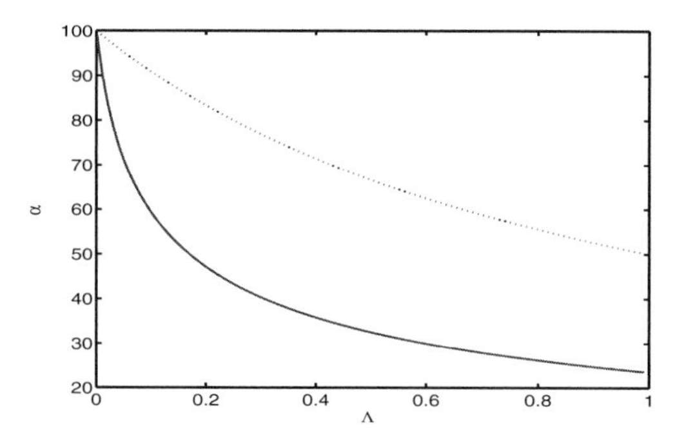

Fig. 4.5. α vs. Λ for $c = 0, r = 0$ (dotted line, independent of r according to Theorem 1) and $c = 0.1, r = 1$ (solid line).

integrate-and-fire model with $c = 0.1$. It is worth pointing out that the lowest limit of α is about $\alpha = 23$. Finally, we want to emphasize that our results are independent of input distributions. No matter what the input distribution is, as soon as p_c is greater than 40, the input signal can be perfectly separated from an observation of efferent spike trains provided that $\Lambda = 0.5$. The improvement of discrimination capacity from $r = 0$ to $r = 1$ is remarkable, almost halving in most cases.

A universal result as in (4.26) and (4.27) above is illuminating since it is independent of model parameters and then widely applicable. However, the downside of such a result is that neurons modify their connections to improve their performance. Therefore, we would argue that learning plays no role in improving its discrimination capacity, or discrimination tasks are not a primary computational task for a neuronal system. However, we want to point out that (4.26) is obtained for $r = 1$, the case with exactly balanced inhibitory and excitatory inputs. In a biologically realistic situation, neuron systems might operate in a region with $r < 1$. In the circumstances, α could depend on various model parameters and so learning might be important to improve the integrate-and-fire model discrimination capacity. Certainly to find a learning rule to improve neuronal discrimination capacity would be an interesting topic.

Input-Output Relationship

We first want to assess whether $R_{\min}^{(out)}(\lambda^{(2)}) - R_{\max}^{(out)}(\lambda^{(1)}) > 0$ even when $R_{\min}^{(in)}(\lambda^{(2)}) - R_{\max}^{(in)}(\lambda^{(1)}) < 0$, that is, the input signal is mixed, but the output signal is separated. In Figure 4.6 we plot $R_{\min}^{(out)}(\lambda^{(2)}) - R_{\max}^{(out)}(\lambda^{(1)})$ vs $R_{\min}^{(in)}(\lambda^{(2)}) - R_{\max}^{(in)}(\lambda^{(1)})$. It is easily seen that after neuronal transformation, mixed signals are better separated when $c > 0$. For example, when

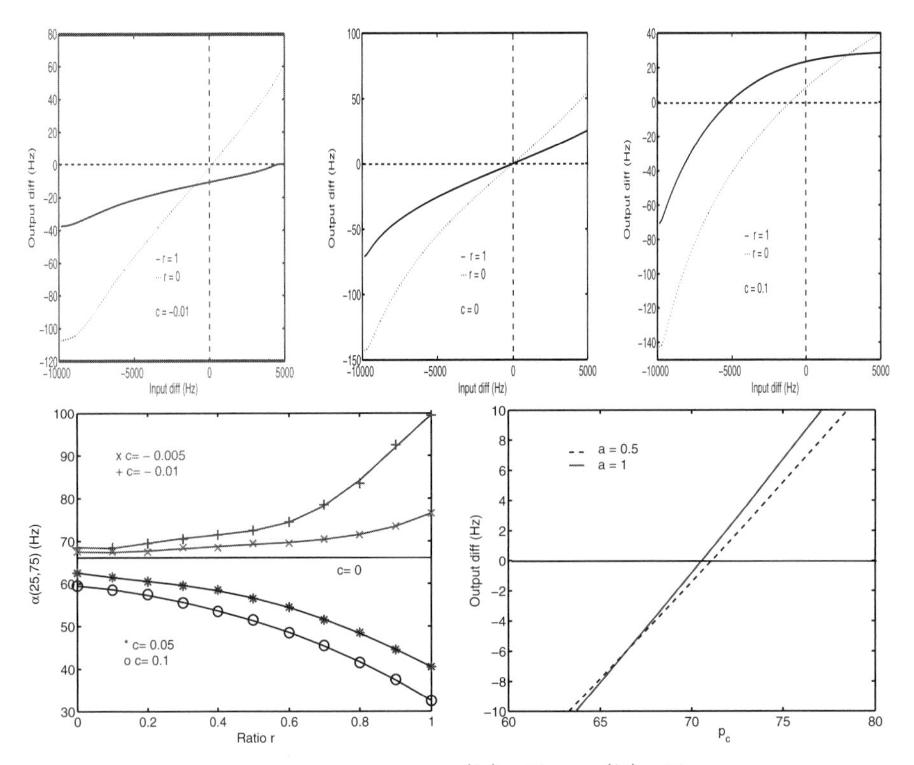

Fig. 4.6. Upper panel, input difference $R^{(in)}_{\min}(\lambda^{(2)}) - R^{(in)}_{\max}(\lambda^{(1)})$ and output difference $R^{(out)}_{\min}(\lambda^{(2)}) - R^{(out)}_{\max}(\lambda^{(1)})$ for $c < 0$ (left), $c = 0$ (middle) and $c > 0$ (right). Output firing rates are equal to $1000/(\langle T^{(k)}\rangle + T_{ref})$ with $T_{ref} = 5$ msec. Bottom panel (left), numerical values are obtained by directly solving (4.24). Bottom panel (right), the dependence of the discrimination capacity on a with $r = 0.4, c = -0.01$.

$c = 0.1, r = 1$ and $R^{(in)}_{\min}(\lambda^{(2)}) - R^{(in)}_{\max}(\lambda^{(1)}) = -5000$ Hz (mixed), but $R^{(out)}_{\min}(\lambda^{(2)}) - R^{(out)}_{\max}(\lambda^{(1)}) > 0$ (separated). The conclusion is not true for $c = 0$, but the separation is not worse after neuronal transformation. In Figure 4.6, it is clearly seen that when $r < 1$ and $c \neq 0$, the discrimination capacity depends on model parameters.

We can prove the following conclusions.

Theorem 3 *For the integrate-and-fire model*

- *if* $c > 0$ *we have*

$$R^{(out)}_{\min}(\lambda^{(2)}) - R^{(out)}_{\max}(\lambda^{(1)}) > 0 \quad \textit{when} \quad R^{(in)}_{\min}(\lambda^{(2)}) - R^{(in)}_{\max}(\lambda^{(1)}) = 0;$$

- *if* $c = 0$ *we have*

$$R^{(out)}_{\min}(\lambda^{(2)}) - R^{(out)}_{\max}(\lambda^{(1)}) = 0 \quad when \quad R^{(in)}_{\min}(\lambda^{(2)}) - R^{(in)}_{\max}(\lambda^{(1)}) = 0;$$

- *if* $c < 0$ *we have*

$$R^{(out)}_{\min}(\lambda^{(2)}) - R^{(out)}_{\max}(\lambda^{(1)}) < 0 \quad when \quad R^{(in)}_{\min}(\lambda^{(2)}) - R^{(in)}_{\max}(\lambda^{(1)}) = 0.$$

The conclusions above completely answer the question raised in the Introduction. When $c > 0$, the output histograms become more separated than input histograms; when $c < 0$, the output histograms are more mixed than input histograms. $c = 0$ is the critical case.

4.4.3 Discrimination Capacity: The Distribution Dependent Case

Consider the random variable $(\sum_{i=p_c+1}^{p} \lambda_i)$ in (4.20) and let us denote $(\sum_{i=p_c+1}^{p} \lambda_i)_j$ as its jth sampling. We see that after m times of sampling, the smallest input signal would be

$$p_c\lambda^{(k)} + \left(\sum_{i=p_c+1}^{p} \lambda_i\right)_{(m)} \tag{4.31}$$

and the largest would be

$$p_c\lambda^{(k)} + \left(\sum_{i=p_c+1}^{p} \lambda_i\right)^{(m)} \tag{4.32}$$

where $(\sum_{i=p_c+1}^{p} \lambda_i)^{(m)}$ and $(\sum_{i=p_c+1}^{p} \lambda_i)_{(m)}$ are the largest and smallest extreme value of the random variable $(\sum_{i=p_c+1}^{p} \lambda_i)$. Note that in subsection 4.4.2, we consider the worst cases and use 0 as its smallest input signal and $(p - p_c)\lambda_{max}$ as its largest input signal.

We can carry out a rigorous analysis on the behaviour of the extreme values of the random variable $(\sum_{i=p_c+1}^{p} \lambda_i)$. However, the conclusion obtained will then depend on the actual distribution of λ_i. To avoid this, we then assume that $p >> p_c$, $\lambda_i, i = p_c + 1, \cdots, p$ are an identically and independently distributed random sequence (only for a technical convenience) and we have

$$\left(\sum_{i=p_c+1}^{p} \lambda_i - (p - p_c)\langle\lambda_p\rangle\right) / (\sqrt{p - p_c}\sigma(\lambda_p)) \sim N(0, 1)$$

where $\sigma(\lambda_p) = \sqrt{\langle\lambda_p^2\rangle - \langle\lambda_p\rangle^2}$.

We need the following lemma [37]:

Lemma 1 *For ξ_n being an identically and independently distributed normal sequence of random variables, then*

$$P(a_m((\xi)^{(m)} - b_m) \leq x) \to \exp(-\exp(-x)) \tag{4.33}$$

where

$$\begin{cases} a_m = (2\log m)^{1/2} \\ b_m = (2\log m)^{1/2} - \frac{1}{2}(2\log m)^{-1/2}(\log\log m + \log(4\pi)). \end{cases} \tag{4.34}$$

Basically Lemma 1 tells us that approximately $(\xi)_{(m)}$ diverges to positive (negative) infinity at a speed of b_m $(-b_m)$. We thus conclude that

$$\left(\sum_{i=p_c+1}^{p} \lambda_i\right)^{(m)} \sim \min\left\{[(p-p_c)\langle\lambda_p\rangle + \sqrt{p-p_c}\sigma(\lambda_p)(b_m)], (p-p_c)\lambda_{max}\right\} \tag{4.35}$$

and

$$\left(\sum_{i=p_c+1}^{p} \lambda_i\right)_{(m)} \sim \max\left\{[(p-p_c)\langle\lambda_p\rangle - \sqrt{p-p_c}\sigma(\lambda_p)(b_m)], 0\right\}. \tag{4.36}$$

We see that when $m \to \infty$, $\left(\sum_{i=p_c+1}^{p} \lambda_i\right)^{(m)} \to (p-p_c)\lambda_{max}$ and $\left(\sum_{i=p_c+1}^{p} \lambda_i\right)_{(m)} \to 0$, which is the worst case we considered in the previous subsection.

For fixed $\lambda^{(1)} < \lambda^{(2)}$ and m, as before, we have two corresponding emperical histograms $p_1^{(out)}(\lambda, m)$ and $p_2^{(out)}(\lambda, m)$ of output firing rates. Let

$$R_{\min}^{(out)}(\lambda^{(2)}, m) = \min\{\lambda, p_2^{(out)}(\lambda, m) > 0\}$$

and

$$R_{\max}^{(out)}(\lambda^{(1)}, m) = \max\{\lambda, p_1^{(out)}(\lambda, m) > 0\}$$

and denote

$$\beta(\lambda^{(1)}, \lambda^{(2)}, c, r, m) = \{p_c : R_{\min}^{(out)}(\lambda^{(2)}, m) = R_{\max}^{(out)}(\lambda^{(1)}, m)\}. \tag{4.37}$$

Hence for fixed $(\lambda^{(1)}, \lambda^{(2)}, c, r, m)$, $\beta(\lambda^{(1)}, \lambda^{(2)}, c, r, m)$ gives us the critical value of p_c and we call it the discrimination capacity of the neuron (under m samplings).

Behaviour of $\beta(\lambda^{(1)}, \lambda^{(2)}, c, r, m)$

As in the previous subsection, we conclude that $\beta(\lambda^{(1)}, \lambda^{(2)}, c, r, m)$ is the solution of the following equation (p_c).

$$\int_0^{V_{thre}L} g\left(\frac{y - a[p_c\lambda^{(1)} + (\sum_{i=p_c+1}^{p}\lambda_i)^{(m)}](1-r)}{a\sqrt{[\lambda^{(1)}p_c(1+c(p_c-1)) + (\sum_{i=p_c+1}^{p}\lambda_i)^{(m)}](1+r)}}\right) dy$$
$$= \frac{\sqrt{[\lambda^{(1)}p_c(1+c(p_c-1)) + (\sum_{i=p_c+1}^{p}\lambda_i)^{(m)}]}}{\sqrt{[\lambda^{(2)}p_c(1+c(p_c-1)) + (\sum_{i=p_c+1}^{p}\lambda_i)_{(m)}]}}$$
$$\cdot \int_0^{V_{thre}L} g\left(\frac{y - a[p_c\lambda^{(2)} + (\sum_{i=p_c+1}^{p}\lambda_i)_{(m)}](1-r)}{a\sqrt{[\lambda^{(2)}p_c(1+c(p_c-1)) + (\sum_{i=p_c+1}^{p}\lambda_i)_{(m)}](1+r)}}\right) dy. \quad (4.38)$$

The critical value $\beta(\lambda^{(1)}, \lambda^{(2)}, c, r, m)$ can be analytically found in the two most interesting cases: $c = 0$ and $r = 1$. Define

$$0 \le \Theta = \frac{\lambda^{(2)} - \lambda^{(1)}}{\sigma(\lambda_p)}.$$

This corresponds to the parameter Λ defined in the previous subsection.

Theorem 4 *For* $\Theta > 0$,

- *when* $c > 0$ *we have*

$$\beta(\lambda^{(1)}, \lambda^{(2)}, c, 1, m) < \beta(\lambda^{(1)}, \lambda^{(2)}, 0, r, m) \quad (4.39)$$

  and furthermore $\beta = \beta(\lambda^{(1)}, \lambda^{(2)}, c, 1, m)$ *is the solution of the following equation:*

$$\Theta\beta(1 + c(\beta - 1)) = 2\sqrt{p - \beta}\, b_m \quad (4.40)$$

  provided that the approximations (4.35) and (4.36) are used.

- *When* $c = 0$ *we have*

$$\beta(\lambda^{(1)}, \lambda^{(2)}, 0, r, m) = \frac{2b_m\left[\sqrt{b_m^2 + p\Theta^2} - b_m\right]}{\Theta^2} \quad (4.41)$$

  independent of r*, provided that the approximations (4.35) and (4.36) are used.*

- *When* $c < 0$ *we have*

$$\beta(\lambda^{(1)}, \lambda^{(2)}, c, 1, m) > \beta(\lambda^{(1)}, \lambda^{(2)}, 0, r, m). \tag{4.42}$$

Again from numerical simulations (see Figure 4.8 and Figure 4.9) we conclude:

- *When* $c > 0$ *we have*

$$\beta(\lambda^{(1)}, \lambda^{(2)}, c, r_2, m) < \beta(\lambda^{(1)}, \lambda^{(2)}, c, r_1, m) \tag{4.43}$$

where $1 \geq r_2 > r_1 > 0$.

- *When* $c < 0$ *we have*

$$\beta(\lambda^{(1)}, \lambda^{(2)}, c, r_2, m) > \beta(\lambda^{(1)}, \lambda^{(2)}, c, r_1, m) \tag{4.44}$$

where $1 \geq r_2 > r_1 > 0$

Again it is interesting to note that (4.41) and (4.40) are independent of a, V_{thre}, and L, three essential parameters in the integrate-and-fire model. In other words, the results of (4.41) and (4.40) of the integrate-and-fire model are *universal*. In Figure 4.7 we plot β vs. Θ for $c = 0$ and $c = 0.1$ according to (4.41) and (4.40). For a given Θ and $c = 0.1$, the solid line in Figure 4.7 gives us the smallest number of coherently synaptic inputs for an integrate-and-fire model to discriminate between input signals for $r \in [0, 1]$. Hence the solid line in Figure 4.7 is the smallest discrimination capacity of an integrate-and-fire model with $c = 0.1$ if we assume that $r \in [0, 1]$. It is worth pointing out that the lowest limit of β is about $\beta = 14$.

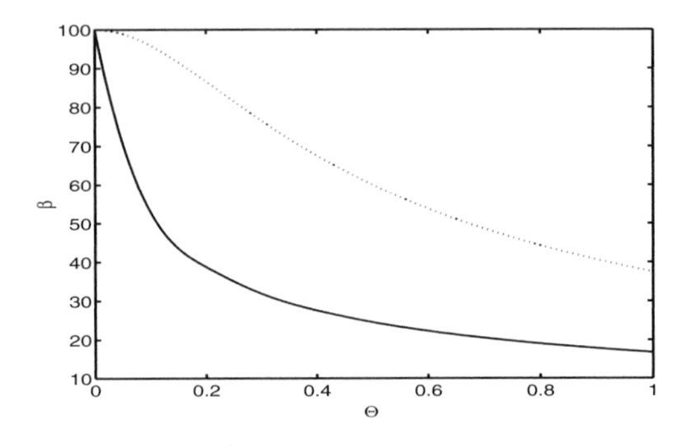

Fig. 4.7. β vs. Θ for $c = 0, r = 0$ (dotted line, independent of r according to Theorem 3) and $c = 0.1, r = 1$ (solid line) with $m = 100$.

Input-Output Relationship

As before, we can assess the relationship between input and output histograms. We can prove the following conclusions:

Theorem 5 *For the integrate-and-fire model,*

- *if* $c > 0$ *we have*

$$R_{\min}^{(out)}(\lambda^{(2)}, m) - R_{\max}^{(out)}(\lambda^{(1)}, m) > 0$$

$$\textit{when} \quad R_{\min}^{(in)}(\lambda^{(2)}, m) - R_{\max}^{(in)}(\lambda^{(1)}, m) = 0;$$

- *if* $c = 0$ *we have*

$$R_{\min}^{(out)}(\lambda^{(2)}, m) - R_{\max}^{(out)}(\lambda^{(1)}, m) = 0$$

$$\textit{when} \quad R_{\min}^{(in)}(\lambda^{(2)}, m) - R_{\max}^{(in)}(\lambda^{(1)}, m) = 0;$$

- *if* $c < 0$ *we have*

$$R_{\min}^{(out)}(\lambda^{(2)}, m) - R_{\max}^{(out)}(\lambda^{(1)}, m) < 0$$

$$\textit{when} \quad R_{\min}^{(in)}(\lambda^{(2)}, m) - R_{\max}^{(in)}(\lambda^{(1)}, m) = 0.$$

4.4.4 Numerical Results

Let us now consider the minimum total probability of misclassification (TPM) defined by

$$\begin{aligned}\text{TPM} =& \frac{1}{2}P(\text{misclassfied as } \lambda^{(2)}|\text{input is } \lambda^{(1)})\\ &+\frac{1}{2}P(\text{misclassfied as } \lambda^{(1)}\ |\text{input is } \lambda^{(2)}).\end{aligned}$$

For example, in Figure 4.8, we see that TPM (in percentile) for the left upper panel is about 13.5% and for the right upper panel is 5.5%. Therefore, adding inhibitory inputs to the neuron considerably improves its discrimination capability, reducing TPM from 13.5% to 5.5%.

The parameters used in simulating the IF model are $V_{thre} = 20mV$, $V_{rest} = 0\,mV, L = 1/20$, $a = b = 1\,mV, p = 100$, $\lambda^{(1)} = 25$ Hz, and $\lambda^{(2)} = 75$ Hz. A refractory period of 5 msec is added for all numerical results of efferent

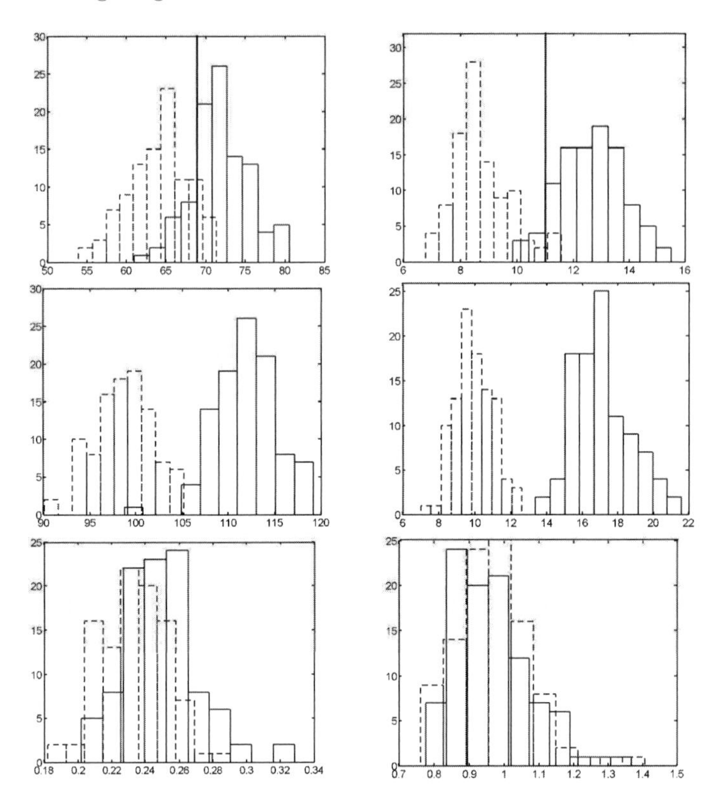

Fig. 4.8. Upper and middle panel, histogram of firing rates (Hz) with $c = 0.1$ for the IF model. Left, exclusively excitatory inputs $r = 0$. Right, $r = 0.95$. Upper panel: $p_c = 15$. The minimum TPM is calculated according to the thick vertical lines (the optimal discrimination line). Middle panel: $p_c = 25$. Bottom panel: histograms of coefficient of variation (CV) for $p_c = 25$ and left with $r = 0$, right with $r = 0.95$, corresponding to the middle panel.

firing frequency. For each fixed set of parameters of the model, 100 spikes are generated to calculate each mean, standard deviation, etc. The histograms are obtained using 100 firing rates, that is, $m = 100$.

It is interesting to compare numerical results with theoretical results in the previous subsections. From previous subsections we have

$$\alpha(25, 75, 0.1, 1) = 32.5133$$

and

$$\beta(25, 75, 0.1, 1, 100) = 12.1251 \qquad \beta(0, 100, 0.1, 1, 100) = 7.8058.$$

From Figure 4.9 (right) we conclude that the discrimination capacity would be between 15 and 20. The discrimination capacity from actual numerical simulations for $r = 1$ closes to $\beta(25, 75, 0.1, 1, 100)$.

Numerical simulations have been extensively carried out for the integrate-and-fire model and the IF-FHN model. All results are in agreement with our theoretical results in the previous subsections and were reported in a meeting [21].

4.4.5 Discussion

We have considered the problem of discriminating between input signals in terms of an observation of efferent spike trains of a single integrate-and-fire neuron. We have demonstrated, both theoretically and numerically, that two key mechanisms to improve the discrimination capability of the model neuron are to increase inhibitory inputs and increase correlation between coherent inputs. Analytical results for the two most interesting cases, $c = 0$ and $r = 1$, are obtained and the results are independent of model parameters.

Our results offer answers to a few issues that were extensively discussed in the literature. We simply summarize two of them.

First increasing inhibitory inputs and increasing correlations between coherent inputs can enhance discrimination capacity of a neuron. However, on the other hand we all know that increasing inhibitory inputs and correlations between inputs increase its output variability of efferent spike trains, which will simply broaden the efferent firing rate histograms and so reduce the discrimination capacity of a neuron. It seems our results here simply contradict this. Nevertheless, we must note that all theoretical results in sections 4.4.3 and 4.4.2 are obtained under the assumption that the efferent firing rates are exactly obtained. Results in section 4.4.4 clearly demonstrate that theoretical results in sections 4.4.3 and 4.4.2 are true even when the number of spikes used to obtained the firing rates histogram is small (100 spikes). In general our results reveal that a neuron system faces two opposite requirements: to obtain the mean firing rates as exactly as possible by reducing the variability of output spike trains (reducing inhibitory inputs and input correlations), and to increase the discrimination capacity by increasing inhibitory inputs and input correlations. To elucidate our points further, in Figure 4.10, we plot the firing rate histograms, using the identical parameters as in Figure 4.8 (middle panel, right), but with 10 spikes to estimate the mean, rather than 100 spikes. It is clearly shown that the firing rate histograms in Figure 4.10 are less widely separated than in Figure 4.8, middle panel (right), and it is impossible to perfectly distinguish between two inputs. A neuronal system must find a compromise way to resolve the issue. How to find an optimal trade-off between the two requirements is an interesting research topic.

Second, many publications argue that there is an optimal value of noise at which a neuronal system can optimally extract information. Nevertheless, our results indicate that the optimal point is simply the one at which the neuron's output is most variable. We thus conclude that the larger the noise, the better

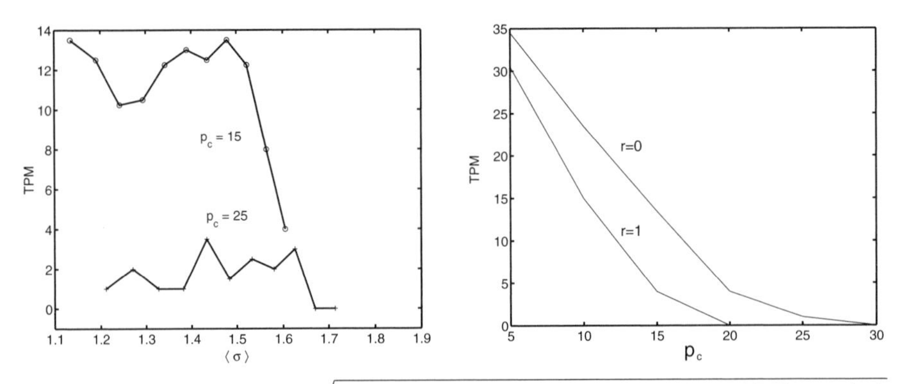

Fig. 4.9. TPM % vs. $\langle\sigma\rangle = \sqrt{a^2(1+r)\lambda^{(1)}p_c(1+c(p_c-1)) + a^2(1+r)\sum_{i=p_c+1}^{p}\langle\lambda_i\rangle}$ with $\langle\lambda_i\rangle = 50$ Hz (left) and TPM vs. p_c (right) for the IF model, $r \in [0,1], c = 0.1, \lambda^{(1)} = 25$ Hz. When $p_c = 15$ (left), it is clearly shown that TPM attains its optimal value at $r = 1$, that is, the larger the noise, the better the discrimination (see the right figure as well). All other parameters are the same as in subsection 4.4.4.

for the neuron system (see the paragraph above and Figure 4.9) to separate masked signals. This confirms the fact that noise is useful in a neural system, but not via the form of stochastic resonance.

The only assumption we introduced in the model is that coherent signals are more correlated than random signals. This seems a quite natural assumption given the structured cortical areas. Figure 4.11 illustrates the point. Coherent signals are transmitted by neurons grouping together (cortical columns) and neurons in the same column are bound to fire with a correlation. In contrast, noisy (distorted) signals are less correlated.

The integrate-and-fire model is the simplest model in theoretical neuro-

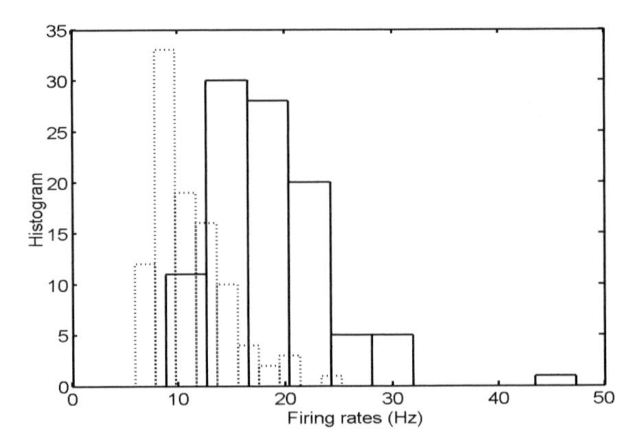

Fig. 4.10. Histogram of firing rates (Hz) with $c = 0.1$ for the IF model, with identical parameters as in Figure 4.8, bottom panel (right), but only 10 spikes are used for estimating mean firing rates.

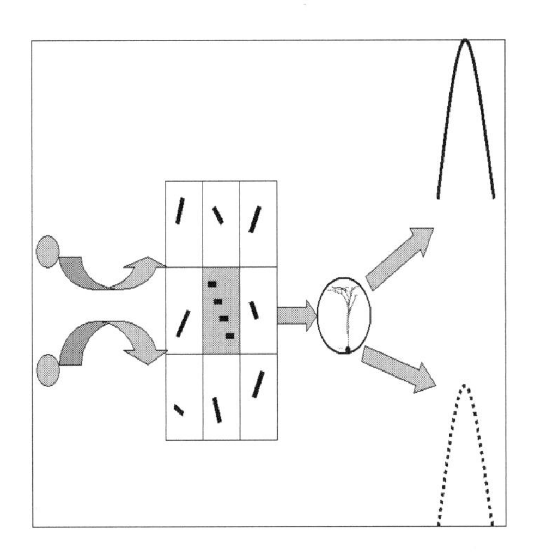

Fig. 4.11. Correlations between coherent signals can be naturally introduced via neuronal pathways.

science. One might easily argue that the model is too simple to be true for a realistic biological neuron. Nevertheless, we have seen in the past few years that the integrate-and-fire model fits well with many experimental data (see for example, [3, 29, 38]) and still serves as a canonical model of neurons. On the other hand, we have tested results here on other models as well ([21]. It would be very interesting to investigate the impact of adaptation and dynamical synapses [29] on discrimination tasks.

Currently we are also working on building a system that mimics experiments carried out in [44, 56] with random dot stimuli and making judgements on dot-moving directions [26].

4.5 Multineuron Activity: Correlation

In the past decade, we have witnessed the prevailing of the synchronization theory in nervous systems. Certainly two (partially) synchronized spike trains are positively correlated. Are two spike trains recorded in *in vivo* experiments exclusively positively correlated? This is a question asked by many experimentalists. Here we first present some of our recent data from *in vivo* multi-electrode array recording in the rat olfactory bulb.

Noise observed in all in vivo experiments in the nervous system is one of the main curbs to determining the computational principles employed by the brain, despite a century of research. If the individual neuron behaves unreliably, how we can recognise a face, react to a stimulus, and catch a ball in such an accurate way? One might argue that the noise in the nervous system can be averaged out if a population coding strategy is implemented. In fact, it is pointed out in the literature [76] that increasing the number of neurons cannot enhance the signal-to-noise ratio without limitations, if neurons are positively correlated. Of course, a pair of positively correlated spike trains implies that they are synchronized in a certain degree, and the synchronization assumption was very popular in the literature in past decades.

However, we have found evidence confirming the existence of a negative correlation rather than positive correlations in *in vivo* experiments on the rat's olfactory bulb [60]. What is the functional meaning of our findings [17, 22]? To answer the question, we run a few simulations here. Assume there are 100 neurons and the ith neuron sends out a spike train with an instantaneous rate

$$\xi_i = \sqrt{\sin(2\pi t) + 1.5}\eta_i + \sin(2\pi t) + 1.5$$

where η_i is noise (normally distributed random variable with mean 0 and variance 1). Note that the firing rate takes the form of Poisson processes: its noise magnitude is proportional to the signal term [57]. In Figure 4.12 (left), we depict the summation of all $N = 65$ neurons firing rates with various correlation between η_i. The correlation between the ith and jth neuron is $-c^n$ (positive) when $i, j < N^p, N^p = 1, \cdots, 65$ and otherwise is negative as specified in the figure. It is clearly seen that the noise level is significantly reduced when the correlation becomes more negative, that is, N^p decreases. When the correlation between neurons is around $c^n = -0.015$ and $N^p = 0$, the obtained signal is almost perfect.

The mechanism behind the observed phenomenon in Figure 4.12 is essentially due to the central limit theorem. Suppose that we have N normally distributed random variables [(instantaneous) firing rate of N neurons] $\xi_i, i = 1, \cdots, N$ with mean and covariance matrix

$$\sigma_{ij} = \begin{cases} (\sin(2\pi t) + 1.5)(\delta_{ij} - c^n(1 - \delta_{ij}), & i, j = 1, \cdots, N^p, \\ (\sin(2\pi t) + 1.5)(\delta_{ij} + c^n(1 - \delta_{ij}), & i, j \notin \{1, , N^p\} \end{cases}$$

From the central limit theorem we know that

$$\frac{\xi_1 + \cdots \xi_N - N(\sin(2\pi t) + 1.5)}{\sqrt{N}} \to N\left(0, \frac{\sum \sigma_{ij}}{N}\right)$$

where $N(\cdot, \cdot)$ is the normal distribution. Note that the variance term can be decomposed into two terms: the first is $i = 1, \cdots, N, \sigma_{ii}$, which is always positive, and the remaining term could be positive or negative depending on the correlation between neuronal activity. With a positive correlation, the variance will increase rather than reduce. In other words, the noise level is increased. But with a negative correlation, the variance could be dramatically reduced. In Figure 4.2, right, we plot the variance term with $N^p = 0, 1, \cdots, 65$. When $N^p = 0$, all neurons are negatively correlated. When $N^p = 65$, all neurons are positively correlated, N^p is around 45, and the variances is 1,

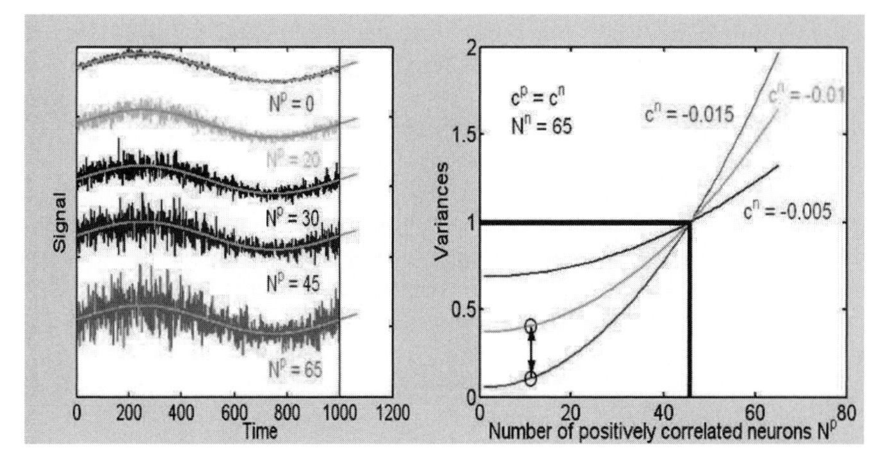

Fig. 4.12. Left, green lines are the true signals, and jagged lines are averaged signals with various correlation levels as indicated in the figure. It is easily seen that an almost perfect signal is obtained when the neuron is negatively correlated with a correlation coefficient of -0.015. Right, theoretically calculated variance vs. number of positively correlated neurons N^p. Curves in the left figure correspond to marked points of the red curve. For example, when $N^p = 0$, it is the lowest point of the red curve, $N^p = 45$ is the point where two thick black lines meet. From our experimental data, if we assume that 10 neurons are positively correlated (indicated by circles), that is, there are 100 positive correlations, the improvement of the signal-to-noise ratio is considerably significant.

that is, all neurons are independent. Corresponding to Figure 4.12 left, we see that when $N^p = 0$, the variance is almost zero with $c^n = -0.015$. From Figure 4.12 left, we see that the difference between $N^p = 30$ and $N^p = 45$ is not easy to visually detect. However, theoretical results (right) tell us that there is a difference around 0.4.

Hence the negatively correlated terms contribute to reduce the variance, while positively correlated terms result in an increase of the variance. Our results above tell us that even a single neuron is very unreliable: it could behave as a Poisson process (the noise magnitude is proportional to the signal as we used here), and the obtained signal in terms of a group of neurons could be almost perfect. In other words, an almost perfect signal can be extracted, as long as the neuronal activity is negatively correlated, as reported in our biological data [60].

In the literature, the redundancy reduction principle is widely regarded as one of the operation principles in the brain [6]. However, recently it is agreed that the principle has difficulty in explaining the experimental data, such as why the number of cells in V1 is much larger than the number of cells in LGN. In fact, the negatively correlated computation principle is a step further than the redundancy reduction principle and is not in contradiction with the known facts; the functional meaning is clear. In Figure 4.13, we plot 65 random numbers generated with mean zero and variance 16 with correlation −0.015 and 0.8. With a negative correlation, the space is more completely and symmetrically explored, resulting in a mean of −0.0080. With a positive correlation, the situation is the worst, as only a portion of the whole space is covered by the points and the mean is −0.72. The figure also demonstrates another possible role of a negative correlated neuronal activity in discrimination tasks. We all know that the further away the distance between patterns to be discriminated, the easier the discrimination task is. Obviously negative correlation is a mechanism to push each pattern to the corner of its feature space and so make the discrimination task easier, in comparison with the independent case, that is, the redundancy reduction principle.

It might not be surprising to observe the negatively correlated activity in the nervous system. It is known that the interaction between neurons in the olfactory bulb is lateral inhibitory, which is the natural cause of negative correlation between neuronal activity.

In the previous sections, we have presented several examples to explain the importance of the second-order statistics—both variances and correlations. Can we develop a general theory to describe network activity including both the first- and second-order statistics? In the next section, we develop such a theory.

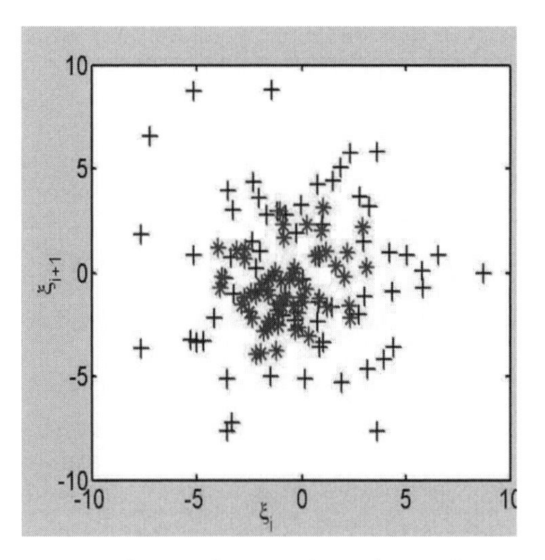

Fig. 4.13. Returning map of 65 random numbers, $(\xi_i, \xi_{i+1})i = 1, 2, \cdots, 64$. +, point with a negative correlation of −0.015; * random number with a positive correlation of 0.8.

4.6 Moment Neural Networks

We confine ourselves to feedforward networks (Figure 4.14) only, although it is readily seen that we can easily generalize our results below to feedback and recurrent networks.

For two given quantities V_{thre} (threshold) $> V_{rest}$ (resting potential) and when $v_i^{(k+1)}(t) < V_{thre}$, the membrane potential of the ith neuron in the $(k+1)$th layer $v_i^{(k+1)}(t)$ satisfies the following dynamics:

$$\begin{cases} dv_i^{(k+1)}(t) = -L(v_i^{(k+1)}(t) - V_{rest})dt + dI_{i,syn}^{(k+1)}(t) \\ v_i^{(k+1)}(0) = V_{rest} \end{cases} \tag{4.45}$$

where L is the decay rate, and $I_{i,syn}^{(k+1)}(t)$ is the synaptic input given by

$$dI_{i,syn}^{(k+1)}(t) = \sum_{j=1}^{p^{(k)}} w_{ij}^{E,(k)} dN_j^{E,(k)}(t) - \sum_{j=1}^{q^{(k)}} w_{ij}^{I,(k)} dN_j^{I,(k)}(t)\,. \tag{4.46}$$

Here $w_{ij}^{E,(k)}$ is the magnitude of EPSPs, $w_{ij}^{I,(k)}$ is the magnitude of IPSPs, $N_i^{E,(k)}$ and $N_j^{I,(k)}$ are renewal processes (EPSPs and IPSPs) arriving from the ith and jth synapse with interspike intervals $T_{im}^{E,(k)}, m = 1, 2, \cdots$, and $T_{jm}^{I,(k)}, m = 1, 2, \cdots$, $p^{(k)}$ and $q^{(k)}$ are the total number of active excitatory and inhibitory synapses in the kth layer. Once $v_i^{(k+1)}(t)$ is greater than V_{thre},

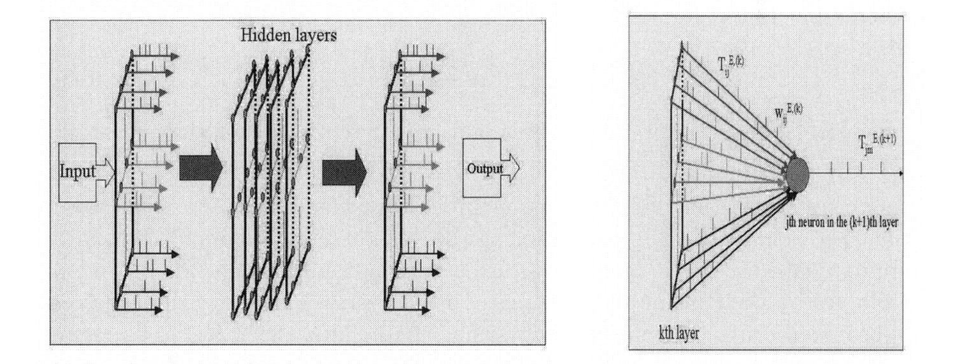

Fig. 4.14. Schematic plot of a feedforward spiking network (left). Right, a single neuron in the $(k+1)$ layer receives inputs from the ith layer with weights (EPSP size) $w_{ij}^{E,(k)}, i = 1, \cdots, p^{(k)}, j = 1, \cdots, p^{(k+1)}, k = 1, 2, \cdots$.

it is reset to V_{rest}. The model defined by (4.45) is called the (leaky) integrate-and-fire model [57]. In the sequel, we define $T_0 = 0$ and

$$T_{ij}^{(k+1)} = \inf\{t : v_i^{(k+1)}(t) \geq V_{thre}, t > T_{i(j-1)}^{(k+1)}\}, \qquad j = 1, 2, \cdots \tag{4.47}$$

as the firing time (interspike intervals), see Figure 4.14 right.

According to the renewal theorem, we have

$$\frac{dN_i^E(t)}{dt} \to N\left(\frac{1}{\langle T_{ij}^{E,(k)}\rangle}, \frac{[\langle (T_{ij}^{E,(k)})^2\rangle - \langle T_{ij}^{E,(k)}\rangle^2]}{\langle T_{ij}^{E,(k)}\rangle^3}\right)$$

where $N(\cdot,\cdot)$ is the normal distribution.

In fact, a more detailed calculation tells us that

$$\begin{cases} \langle N_i^{E,(k)}\rangle = \dfrac{t}{\langle T_{ij}^{E,(k)}\rangle} + \dfrac{\mathrm{var}(T_{ij}^{E,(k)}) - (\langle T_{ij}^{E,(k)}\rangle)^2}{2(\langle T_{ij}^{E,(k)}\rangle)^2} + o(1) \\ \mathrm{var}(T_{ij}^{E,(k)}) = \dfrac{[\langle (T_{ij}^{E,(k)})^2\rangle - \langle T_{ij}^{E,(k)}\rangle^2]t}{\langle T_{ij}^{E,(k)}\rangle^3} + c \end{cases} \tag{4.48}$$

where c is a constant depending on the first-, second- and third-order statistics of $T_{ij}^{E,(k)}$. Hence N_t^E is in general not a Markov process due to the constant term independent of t in (4.48).

Let us introduce more notation here. Define

$$\mu_{ij}^{E,(k)} = \frac{1}{\langle T_{ij}^{E,(k)}\rangle + T_{ref}}, \qquad (\sigma_{ij}^{E,(k)})^2 = \frac{[\langle (T_{ij}^{E,(k)})^2\rangle - \langle T_{ij}^{E,(k)}\rangle^2]}{\langle T_{ij}^{E,(k)}\rangle^3}$$

where T_{ref} is the refractory period. For the ith neuron in the kth layer, we have

$$dN_i^{E,(k)}(t) \sim \mu_{i1}^{E,(k)} dt + \sigma_{i1}^{E,(k)} dB_i^{E,(k)}(t) \tag{4.49}$$

where $B_i^{E,(k)}(t)$ is the standard Brownian motion with a correlation coefficient $\rho_{ij}^{(k)}, i, j = 1, \cdots, N^{(k)}$. Of course, (4.54) is an approximation: the so-called usual approximation [57] to Poisson process is a special case[3]. The essential idea behind the approximation is the central limit theorem, that is, to approximate the input pulses that are a summation of independent random variables by their mean and variance. To the best of our knowledge, such an approximation to the renewal process has not be introduced in the literature and we expect our approach could be quite significant and open up many new

[3] To have a more accurate approximation, we have to resort to the similar idea employed in [20]. For the IF model, which is linear before resetting, the approximation developed above without the constant term is quite accurate already (not shown here).

problems to be further addressed. We also stress here that in the literature there are many papers devoted to investigating the network activity of spiking neurons. However, all of them are confined to the case of Poisson inputs, which is of course not satisfactory since the output of an integrate-and-fire model is no longer a Poisson process (hence the self-consistency issue has to be checked) but is surely a renewal process. Summarizing the results above, we obtain

$$\sum_{j=1}^{p^{(k)}} w_{ij}^{E,(k)} dN_j^{E,(k)}(t) = \sum_j w_{ij}^{E,(k)} \mu_{j1}^{E,(k)} dt + \sum_j w_{ij}^{E,(k)}(t) \sigma_{j1}^{E,(k)} dB_j^{E,(k)}(t) .$$

Let us further suppose that $p^{(k)} = q^{(k)}$ and for the simplicity of notation

$$\begin{aligned} \mu_j^{E,(k)} &= \mu_{j1}^{E,(k)}, & \sigma_j^{E,(k)} &= \sigma_{j1}^{E,(k)}, \\ r\mu_i^{(k)} &= r\mu_i^{E,(k)} = \mu_i^{I,(k)}, & w_{ij}^{(k)} &= w_{ij}^{E,(k)} = w_{ij}^{I,(k)} \end{aligned} \tag{4.50}$$

and

$$r\sigma_i^{(k)} = r\sigma_i^{E,(k)} = \sigma_i^{I,(k)} \tag{4.51}$$

where $i = 1, \cdots, p^{(k+1)}, j = 1, \cdots, p^{(k)}$, and r is the ratio between inhibitory inputs and excitatory inputs. In particular, when $r = 0$ the neuron exclusively receives excitatory inputs; when $r = 1$ the inhibitory and excitatory input is exactly balanced. Note that (4.51) is a quite strong assumption. Now we have

$$dI_{i,syn}^{(k+1)} = \sum_j w_{ij}^{(k)} \mu_j^{(k)} (1 - r) dt + \sum_j w_{ij}^{(k)}(t) \sigma_j^{(k)} \sqrt{1 + r} dB_j^{(k)}(t) \tag{4.52}$$

where $B_i^{(k)}(t), i = 1, \cdots, p^{(k)}$ are correlated Brownian motion with correlation coefficient $\rho_{ij}^{(k)}$.

In terms of Siegert's expression, we have the expression of all moments of the output firing rate of the model. For the mean we have

$$\langle T_i^{(k+1)} \rangle = \frac{2}{L} \int_{A_i^{(k)}(0))}^{A_i^{(k)}(V_{thre})} g(x) dx, \qquad A_i^{(k)}(y) = \frac{yL - \bar{\mu}_i^{(k)}}{\bar{\sigma}_i^{(k)} \sqrt{L}} \tag{4.53}$$

where

$$\bar{\mu}_i^{(k)} = \sum_j w_{ij} \mu_j^{(k)} (1 - r), \qquad (\bar{\sigma}_i^{(k)})^2 = \sum_{m,n} w_{im} w_{in} \sigma_m^{(k)} \sigma_n^{(k)} \rho_{mn}^{(k)} (1 + r) \tag{4.54}$$

and

$$g(x) = \left[\exp(x^2) \int_{-\infty}^{x} \exp(-u^2) du \right] = \begin{cases} D_-(x) + \exp(x^2) \dfrac{\sqrt{\pi}}{2} & \text{if } x \geq 0 \\ -D_-(-x) + \exp(x^2) \dfrac{\sqrt{\pi}}{2} & \text{otherwise} \end{cases}$$

with Dawson integral $D_{-}(x), x \geq 0$.

For the second-order moment variance, we know that

$$\mathrm{Var}(T_i^{(k+1)}) = \frac{4}{L^2} \int_{A_i^{(k)}(0)}^{A_i^{(k)}(V_{thre})} \exp(x^2) \cdot \left\{ \int_{-\infty}^{x} \exp(-u^2) g^2(u) du \right\} dx. \tag{4.55}$$

Finally we have to consider the correlation relationships between neuronal activity. In the literature, many publications investigated the functional role of correlations. Unfortunately, it turns out that to find a rigorous relationship between input correlation and output correlation is not an easy task. To this end, we simulate two IF neurons with synaptic inputs

$$\mu(1-r) + \sqrt{\mu(1+r)} dB^{(i)}, i = 1, 2$$

where $B^{(i)}$ are correlated Brownian motion with a correlation coefficient ρ_{in}, and μ is a constant. It is found from Figure 4.15 that the input correlation ρ_{in} and output correlation ρ_{out} are almost identical, independent of r and μ.

Hence we have the following *heuristic* relationship between the input and output correlation in the network:

$$\rho_{ij}^{(k+1)} = \frac{\sum_{m,n} w_{im}^{(k)} \sigma_m^{(k)} w_{jn}^{(k)} \sigma_n^{(k)} \rho_{mn}^{(k)}}{\sqrt{\sum_{m,n} w_{im}^{(k)} \sigma_m^{(k)} w_{in}^{(k)} \sigma_n^{(k)} \rho_{mn}^{(k)}} \cdot \sqrt{\sum_{m,n} w_{jm}^{(k)} \sigma_m^{(k)} w_{jn}^{(k)} \sigma_n^{(k)} \rho_{mn}^{(k)}}}. \tag{4.56}$$

Note that the right-hand side of (4.56) is the input correlation to the ith and the jth neuron in the $(k+1)$th layer.

Let us address the implications of (4.56). Assume that $w_{im}^{(k)} = w > 0$, $\sigma_m^{(k)} = \sigma > 0$ and $\rho_{mn}(k) = \rho, m \neq n$, then we have

$$\rho_{ij}^{(k+1)} = \frac{p^{(k)} + (p^{(k)} - 1)^2 \rho}{p^{(k)} + (p^{(k)} - 1)^2 \rho} = 1 .$$

In other words, the neuronal activity is correlated (synchronized). The synchronized spike trains in a feedforward network as we discussed here have been observed early in the literature for a feedforward spiking network. In the general case where $w_{ij}^{(k)}, \sigma_m^{(k)}$ and $\rho_{ij}^{(k)}$ are not homogeneous and the output correlation is slightly higher or lower than the input correlation, $\rho_{ij}^{(k)}$ will not be unity.

Another extreme case is that we have $w_i^{(k)} = \{w_{im}^{(k)}, m = 1, \cdots, p^{(k)}\} \perp w_j^{(k)} = \{w_{jm}^{(k)}, m = 1, \cdots, p^{(k)}\}$ (orthogonal). Since we require that $w_{im}^{(k)} \geq 0$, we conclude that $\sum_{m,n} w_{im}^{(k)} \sigma_m^{(k)} w_{jn}^{(k)} \sigma_n^{(k)} \rho_{mn}^{(k)} = 0$ and therefore $\rho_{ij}^{(k)} = 0$. For $A, B \subset \{1, \cdots, p^{(k+1)}\}$, $A \cap B = \phi$, and $w_i^{(k)} \perp w_j^{(k)}$ for $i \in A, j \in B$, then

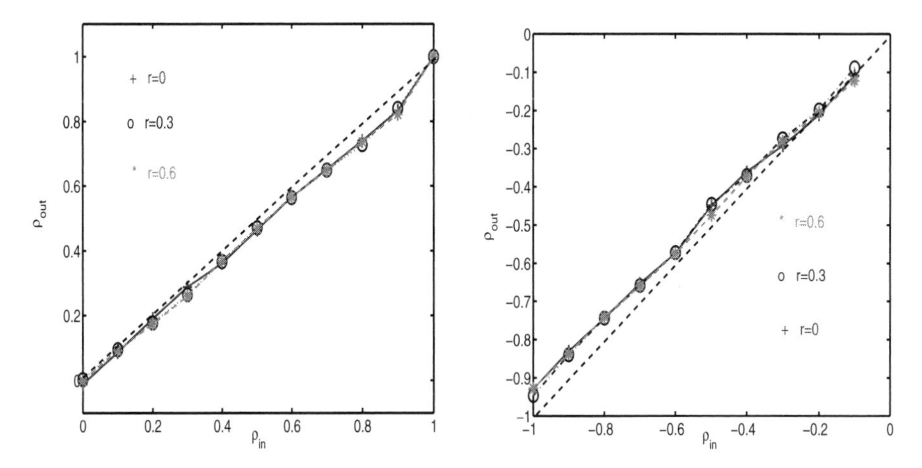

Fig. 4.15. Left, the curve of ρ_{out} vs. ρ_{in} for $r = 0$ (+), $r = 0.3$ (o) and $r = 0.6$ (*) with 5000 interspike intervals. The relationship between the input correlation coefficient and the output correlation coefficient is almost linear and independent of r. Right, ρ_{out} vs. $\rho_{in} \leq 0$. Each point is obtained with 5000 spikes.

the neuronal activity in A and B is independent; that is, patched neuronal activity is observed in this case.

In summary, from (4.22), (4.55), and (4.56), we have the following relationship between inputs and outputs:

$$(\boldsymbol{\mu}^{(k+1)}, \boldsymbol{\sigma}^{(k+1)}, \rho^{(k+1)}) = \mathcal{M}(\boldsymbol{\mu}^{(k)}, \boldsymbol{\sigma}^{(k)}, \rho^{(k)}) \tag{4.57}$$

where $\boldsymbol{\mu}^{(k)} = \{\mu_i^{(k)}, i = 1, \cdots, p^{(k)}\}, \boldsymbol{\sigma}^{(k)} = \{\sigma_i^{(k)} i = 1, \cdots, p^{(k)}\}$ and $\mathcal{M}$ is the mapping defined by (4.22), (4.55), and (4.56), (4.57) gives us the relationship of the first- and second-order moments in a spiking neuronal network and is called *moment mapping*, which is one of the central results in MNN.

4.6.1 Application: Spontaneous Activity

The first question we address here is: How can spontaneous activity be maintained in a feedforward network? In our simulation typical parameters used are $V_{rest} = 0\,mV, L = 1/20\,msec^{-1}$, and $V_{th} = 20\,mV$, in agreement with most published results [39, 55]. The simulations are carried out with Matlab [61].

Clamped Correlation

We first consider an ideal case: all weights, afferent mean, and variance are identical, that is, the network is homogeneous; (4.57) is now reduced to

$$(\boldsymbol{\mu}^{(k+1)}, \boldsymbol{\sigma}^{(k+1)}, \rho^{(k+1)}) = \mathcal{M}(\boldsymbol{\mu}^{(k)}, \boldsymbol{\sigma}^{(k)}, \rho^{(k)}) \tag{4.58}$$

with $\mu_1^{(k)} = \mu_j^{(k)}, \sigma_1^{(k)} = \sigma_j^{(k)}, \rho_{12}^{(k)} = \rho_{mn}^{(k)}, m \neq n$. As we discussed before, now the propagation of correlation becomes trivial: it will simply reach 1. To prevent it from happening, we clamp the correlation coefficient in the current subsection. From experimental data, we know that correlation coefficient has a mean around 0.1. In all simulation below, we also fixed $w = w_i = 0.5$.

In Figure 4.16, we plot simulation results for various $\mu^{(1)}$ and $\sigma^{(1)}$ [$C_i^{(k)} = \sigma_i^{(k)}/\mu_i^{(k)}$, the coefficient of variation (CV), is used] with $p^{(k)} = 100, \rho_{ij}^{(k)} = 0, 0.1, i \neq j$. For example, in the upper panel left, for the left branch, we simulate the model with initial state as specified in the figure. From left to right, the ratio r is 0.4, 0.3, 0.2, 0.1, 0.0. Each data point is $(\mu^{(k)}, \sigma^{(k)})$ and we connect $(\mu^{(k)}, \sigma^{(k)})$ with $(\mu^{(k+1)}, \sigma^{(k+1)})$ by a straight line. From Figure 4.16, upper panel and bottom left, $\rho_{ij}^{(k)} = 0$, and we see that there are only two possibilities: either the cell becomes completely silent[4] or fires with a relatively high frequency. In all of our simulations, we have never found that values of r between 0 and 1 lead to a system that is stable with a firing rate in accordance with experimental observations, that is, with a firing rate below 10 Hz. To confirm our conclusions, in Figure 4.16 bottom right, we carry out a more detailed simulation with $\rho_{ij}^{(k)} = 0.1, r = 0.25, 0.26, 0.27, 0.29, 0.3$. It is easily see that the network activity converges to either a highly firing frequency or completely silent state.

Having observed this difficulty: the network becomes either too active or completely silent, we then try to explore wider parameter regions. From Figure 4.17, upper panel left, it is concluded that the ratio r must be significantly higher than for exactly balanced input to make the network become stable. When $r = 2.3$, the network settles down to a state with a firing rate below 10 Hz and a coefficient of variation around 1.2. This also validates our approach since the output process is surely not a Poisson process, but a renewal process.

The reason why such a high inhibitory is needed in a network is clear. The input-output firing rate relationship of a neuron takes the shape of a sigmoidal function or a skew-bell shape function starting from the origin (see Figure 8.13 at page 310, middle panel in [28]). Hence there are always three or one fixed point when the output is sigmoidal. When there are three fixed points, two of them are stable. When the initial input is high, the firing rate

[4] We stop the simulation if the simulation time is exceedingly long, which implies that the cell is silent. For example, in Figure 4.16 upper trace left, for the left branch, with $r = 0.3$, the next point should have a firing rate of zero.

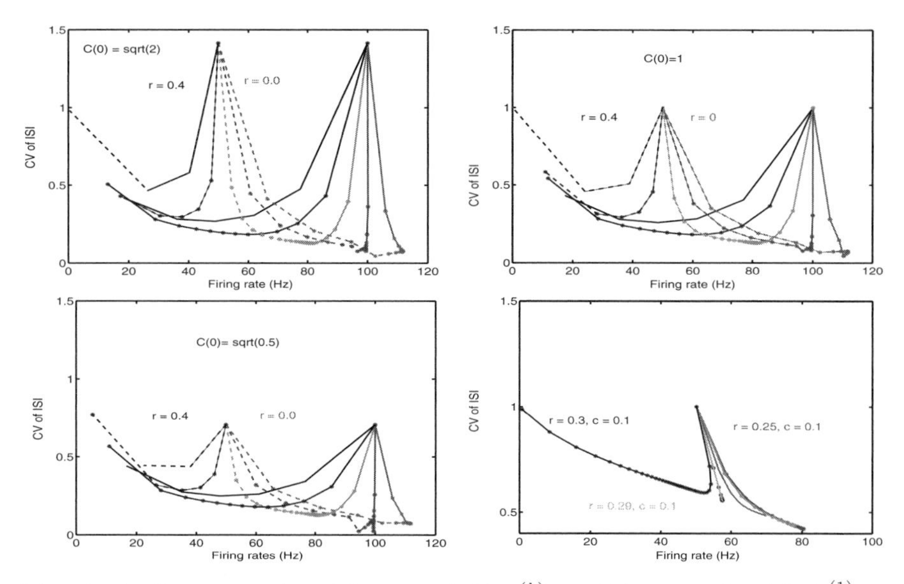

Fig. 4.16. Upper panel and bottom (left) $p^{(k)} = 100, \rho_{ij} = 0$, with $C_1^{(1)} = \sqrt{0.5}, 1, \sqrt{2}$ and $\mu_1^{(1)} = 100$ Hz (solid lines), 50 Hz (dashed lines). When $r \geq 0.3$, the neuron becomes silent (simulations stop when $\langle T \rangle$ is too large). Bottom panel (right) with $r = 0.25, 0.26, 0.27, 0.29, 0.3$, $\rho_{ij}^{(k)} = 0.1$, and $C_i^{(1)} = 1, \mu_i^{(1)} = 50$ Hz.

will become stable at a point with a positive firing rate. Otherwise, it becomes silent. To make sure that the network operates at a low but positive firing rate region, we need a strong excitatory input (see Figure 4.17 bottom panels). A strong excitatory input implies a high frequency of stable point (see Figure 4.17, bottom panel left with $r = 0$ and $r = 1$). To get a low frequency of stable firing, a significant inhibitory input is needed. In fact, from Figure 4.17, we conclude that to obtain a stable, low firing rate the input-output relationship should be a skewed bell shape, rather than the usual sigmoidal function. When $r = 1$, we see that the output firing rate is an increasing function of input firing rate and hence the input-output relationship is still sigmoidal. To have a skewed bell shape input-output relationship requires $r > 1$ since the output firing rate is no longer an increasing function of the input firing rate. For a bell shape input-output function, we know that there might be two fixed points and the positive fixed point could be stable. Of course, a bell-shaped input-output relationship possibly leads to more complicated dynamics such as a limit cycle or chaotic activity.

Furthermore, with a larger $p^{(k)}$ (for example, $p^{(k)} > 250$ in Figure 4.17), we can have a wider encoding and decoding region. When $p^{(k)} = 200$, the output firing rate is either greater than 40 Hz or silent, but when $p^{(k)} > 250$, the output firing rate is continuous. We have also tested our results for extreme parameters, for example, $V_{thre} = 10$ [39] and with refractory periods. Our conclusions all remain true (not shown).

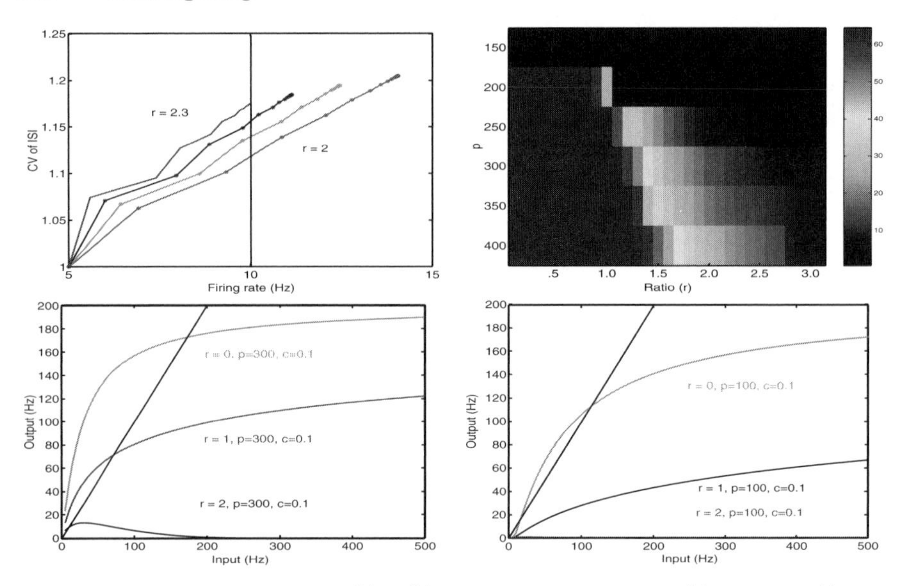

Fig. 4.17. Upper panel, left, $(\mu_i^{(k)}, C_i^{(k)})$ for $k = 1, 2, \cdots$ with $\mu_i^{(1)} = 5\text{Hz}$, $C_i^{(i)} = 1$ and $p^{(k)} = 300, \rho_{ij} = 0.1, r = 2., 2.1, 2.2, 2.3$. Upper panel, right, output firing rate (Hz) indicated by color bar with $\mu_i^{(1)} = 5$ Hz and $C_i^{(1)} = 1$. Bottom panel, input-output relationship for $p^{(k)} = 300$ and $p^{(k)} = 100$.

In the literature, the exactly balanced spiking neuronal network assumption is widely accepted (see for example [39, 55, 59]). However, with biologically reasonable parameters [55], we found that a feedforward spiking neuronal network cannot maintain a low-frequency spontaneous activity. Hence a *much* strong inhibitory input is needed. It was recently estimated that the magnitude of IPSP is around fives times larger than EPSP, that is, $b/a = 5$ (see also [28], page 239).

Finally, we have to emphasize here that all results are obtained for homogeneous networks.

Due to space limitations we will present our detailed results elsewhere [24].

4.7 Conclusion

We first presented some examples to show the importance of high-order statistics in neuronal computations. The examples include simple input and output surfaces, optimal control models, noise reduction, and storage capacity. Moment neuronal networks (first- and second-order statistics) are developed and some simple applications are included.

Our approach is reminiscent of the hydrodynamics limit in dealing with fluid dynamics, where the Navier-Stokes equation is the ultimate description of fluid dynamics. Of course, our development is more complex than the derivation of the Navier-Stokes equation where only the first-order statistics are considered.

Acknowledgement. Partially supported by grants from UK EPSRC (GR/R 54569), (GR/S 20574), and (GR/S 30443).

References

1. Abbott LF, Dayan P (1999) The effect of correlated variability on the accuracy of a population code. Neural Comput 11:91–101.
2. Abbott LF, van Vreeswijk C (1993) Asynchronous states in networks of pulse-coupled oscillators. Phys Rev E 48:1483–1490.
3. Abbott LF, Varela JA, Sen K, Nelson SB (1997) Synaptic depression and cortical gain control. Science 275:220–223.
4. Amit DJ, Tsodyks MV (1991) Quantitative study of attractor neural networks retrieving at low spike rates. Network 2:259–274.
5. Amit DJ, Brunel N (1997) Model of global spontaneous activity and local structured activity during delay periods in the cerebral cortex. Cerebral Cortex 7:237–252.
6. Barlow H (2001) Redundancy reduction revisited. Network-Comp Neural 12:241–253.
7. Bell AJ, Sejnowski TJ (1995) An information maximization algorithm that performs blind separation. Adv Neural Inform Process Sys, MIT Press, 7:456–474.
8. Bezzi M, Diamond ME, Treves A (2002) Redundancy and synergy arising from pairwise correlations in neuronal ensembles. J Comput Neurosci 12:165–174.
9. Britten KH, Shadlen MN, Newsome WT, Celebrini S, Movshon JA (1992) The analysis of visual motion: a comparison of neuronal and psychophysical performance. J Neurosci 12:4745–4765.
10. Brown D, Feng J, Feerick S (1999) Variability of firing of Hodgkin-Huxley and FitzHugh-Nagumo neurons with stochastic synaptic input. Phys Rev Lett 82:4731–4734.
11. Brunel N (2000) Dynamics of sparsely connected networks of excitatory and inhibitory spiking neurons. J Comput Neurosci 8:183–208.
12. Brunel N, Chance F, Fourcaud N, Abbott L (2001) Effects of synaptic noise and filtering on the frequency response of spiking neurons. Phys Rev Lett 86:2186–2189.
13. Brunel N, Hakim V (1999) Fast global oscillations in networks of integrate-and-fire neurons with low firing rates. Neural Comput 11:1621–1671.
14. Deng YC, Williams P, Liu F et al (2003) Neuronal discrimination capacity. J Phys A: Math Gen 36: 12379–12398.

15. Destexhe A, Rudolph M, Fellous JM, Sejnowski TJ (2001) Fluctuating synaptic conductances recreate in-vivo-like activity in neocortical neurons. Neuroscience 107:13–24.
16. Fairhall AL, Lewen GD, Bialek W et al (2002) Efficiency and ambiguity in an adaptive neural code. Nature 412:787–792.
17. Feng JF (2002) Correlated neuronal comput. In: Unsolved Problems Of Noise and Fluctuations. Ed.: Bezrukov SM, Springer, 208–215.
18. Feng J (1997) Behaviours of spike output jitter in the integrate-and-fire model. Phys Rev Lett 79:4505–4508.
19. Feng J (2001) Is the integrate-and-fire model good enough? —a review. Neural Networks 14:955–975.
20. Feng JF, Li GB (2001) Integrate-and-fire and Hodgkin-Huxley models with current inputs. J Phys A: Math Gen 34 (8):1649–1664.
21. Feng J, Liu F (2002) A modelling study on discrimination tasks. Biosystems 67:67–73.
22. Feng J, Tirozzi B (2000) Stochastic resonance tuned by correlations in neuronal models. Phys Rev E 61:4207–4211.
23. Feng JF, Zhang P (2001) Behavior of integrate-and-fire and Hodgkin-Huxley models with correlated inputs. Phys Rev E 63:051902.
24. Feng JF, Deng YC, Rossoni E (2006) Dynamics of moment neuronal networks. Phys Rev E, in press.
25. Gammaitoni L, Hänggi P, Jung P, Marchesoni F (1998) Stochastic resonance. Rev Mod Physics 70: 224–287.
26. Gaillared B (2002) Moving dots discrimination. Master's thesis, Sussex University, Sussex, England.
27. Gerstner W (2000) Population dynamics of spiking neurons: fast transients, asynchronous states, and locking. Neural Comput 12:43–89.
28. Gerstner W, Kistler W (2002) Spiking Neuron Models: Single Neurons, Populations, Plasticity. Cambridge University Press, Cambridge, UK.
29. Goldman MS, Maldonado P, Abbott LF (2002) Redundancy reduction and sustained firing with stochastic depressing synapses. J Neurosci 22:584–591.
30. Gray CM, Koenig P, Engel AK, Singer W (1989) Oscillatory responses in cat visual cortex exhibit inter-columnar synchronization which reflects global stimulus properties. Nature 338:334–337.
31. Hyvärinen A, Karhunen J, Oja E (2001) Independent Component Analysis. Wiley, New York.
32. Hyvärinen A, Hoyer PO (2000) Emergence of phase- and shift- invariant features by decomposition of natural images into independent feature subspaces. Neural Comput 12 (23):1705–1720.
33. Hyvärinen A, Oja E (1997) A fast fixed-point algorithm for independent component analysis. Neural Comput 9 (7):1483–1492.
34. Kast B (2001) Decisions, decisions... Nature 411:126–128.
35. Kreiter AK, Singer W (1996) Stimulus-dependent synchronization of neuronal responses in the visual cortex of the awake macaque monkey. J Neurosci 16:2381–2396.
36. Laughlin SB (1981) A simple coding procedure enhances a neuron's information capacity. Z Naturforsch 36c:910–912.
37. Leadbetter MR, Lindgren G, Rootzén H (1983) Extremes and Related Properties of Random Sequences and Processes. Springer, New York.

38. Leng G et al (2001) Responses of magnocellular neurons to osmotic stimulation involves co-activation of excitatory and inhibitory input: an experimental and theoretical analysis. J Neurosci 21:6967–6977.
39. Litvak V, Sompolinsky H, Segev I, Abeles M (2003) On the transmission of rate code in long feedforward networks with excitatory-inhibitory balance. J Neurosci 23:3006–3015.
40. Longtin A, Moss F, Bulsara A (1991) Time interval sequences in bistable systems and noise induced transmission of neural information. Phys Rev Lett 67:656–659.
41. Mattia M, Del Giudice P (2002) Population dynamics of interacting spiking neurons. Phys Rev E 66:051917.
42. Nykamp DQ, Tranchina D (2000) A population density approach that facilitates large-scale modeling of neural networks: analysis and an application to orientation tuning. J Comp Neurosci 8:19–30.
43. Olshausen BA, Field DJ (1997) Sparse coding with an overcomplete basis set: a strategy employed by V1? Vis Res 37:3311–3325.
44. Parker AJ, Newsome WT (1998) Sense and the single neuron: probing the physiology of perception. Annu Rev Neurosci 21:227–277.
45. Rudolph M, Destexhe A (2001) Do neocortical pyramidal neurons display stochastic resonance? J Comput Neurosci 11:19–42.
46. Rudolph M, Destexhe A (2001) Correlation detection and resonance in neural systems with distributed noise sources. Phys Rev Lett 86:3662–3665.
47. Petersen RS, Panzeri S, Diamond ME (2001) Population coding of stimulus location in rat somatosensory cortex. Neuron 32:503–514.
48. Petersen RS, Panzeri S, Diamond ME (2002) Population coding in somatosensory cortex. Curr Opin Neurobiol.
49. Rieke F, Warland D, de Ruyter van Steveninck R, Bialek W (1997) Spikes. MIT Press, Cambridge, MA.
50. Romo R, Salinas E (2001) Touch and go: decision mechanisms in somatosensation. Annu Rev Neurosci 24:107–137.
51. Sompolinsky H, Yoon H, Kang KJ et al (2001) Population coding in neuronal systems with correlated noise. Phys Rev E 64:051904.
52. Stevens CF, Zador AM (1998) Input synchrony and the irregular firing of cortical neurons. Nat Neurosci 1:210–217.
53. Salinas E, Sejnowski TJ (2001) Correlated neuronal activity and the flow of neural information. Nat Rev Neurosci 2:539–550.
54. Salinas E, Sejnowski TJ (2000) Impact of correlated synaptic input on output firing rate and variability in simple neuronal models. J Neurosci 20:6193–6209.
55. Shadlen MN, Newsome WT (1994) Noise, neural codes and cortical organization. Curr Opin Neurobiol 4:569–579.
56. Shadlen MN, Newsome WT (1996) Motion perception: seeing and deciding. Proc Natl Acad Sci 93:628–633.
57. Tuckwell HC (1988) Introduction to Theoretical Neurobiology, vol 2. Cambridge University Press, Cambridge, England
58. van Vreeswijk C, Abbott LF, Ermentrout GB (1994) When inhibition not excitation synchronizes neural firing. J Comput Neurosci 1:313–321.
59. van Vreeswijk C, Sompolinsky H (1996) Chaos in neuronal networks with balanced excitatory and inhibitory activity. Science 274 (5293):1724–1726.
60. Nicol A, Feng J, Kendrick K: Negative correlation yields computal vigour in a mammalian sensory system. (in preparation)

61. Available at http://www.cogs.susx.ac.uk/ users/jianfeng/publications, and we thank to Nicolas Brunel for providing us the matlab code.
62. Gaussel N, Legra J (1999) Black-Sholes... What's next? Quants No. 35.
63. Dupuis P, Kushner, HJ (2002) Stochastic Control Problems in Mobile Communications. U.S. gov. rep. no. A071214.
64. Harris CM, Wolpert DM (1998) Signal-dependent noise determines motor planning. Nature 394:780–784.
65. Kistler W, Gerstner W, van Hemmen JL (1997) Reduction of the Hodgkin-Huxley equations to a single-variable threshold model. Neural Comput 9:1069–1110.
66. Oksendal B (1989) Stochastic Differential Equations, 2nd ed., Springer, Berlin.
67. Renart A et al (2003) Ch. 15, J Comp Neurosci. Feng JF (ed.). Shadlen MN, Newsome WT (1994) Curr Opin Neurobiol 4:569. Livak V et al (2003) J Neurosci 3006. Leng G et al (2002) J Neurosci 21:6967. Matthews PBC (1996) J. Physiology 492:597.
68. Fitts PM (1954) J Exp Psychol 47:381.
69. Feng JF (ed) Computational Neuroscience: A Comprehensive Approach. Chapman and Hall/CRC Press, London.
70. Morasso PG, Sangujineti V (2003) In Feng JF (ed) Computational Neuroscience: A Comprehensive Approach. Chapman and Hall/CRC Press, London.
71. Korman M et al (2003) P Natl Acad Sci USA 100 (21):12492. Richardson MJE, Flash T (2002) J Neurosci 22:8201. Vetter P et. al. (2002) Curr Biol 12:488. Feng JF et al (2003) J Phys A 36:7469. Feng JF et al (2002) J Phys A 35:7287.
72. Fleming WH, Rishel RW (1975) Deterministic and Stochastic Optimal Control. Springer, New York. Bensoussan A. (1992) Stochastic Control of Partially Observable Systems. Cambridge University Press, Cambridge, UK.
73. Csete ME, Doyle J (2002) Science 295:1664. Todorov E. (2002) Neural Comput 14:1233.
74. Joshi P, Maass W (2003) Movement generation and control with generic neural microcircuits (submitted). Kawato M. (1999) Curr Opin Neurobiol 9:718.
75. Usrey WM, Reid RC (1999) Synchronous activity in the visual system. Annu Rev Physiol 61:435–456.
76. Zohary E, Shadlen MN, Newsome WT (1994) Correlated neuronal discharge rate and its implications for psychophysical performance. Nature 370:140–143.

5

Neuronal Model of Decision Making

Benoit Gaillard, Jianfeng Feng, and Hilary Buxton

Summary. We have built a neuronal model of decision making. Our model performs a decision based on an imperfect discrimination between highly mixed stimuli, and expresses it with a saccadic eye movement, like real living beings. We use populations of integrate-and-fire neurons.

To take a decision that depends on imperfectly separated stimuli, we use a model inspired by the principle of accumulation of evidence. More precisely, this accumulation of evidence is performed by a competition between groups of neurons that model a visual column in the lateral intra-parietal area (area LIP) of the brain. In this column, we have groups of neurons that are sensitive to specific stimuli on which the decision is based. They inhibit each other through groups of inhibitory neurons. Simultaneously, they recursively excite themselves through recurrent synaptic excitation, and all the neurons receive a steady low-level excitation from the rest of the brain. The competition is generated by these recurrent inhibitory and excitatory loops. We study this structure within the framework of dynamical systems. The variables we use are the activities of each group of neurons. This dynamical system has several stable states: one of them occurs when all activities are weak, and other ones when one local group of neurons has a higher activity and dominates the others through lateral inhibition. The convergence to one of these stable states models decision making, guided by sensory evidence. The group of neurons sensitive to the specific stimulus has a comparative advantage on the others during the competition. This structure is not a new idea, but we use it to test our hypothesis on the way the brain controls the dynamics of decision making. Our hypothesis is that the statistical signature of the low-level activity of the brain modifies the stability of the attractors of our model, and thus changes the dynamics of the competition that models decision making.

The criterion by which we judge that a decision is taken is more realistic than just looking at the decisive neurons' activities. We model a saccadic eye movement directed by the activities of our LIP neurons, and we read the decision from the position of the eye. This experimental setup is comparable to biophysical experiments in which living beings express their decisions by saccadic eye movements.

The neurons of the LIP column in which the decisions take place are modeled as neurons in a real brain: besides the stimuli and the recurrent interactions, they receive significant inputs from the rest of the brain. It is well known that neurons in the brain are highly influenced by this activity, called low-level background activity.

We study how the dynamics of the decision making change as a function of the first-order statistics (mean) and second-order statistics (variance) of this global low-level background activity or noise. By studying its influence on the reaction time and error rate, we show that this background activity may be used to control the dynamics of decision making. We compare the performance of such a model (error rate as a function of reaction time) to the performance of living beings during psychophysical experiments. By doing so, we assess the plausibility of the hypothesis that decisions be controlled by the statistical signature of the low-level background activity of the brain.

5.1 Introduction

The study of the reaction time (RT) and error rate (ER) of animals and humans performing a simple two-choice perceptual discrimination task has a long history. One of the first approaches to model such decision processes was developed by psychologists and is known as the accumulation of evidence models. Typically, these models consider two variables that measure the quantity of evidence for each of the two decisions, and the first variable that reaches a decision boundary determines the decision. Those models were introduced by, for example Townsend and Ashby [17]. The diffusion model is often used to explain relations among RT, accuracy, error trials, and clarity of the input. Thanks to successful neural recording during decision making (for example, Shadlen and Gold [14], Platt and Glimcher [12]), experimental evidence confirmed that the diffusion model can account for the accumulation of evidence, but also inspired more precise models of the neural dynamics involved. In particular, accumulation of evidence through neural competition has been modeled by Wang [20] and by Brunel (Amit and Brunel [1]). Glimcher [9] reviews stochastic decision making. According to Glimcher, who tested this claim by experimenting on monkeys, the tendency to make a choice is implemented in the firing rate (FR) of area LIP, and the uncertainty about this decision could be implemented by what we call input noise in area LIP. However, decision making is also controlled by our context, our motivations, and our history. This idea has been addressed by Salinas [13]. He uses, as well, background synaptic activity as a switch between dynamical states in a network. We propose a model in which sensory inputs do not automatically trigger a decision. The decision making is dependent on the characteristics of the low-level background neuronal activity of the brain. This background activity does not only control if a decision is taken, but controls the trade-off between speed and accuracy as well: in urgent situations, we tend to take less accurate decisions.

We restrict ourselves to a very simple decision in the case of very simple and well-studied sensory evidence setup. We use moving-dot kinematograms as stimuli, and the model has to decide in which direction to move its eye, signifying if it thinks that the dots generally move downward or upward. This experimental setup has been extensively studied on alive monkeys and on humans by Newsome and colleagues [2, 15, 16, 21]. Our moving-dot kinematograms are composed of a hundred dots. A percentage of them move coherently in one direction. This percentage is called coherence in this chapter. The rest of them move randomly in any direction. In our experimental setups, they don't even have consistent directions. This setup is the same as in previous studies [3, 5, 6], and as in many of Newsome's experimental setups.

We currently treat these stimuli in a very crude way. We suppose that the retina and the early parts of the visual system evaluate the direction of each dot during successive short time steps. Authors have argued for such a discretization of time in the visual system (VanRullen and Koch [19]). We suppose that each dot moving in a given direction triggers the activity of specific motion detectors. There are more detailed models that support this idea: It has been proved that in the visual system, we find columns of neurons that detect specific directions of movement (Mountcastle [11] Hubel and Wiesel [10]). So we suppose that the first detailed neurons of our model receive a hundred different synaptic inputs, each of these inputs corresponding to the direction of one dot. These neurons are detectors of the global direction of the dots in the kinematogram: they react strongly to a given general direction and weakly to its opposite direction. To implement this neural behaviour, we set their synaptic input to be inversely proportional to the difference between their preferred direction and the direction of the corresponding dot.

These global direction detectors and the rest of the model is described more precisely in this chapter.

5.2 Model

5.2.1 Overview

Our model can be divided into three parts, illustrated in Figure 5.1. First we have the global direction detectors. They are made of populations of integrate-and-fire (IF) neurons whose firing rates (FRs) depend on the global direction of the dots in the stimulus. They are a very simple generalisation of the model we studied in earlier publications [6, 7]. Second; using the output of the direction detectors, we have our LIP decisive column. Internal recursive loops and global inputs from the brain determine the activities of the neurons in the column. Specific patterns of activity of this column can generate an eye movement. The generation of this eye movement is the third part of the model.

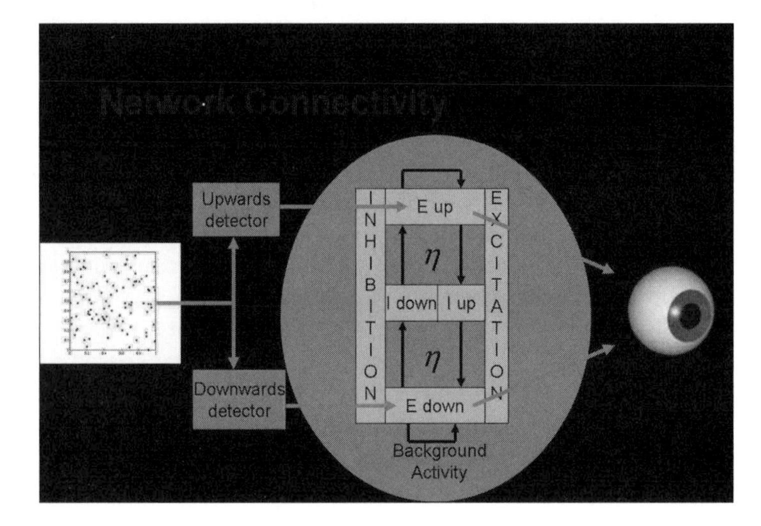

Fig. 5.1. The left panel represents the stimulus: Normally all those dots are moving, most of them randomly. The middle panel is an illustration of the neuronal activities in the decision column of area LIP. In the middle panel, we see the low-level background activity of the brain (represented by the green colour) that surrounds the column. All the represented subpopulations of neurons are connected to themselves and to each other, but the arrows represent Hebbian learning-potentiated connections. When the eye on the right has turned to one side, a decision has been taken.

The neuron model used here is the classic leaky integrate-and-fire (IF) model [4, 8, 18]. The dynamics of the membrane potential below threshold are defined as follows:

$$dV = -\frac{1}{\tau}(V - V_{rest})dt + dI_{syn}(t) \tag{5.1}$$

where τ is the time constant of the neuron and the synaptic input is

$$I_{syn}(t) = \sum_{i=1}^{N_s} a_i E_i(t) \tag{5.2}$$

where $E_i(t)$ is the instantaneous frequency of incoming spikes to the ith synapse, a_i is the positive or negative magnitudes of the excitatory postsynaptic potential (EPSP) and the inhibitory postsynaptic potential (IPSP), respectively, and N_s is the number of synapses. When the membrane potential reaches V_θ, the threshold potential, the neuron emits a spike and the potential is reset to V_{rest}. Then, the neuron goes through a refractory period τ_{ref} during which its membrane potential remains close to V_{rest}. We assume that large groups of incoming synapses receive the activity of large populations of neurons that have a similar activity, and that this activity is small. In that case, as shown by Tuckwell [18], we can use the diffusion approximation. So we can rewrite (5.2) as

$$I_{syn}(t) = J \sum_{i=1,w_i>0}^{N_g} w_i \nu_i(t) - J_{inh} \sum_{i=1,w_i<0}^{N_g} w_i \nu_i(t)$$

- N_g is the number of groups of similar incoming synapses to the neuron.
- w_i is the normalized synaptic weight. It is negative for inhibitory synapses.
- ν_i is the rate of the Poisson process coming to the synapses of the group i.
- J is the global excitatory synaptic strength.
- J_{inh} is the global inhibitory synaptic strength.

Then we reach the following expression for the output FR:

$$\nu_{out} = \left[\tau_{ref} + \tau \int_{\frac{V_{rest}-\mu}{\sigma}}^{\frac{V_\theta-\mu}{\sigma}} \phi(u) du \right]^{-1} \tag{5.3}$$

where

$$\phi(u) = \sqrt{\pi} e^{u^2} (1 + erf(u))$$

$$\mu = J \sum_{i=1}^{N_g} w_i \nu_i$$

$$\sigma^2 = J^2 \sum_{i=1}^{N_g} w_i^2 \nu_i + J^2 \sum_{i,j=1,i\neq j}^{N_g} c_{i,j} w_i w_j \sqrt{\nu_i \nu_j}$$

- $c_{i,j}$ is the correlation coefficient between synapse group i and synapse group j. We assume that the correlation coefficient between the inhibitory and excitatory synapse is zero.
- The resting membrane potential: $V_{rest} = 0$ mV.
- The threshold membrane potential: $V_\theta = 20$ mV.
- τ_{ref} is the refractory period of the leaky integrate-and-fire neuron.
- τ is the time constant of the leaky integrate-and-fire neuron.

5.2.2 Global Direction Detectors

The global direction detectors are constituted of populations of leaky integrate-and-fire neurons described in the previous section. Each detector is made of 100 such neurons that are not laterally connected and that receive the same input. The discrimination of such a population is quicker and more accurate, as shown in previous publications (Gaillard et al. [6, 7]). We do not evaluate statistically the output FR of the neuron, as in (5.3), but we simulate the production of spikes. For one time window t, the population produces N spikes, and the firing rate is $\frac{N}{100t}$.

The synaptic excitation is basically inversely proportional to the distance between the preferred angle of the group and the stimulus angle. Keeping the paradigm of one group of synapses for one dot, synaptic excitation for the upward detector will be π for a dot moving upward, 0 for a dot moving downward, and $\frac{\pi}{2}$ for a dot moving horizontally.

$$\nu_i = k \, \|(\|Angle - Direction_i\| + \pi)[2\pi] - pi\|$$

where ν_i is the incoming rate corresponding to ith dot number, *angle* is the preferred direction of the detector, $direction_i$ is the moving dot's direction, and k is a normalizing parameter, so that the excitation stays in the range of rates used in our experimental models. This equation is justified by the idea that the direction detector has more synaptic connections to the motion detector of its preferred direction. We assume as well that the incoming activity corresponding to dots having a coherent direction is correlated. n_c is the number of dots moving coherently, and c is the correlation coefficient. For simplicity, we assume that the detector's synaptic characteristics are as follows: the magnitude (a_i) of an EPSP is the exact opposite of the magnitude of an IPSP, and has the same absolute value for all synapses: a. r is the ratio between the number of excitatory and inhibitory synapses of the direction detector. Thus, as shown by Tuckwell [18], we can simplify (5.1) into

$$dV = -\frac{1}{\tau}(V - Vrest)dt + \mu dt + \mathcal{N}\sigma\sqrt{dt}$$

where

$$\mu = a\sum_{j=1}^{p}(1-r)\nu_j;$$

$$\sigma^2 = a^2 \left[\sum_{j=1}^{p} (1+r)\nu_j + \sum_{i,j=1,i\neq j}^{n_c} c(1+r)\sqrt{\nu_i \nu_j} \right] \tag{5.4}$$

- r is the ratio between inhibitory synapses and excitatory synapses.
- $p = 100$ is the number of groups of synapses that receive inputs corresponding to the direction of one dot.
- $\nu(j)$ is the incoming rate corresponding to the direction of the jth dot.
- The time constant of the neuron: $\tau = 20$ ms.
- The time step for the integration: $dt = 0.01$ ms.
- The correlation coefficient between inputs from dots that have a coherent motion: $c = 0.1$.
- The number of coherent inputs: $n_c \leq p$. Coherent inputs are dots that move consistently in one direction.
- The resting membrane potential: $V_{rest} = 0$ mV.
- The threshold membrane potential: $V_{threshold} = 20$ mV.
- $\mathcal{N}$ is a normally distributed random variable (mean 0, variance 1); in the formal IF model, $\mathcal{N}\sqrt{dt}$ is the standard Brownian motion.

Characteristics of This Model

We previously studied this model (Gaillard et al. [6, 7]). We showed that the discrimination accuracy is better when the ratio r between excitatory and inhibitory inputs is closer to $r = 1$. We showed that the population coding reduces considerably the time needed to evaluate the firing rate, and increases the discrimination accuracy. The FR decreases with r. However, to obtain a reliable measure of the FR, we need to produce at least 100 spikes. Thus, in our model, in order to take decisions faster, we use $r = 0$.

The output FR of the direction detector is used as the specific inputs of the competing groups of neurons in area LIP. These competing groups of neurons form the column that is described in the next subsection.

Figure 5.2 shows that the mean FR increases linearly with the coherence, as assumed in Wang [20]. The output of our detectors actually fits his assumption of a Gaussian distribution of rates very well. We used this approximation to measure the TPM in Chapter 4.

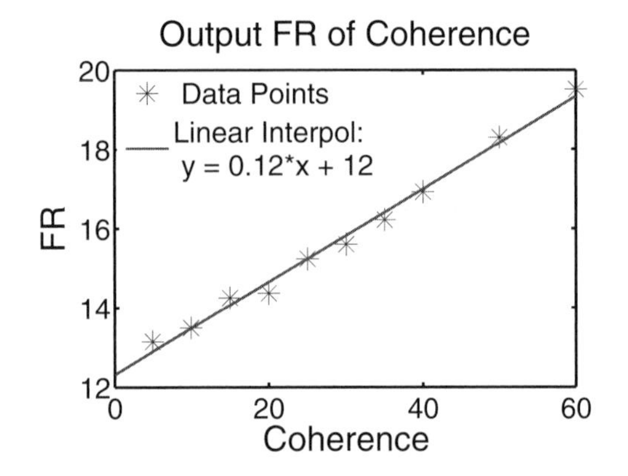

Fig. 5.2. The firing rate of a motion detector as a function of the coherence of the stimulus. The global direction of the stimulus is the preferred one of the detector. It is a linear relationship.

5.2.3 LIP Column: Where the Decision Is Made

Column Organisation

If we have p specific stimuli, then our column is described by $2p+2$ subpopulations of neurons:

- p subpopulations of excitatory neurons specifically sensitive to the specific stimuli
- one subpopulation of nonspecific excitatory neurons
- one subpopulation of nonspecific inhibitory neurons
- p subpopulations of specific inhibitory neurons. They are specifically connected to the specific excitatory neurons, and specifically inhibit populations of neurons.

The specific connections between neurons are modeled to have arisen through Hebbian learning, η being the learning coefficient, or long-term potentiation coefficient. The rest of the synapses are depressed, to keep a normalized synaptic strength.

This leads us to the following parameters:

- $x(= 0.8)$: the fraction of local module synapses relatively to external input.
- $J(= 11)$: synaptic excitatory unstructured strength to one neuron.
- $C(= 1.01)$: inhibitory synaptic strength is slightly stronger than excitatory; this is a stability condition. (I have the formal proof of it in one dimension. Amit and Brunel [1] showed it as well).
- $\eta(= 9)$: learning coefficient.
- $f(= 0.01)$: proportion of neurons specifically sensitive to a specific stimulus.
- $p(= 2)$: number of specific stimuli.
- $f_i(= 0.1/p)$: proportion of inhibitory neurons specifically sensitive to a subpopulation of excitatory neurons.
- $\eta^- = max(0, \frac{1-\eta\times f}{1-f})$: synaptic depression.
- $C^- = max(0, \frac{C*(1-\eta\times f_i)}{1-f_i})$: synaptic depression of inhibitory synapses to an excitatory neuron.
- ξ: standard deviation of the external background activity. This standard deviation varies spatially and temporally.
- $f_b(= 2/3)$: relative strength of the background activity in comparison to the whole external input.
- $f_1 = f_2 = ...f_p = \frac{1-f_b}{p}$: relative strength of each specific input in comparison to the whole external input.

The learning rule is expressed as follows: Since neurons that are more sensitive to a specific stimulus will more often fire together, their connection will be strengthened. We express it: $J^+ = \eta \times J$, where J is the average unstructured synaptic strength. The rest of the synapses will be depressed, so that the mean synaptic strength remains J. Thus, in our case, we have: $\eta^- = max(0, \frac{1-\eta\times f}{1-f})$. We truncated the synaptic strength so that it cannot be a negative strength, as seen in biology. We also suppose learning between inhibitory neurons. Our assumption is that some inhibitory neurons will show self-organisation properties, strengthen their incoming synapses from subpopulations of excitatory neurons, and fire correlatively. This learning parameter will be the same, η, and the synaptic depression will be similar.

We can concisely represent the structure of the column with the following matrix A. Each element $A_{i,j}$ of the matrix represents the strength of the influence of the neurons of group j on the neurons of group i. This is the synaptic strength of each synapse, multiplied by the number of synapses.

$$A = (M, N)$$

Where
$M =$

$$\begin{pmatrix} xf\eta & xf\eta^- & -xf_iC^- & -xf_i\eta C & -x(1-pf_i)C & x(1-pf)\eta^- \\ xf\eta^- & xf\eta & -xf_i\eta C & -xf_iC^- & -x(1-pf_i)C & x(1-pf)\eta^- \\ xf\eta & xf\eta^- & -xf_iC & -xf_iC & -x(1-pf_i)C & x(1-pf)\eta^- \\ xf\eta^- & xf\eta & -xf_iC & -xf_iC & -x(1-pf_i)C & x(1-pf)\eta^- \\ xf & xf & -xf_iC & -xf_iC & -x(1-pf_i)C & x(1-pf) \\ xf & xf & -xf_iC & -xf_iC & -x(1-pf_i)C & x(1-pf) \end{pmatrix}$$

$$N = \begin{pmatrix} (1-x)f_b(1+\xi\mathcal{N}) & (1-x)f_1 & 0 \\ (1-x)f_b(1+\xi\mathcal{N}) & 0 & (1-x)f_2 \\ (1-x)f_b(1+\xi\mathcal{N}) & 0 & 0 \\ (1-x)f_b(1+\xi\mathcal{N}) & 0 & 0 \\ (1-x)f_b(1+\xi\mathcal{N}) & 0 & 0 \\ (1-x)f_b(1+\xi\mathcal{N}) & 0 & 0 \end{pmatrix}$$

System Equations

We use a discrete formalism. At each time step, we reevaluate the variables of the system (rates of the output Poisson processes), as a function of their value at the previous time step, and of the value of the contextual variables. The contextual variables are the external inputs: global activity of the brain, and direction detector activity.

$$\nu_{n+1}^{int} = F(\nu_n^{int}, \nu_n^{ext}) \tag{5.5}$$

where

$$F(\nu_n^{int}, \nu_n^{ext}) = \left[\tau_{ref} + \tau \int_{\frac{V_{rest}-\mu}{\sigma}}^{\frac{V_\theta-\mu}{\sigma}} \phi(u)du \right]^{-1}$$

where

- $\phi(u) = \sqrt{\pi}e^{u^2}(1 + erf(u))$.
- $\mu = \mu(\nu_n^{int}, \nu_n^{ext})$is described in the next subsections.

- $\sigma = \sigma(\nu_n^{int}, \nu_n^{ext})$ is described in the next subsections.
- ν^{int} is the vector representing the output rates of the neurons in the column.

$$\nu^{int} = \begin{pmatrix} \nu_{up} \\ \nu_{down} \\ \nu_{i,up} \\ \nu_{i,down} \\ \nu_i \\ \nu_0 \end{pmatrix}$$

- ν^{ext} is the vector representing the output rates of the neurons in the brain and the direction detectors.

$$\nu^{ext} = \begin{pmatrix} \nu_{global} \\ \nu_{ext,up} \\ \nu_{ext,down} \end{pmatrix}$$

- $\tau_{ref} = 0.0005$ refractory period in seconds.
- $\tau = 0.05$ neuronal time constant in seconds.

The Variable μ

In (5.2), the relation between μ and the incoming Poisson rate is linear. So we can write this $(2p + 5)$ linear equation:

$$\mu = J \times A \times \nu \tag{5.6}$$

where μ is a vector of size $(2p+2)$, representing the various subpopulations of neurons in the LIP column. ν_n is a vector of size $(2p+5)$, representing the rate of the Poisson processes received from the various subpopulations of neurons. ν is the combination of ν^{ext} and ν^{int}. A is a matrix of size $(2p+2) \times (2p+5)$. If we take the example of $p = 2$, for the two choices of the dot direction decision task (upward or downward), we have: $\mu = \begin{pmatrix} \mu_{up} \\ \mu_{down} \\ \mu_{i,up} \\ \mu_{i,down} \\ \mu_i \\ \mu_0 \end{pmatrix}$, $\nu = \begin{pmatrix} \nu_{up} \\ \nu_{down} \\ \nu_{i,up} \\ \nu_{i,down} \\ \nu_i \\ \nu_0 \\ \nu_{global} \\ \nu_{ext,up} \\ \nu_{ext,down} \end{pmatrix}$

The Variable σ

Currently, we suppose that the correlation between neurons that belong to the same group (for example, the group of excitatory neurons that is especially sensitive to the upward direction detector) is $c = 0.1$. The correlation between neurons of two different groups is $c = 0.01$. The correlation between inhibitory and excitatory neurons is $c = 0$. So, in the particular case that we study here ($p = 2$), we obtain the following correlation matrix:

$$c = \begin{pmatrix} 0.1 & 0.01 & 0 & 0 & 0 & 0.01 & 0.01 & 0.01 & 0.01 \\ 0.01 & 0.1 & 0 & 0 & 0 & 0.01 & 0.01 & 0.01 & 0.01 \\ 0 & 0 & 0.1 & 0.01 & 0.01 & 0 & 0 & 0 & 0 \\ 0 & 0 & 0.01 & 0.1 & 0.01 & 0 & 0 & 0 & 0 \\ 0 & 0 & 0.01 & 0.01 & 0.1 & 0 & 0 & 0 & 0 \\ 0.01 & 0.01 & 0 & 0 & 0 & 0.1 & 0.01 & 0.01 & 0.01 \\ 0.01 & 0.01 & 0 & 0 & 0 & 0.01 & 0.1 & 0.01 & 0.01 \\ 0.01 & 0.01 & 0 & 0 & 0 & 0.01 & 0.01 & 0.1 & 0.01 \\ 0.01 & 0.01 & 0 & 0 & 0 & 0.01 & 0.01 & 0.01 & 0.1 \end{pmatrix}$$

Consequently, for a neuron belonging to group n, we have

$$\sigma_n^2 = \frac{1}{2} J^2 \left(\sum_{i=1}^{N_g} A_{n,i}^2 \nu_i + \sum_{i,j=1, i \neq j}^{N_g} c_{i,j} A_{n,i} A_{n,j} \sqrt{\nu_i \nu_j} \right) \tag{5.7}$$

Where

- N_g is the number of group of neurons.

- A is the matrix seen before.
- $c_{i,j}$ is the correlation coefficient between the ith and the jth synapses.

5.2.4 Saccadic Eye Movement

To express the decision, we model a saccadic eye movement (SEM). This is the paradigm that is used in Newsome's lab to evaluate the decisions of monkeys during the same task. In our model, the eye position is governed by the following equation:

$$\frac{\partial x}{\partial t} = k \times R - l \times tan(x)$$

- x represents the angular position of the eye.
- k and l are parameters. In our recent experiments: $k = l = 3$.
- R is the input that comes from the LIP column. In the two-dimensional case ($p = 2$), $R = \nu_{up} - \nu_{down}$.

The term $tan(x)$ models an elastic force that tends to bring the eye in its central position if no specific command is sent, and that also prevents the eye's angular position from diverging. In fact, when x tends to $\frac{pi}{2}$, the force modeled by $tan(x)$ tends to ∞, thus the moving span of the eye is limited. The position of the eye gives us a natural criterion to measure ER and RT. At the moment, we consider that a decision is made when $x = 0.95\frac{\pi}{2}$.

5.3 Results/Predictions

5.3.1 Methods

We ran a set of experiments, covering the parameter space that follows: The mean of the global background activity varied from 3 to 10 Hz, and the standard deviation varied from 0 to 5 Hz. We measured the reaction time and error rate of the model. For each parameter combination, we averaged the results over 10 repetitions. All simulations were conducted with stimuli that had 5% coherence.

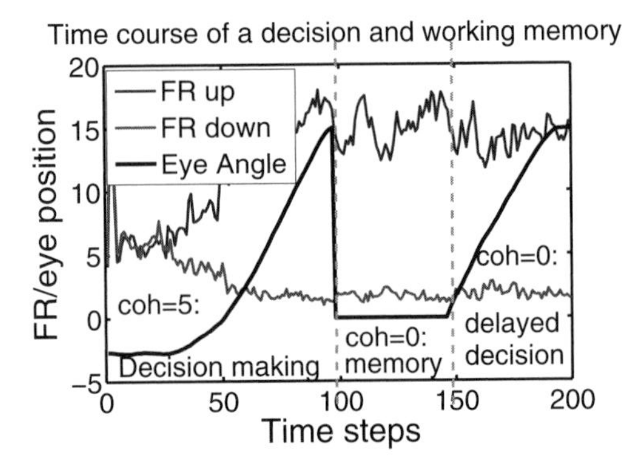

Fig. 5.3. Illustration of decision making and working memory. The stimulus is presented during the first 100 time steps. Then, the eye is reset to the middle position and the stimulus disappears (100 to 150). However, we see that the decision is maintained when we free the eye and the system expresses a delayed decision.

5.3.2 Time Course of Decision: Convergence to a Local Attractor

Figure 5.3 illustrates the time course of decision making. We can see the competition that leads to the convergence to a local attractor, along with the eye position over time. We see as well that the recurrent loops enable the model to keep in mind the decision even when the stimulus disappears. We can argue that we are implementing here a kind of working memory. In fact this decision is not expressed by the model until the difference of activity between the two populations has been consistent enough for the eye to have accomplished its saccadic movement. We see in Figure 5.3 that the decision is kept in memory, because we have artificially reset the eye in the middle position when the stimulus stopped. However, the system stays in the attractor, and when the eye is released it expresses the delayed decision.

5.3.3 Background Activity Controls Decision Making

Mean Intensity of the Background Activity

Reaction Time

We clearly see in Figure 5.4 that increasing the intensity of the background activity reduces the reaction time. We can still see the "switch" effect described by Salinas [13]: If the background activity tends to zero, then the reaction time tends to infinity. This switch effect is part of a more global control exercised by the external activity on the decision process.

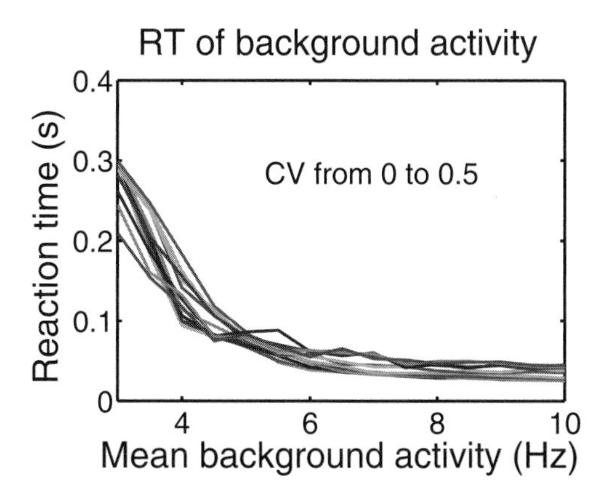

Fig. 5.4. Reaction time as a function of the mean intensity of the background activity of the brain. The decision processes were stopped after 0.3 seconds, to reduce computational time. In fact, if no decision is made, when the background activity is too weak, RT tends to infinity.

Error Rate

The error rate increases with the intensity of the background activity following a sigmoid function. In Figure 5.5, we can see a four parameter exponential approximation of this increase.

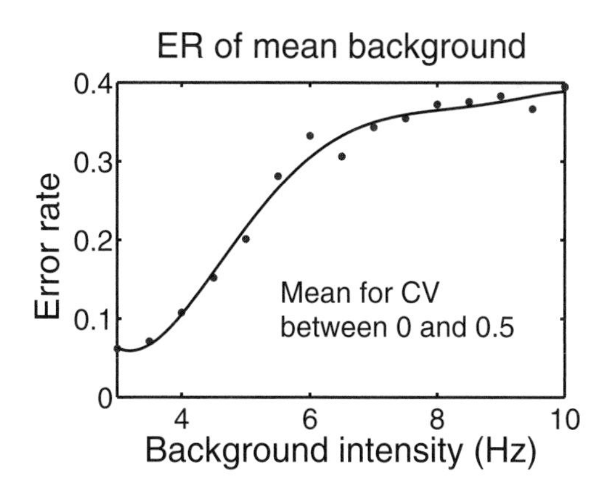

Fig. 5.5. Error rate as a function of the mean intensity of the background activity of the brain. This is an average over all the values of the standard deviation, between 0 and 5.

Error Rate=F(Reaction Time)

Figure 5.6 shows the characteristic of ER as a function of RT, deducted from the previous results. This was possible because RT is a monotonically increasing function of background intensity. This result is important because it is prediction of the model and gives us a tool to evaluate our hypothesis that background activity controls the speed/accuracy trade-off in decision making: If, during psychological experiments, humans show a similar dependency between ER and RT, then our model for the underlying process is likely to be plausible.

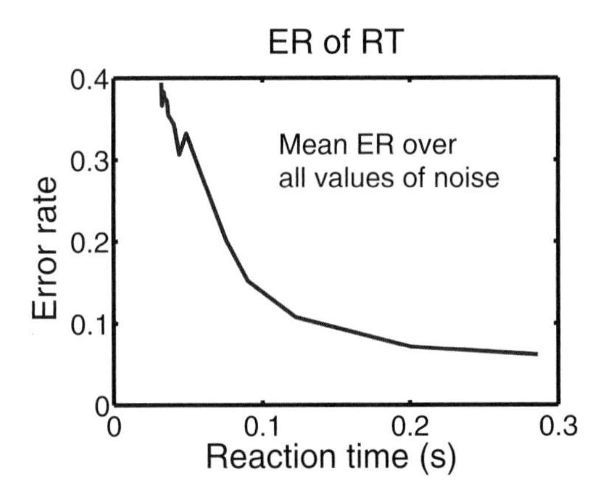

Fig. 5.6. ER as a function of the RT. Our method is basically a change of variables from the mean background activity to the RT. This works because we have seen previously that the RT is a monotonic function of the background intensity. Here, we have taken the mean over all the values of the standard deviation of the background activity.

Standard Deviation of the Background Activity

Reaction Time

Figure 5.7 shows that the real noise also speeds up decision making. In this simulation, the mean of the low-level background activity is constant. Its standard deviation is the only parameter that varies. This is a new result, purely based on the second-order moment of the brain's stochastic neural activity.

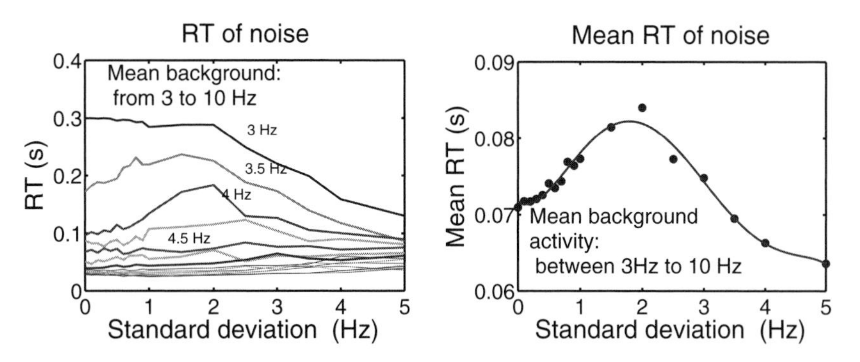

Fig. 5.7. Reaction time as a function of the standard deviation of the background activity. In the first panel, the mean activity varies between 3 and 10 Hz. The second panel is the mean function, averaged over the whole range of background mean intensities (it is the average of the curves presented in the first panel).

Error Rate

Figure 5.8 shows a significant influence of noise on the error rate. For each value of the standard deviation, we have measured the ER for the range of mean background intensities, and then taken the mean over all these simulations.

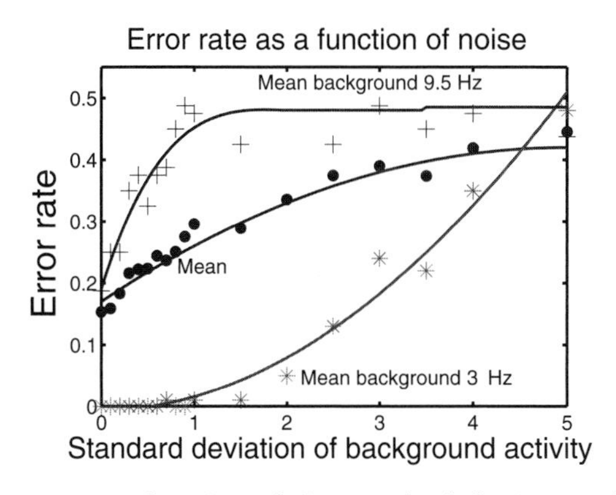

Fig. 5.8. Error rate as a function of the standard deviation of the background activity of the brain. There are three different behaviours for three different mean intensities of the background activity.

5.4 Discussion

5.4.1 Summary of the Results

We have implemented a model that performs decision making and expresses it through a saccadic eye movement. This is a step forward from the models in which the external observer has to watch a neural variable, because we can use measures such as the reaction time and error rate of our model in order to compare it to a living being's performance. We have shown that the low-level noisy background activity of the brain can be beneficial to decision making, in the case of very ambiguous sensory evidence. We have reproduced the results of Salinas that an increase in the mean of the low-level background activity can trigger decision making. We went much further than this intuitive result (if you add more energy in a competitive system, the convergence to the outcome will be quicker). We explored quantitatively the influence of the background intensity on the reaction time and on the error rate, and deduced a relationship between error rate and reaction time. Furthermore, we showed that an increase in the actual noisy property of the background activity, without an increase of its mean, also speeds up decision making at the cost of accuracy. This is a new result.

5.4.2 Limitations

The parameter space is huge. We only explored a small region. The results are dependent on the SEM model, which is currently over simplistic and has not been compared to biological evidence. We only have simulations.

5.4.3 Developments

Moment Mapping

The second-order statistics are fundamental to the results we presented here. However, we have assumed that the spikes are emitted according to Poisson processes. That means that the presynaptic noise is proportional to its mean, and enables us to write (5.4) and gives us (5.5) as an equation for the propagation of the first moment of the interspike interval (ISI) distribution. However, this assumption does not hold in the general case, and furthermore there is little biological evidence for LIP neuronal activity to be Poissonian. So, we have a more general expression than (5.5). In the general case:

$$(\nu_{n+1}^1, \nu_{n+1}^2) = G(\nu_n^1, \nu_n^2) \tag{5.8}$$

where ν_{n+1}^1 and ν_{n+1}^2 are the first and second moment of the ISI distribution at time $n+1$ or at the next neuron layer, and ν_{n+1}^1 and ν_{n+1}^2 are the first and second moment of the ISI distribution at time n or at the current neuron layer. We currently have an analytical expression for G that has been derived from the IF equation, and we will apply it to our setup in subsequent work.

Formal Analysis

Our results are based on numerical simulations. We will try to develop an analytical understanding of a simplified version of this model. We will characterise the attractors of the system defined by (5.8), their domain of attraction, their stability, and the speed of convergence to these attractors as a function of various system parameters.

Hypothesis

We deduced a relation between error rate and reaction time. This relation can be used to test the hypothesis that the background noise activity of the brain controls the speed-accuracy trade-off of decision making. Comparing our model characteristic of ER(RT) to the one of living beings, measured via psychophysical experiments, could support or invalidate our hypothesis.

5.4.4 Conclusion

Our model is a step toward the construction of a bridge between detailed neural models and real living behaviour such as sensory-based decision making, or between neurology and psychology. Computing with noise is also a new step toward less simplistic neural modeling, which takes into account the second-order statistics of spike trains. This more refined approach should give us insights into domains unreachable by the classic mean firing rate approaches.

References

1. Amit N Brunel DJ (1997) Model of global spontaneous activity and local structured activity during delay periods in the cerebral cortex. Cereb Cortex 7:237–252.

2. Britten KH, Shadlen MN, Newsome WT, Movshon JA (1992) The analysis of visual motion : a comparison of neuronal and psychophysical performance. J Neurosci 12:2331–2355.
3. Deng Y, Williams P, Liu F, Feng J (2003) Discriminating between different input signals via single neuron activity. J Physics A: Math and Gen 36(50):12379–12398.
4. Feng J (2001) Is the integrate-and-fire model good enough?—A review. Neural Networks 14:955–975.
5. Feng J, Liu F (2002) A modeling study on discrimination tasks. Biosystems 67:67–73.
6. Gaillard B, Feng J (2005) Modelling a visual discrimination task. Neurocomputing 65-66:203–209.
7. Gaillard B, Feng J, Buxton H (2005) Population approach to a neural discrimination task. Biol Cybernet 94(3):180–191.
8. Gerstner W, Kistler W (2002) Spiking Neuron Models, Single Neurons, Populations, Plasticity. Cambridge University Press, Cambridge, UK.
9. Glimcher PW (2003) The neurobiology of visual-saccadic decision making. Annu Rev Neurosci 26:133–179.
10. Hubel DH, Wiesel TN (1962) Receptive fields, binocular interaction and functional architecture in the cat's visual cortex. J Physiol (Lond) 148:574–591.
11. Mountcastle VB (1957) Modality and topographic properties of single neurons of cat's somatosensory cortex. J Neurophysiol 20:408–434.
12. Platt ML, Glimcher PW (1999) Neural correlates of decision variables in parietal cortex. Nature 400:233–238.
13. Salinas E (2003) Background synaptic activity as a switch between dynamical states in a network. Neural Comput 15:1439–1475.
14. Shadlen MN, Gold JI (2004) The neurophysiology of decision making as a window on cognition. In: Gazzaniga (ed), The Cognitive Neurosciences, 3rd ed. MIT Press, Cambridge, MA.
15. Shadlen MN, Newsome WT (2001) Neural basis of a perceptual decision in the parietal cortex (area LIP) of the rhesus monkey. J Neurophysiol 86:1916–1935.
16. Shadlen MN, Newsome WT (1996) Motion perception : seeing and deciding. Proc Nat Acad Sci 93:628–633.
17. Townsend JT, Ashby F (1983) The stochastic modeling of elementary psychological processes. Cambridge University Press, Cambridge, MA.
18. Tuckwell H (1988) Introduction to Theoretical Neurobiology, vol. 2. Cambridge University Press, Cambridge, MA.
19. VanRullen R, Koch C (2003) Is perception discrete or continuous? Trends Cog Sci 7(5):207–213.
20. Wang XJ (2002) Probabilistic decision making by slow reverberation in cortical circuits. Neuron 36:955–968.
21. Zohary E, Shadlen MN, Newsome WT (1994) Correlated neuronal discharge rate and its implications for psychophysical performance. Nature 370:140–143.

6

Estimation of Spike Train Statistics in Spontaneously Active Biological Neural Networks

Henry C. Tuckwell and Jianfeng Feng

Summary. We consider the theoretical determination of firing rates in some biological neural networks that consist of synaptically connected excitatory and inhibitory elements. A self-consistent argument is employed to obtain equations satisfied by the moments of the firing times of the various cells in the network. We first present results for networks composed of leaky integrate-and-fire model neurons in the case of impulsive currents representing synaptic inputs and an imposed threshold for firing. Solving a differential-difference equation with specified boundary conditions yields an estimate of the mean interspike interval of neurons in the network. We gaphically demonstrate that there may be a critical number of connections $n = n_c$ such that for $n < n_c$ there is no nontrivial solution, whereas for $n > n_c$ there are three solutions. Of these, one is at baseline activity, one is unstable, and one is asymptotically stable. Simulation results are reported which demonstrate that sustained activity is possible even without external afferent input and that the analytical method may yield accurate estimates of the firing rate. We also consider a network of generalized Hodgkin-Huxley model neurons. Assuming a voltage threshold, which is a useful representation for slowly firing such nerve cells, a functional differential equation is obtained whose solution affords an estimate of the mean network firing rate. Related equations enable one to estimate the second- and higher-order moments of the interspike interval.

6.1 Introduction

Investigations and theories of information processing in the nervous system have recently addressed the role of stochasticity in the activity of both single neurons [1–4] and neural networks [5–8]. One source of variability that occurs in cortical (and other) neurons is the "spontaneous" spiking activity in the absence of deliberate or known external input. The role of such spontaneous activity in relation to the processing of afferent input is not yet fully understood. However, spontaneous activity is now not considered to be simply noisy background activity. According to [9], the spontaneous activity of many brain neuronal networks determines the nature of the response of the brain to external stimulation and may contribute to the quality of perception. In visual information processing, it has been found that spontaneous activity does indeed create extra variability in nervous system responses to visual stimuli [10]. Furthermore, some or all such spontaneous activity has been hypothesized to play an important role in visual information processing as it tends to provide a substrate for modulation by afferent input.

Spontaneous neural activity arises because the brain contains a large number of neurons with very dense patterns of connections between cells. Sporadic or sustained random inputs from the thalamus may occur at various parts of the cerebral cortex [9]. Small trains or bursts of activity in one region, possibly due to spontaneous transmitter release at synaptic terminals, are propagated to other regions and may circulate in recurrent loops. Some of the details of brain connection patterns have recently been described in detail for neurons of the mammalian neocortex [11]. However, in most studies, as for example in [5–8], and the current chapter, the precise details of the connections are omitted in favour of randomly assigned connections or fully connected networks.

In a sense, spontaneous neural activity represents the simplest electrophysiological state or ground state for populations of neurons in the living brain, even though the underlying circuits are possibly as complex as for networks involved with perception or cognition. It is clear that by virtue of the complexity of the spatial and temporal aspects of the dynamics of individual neurons as well as the enormous complexity of the underlying biochemical processes, there are difficulties with analytical approaches to finding the quantitative features of spiking or nonspiking activity in neural networks with any degree of realism for the physiological properties of the composite single neurons. However, they may be useful for networks of artificial neurons, which are not usually endowed with a complex biochemical environment. Analytical results concerning dynamical neuronal behaviour are few even for single neurons [2, 12–15] and very sparse for neural networks composed of nonlinear elements, so that efficient simulation and other techniques have been developed for networks with stochastic dynamics [7, 16].

In many network models the neurons are considered to be weakly connected so that they are treated as independent and a mean field approach is adopted. It is then possible to use solutions of a single-neuron Fokker-Planck (or Kolmogorov) equation for the neuronal depolarization [5, 12–15], with suitable boundary conditions, to provide an estimate of the fraction of cells in the network at various levels of excitation. Background noise, usually represented by Poisson processes or white noise, has been found to play an important role in the propagation of signals through model networks [6] and in understanding the behaviour of cortical and subcortical neurons involved in cognitive information processes such as working memory [17–21].

It is of interest to attempt to determine as far as possible with analytical techniques the relationships between the microscopic physiological variables associated with single nerve cells, their patterns of connectivity, and global spiking activity. Such relationships are expected to assist in casting light on computational aspects of neuronal information processing and cognitive activity. Here we describe a method for analytically estimating the firing rates of neurons in stochastic networks of an arbitrary number of elements using Markov process theory. We consider, in section 6.2, networks of generalized integrate-and-fire neurons and present some exact results for networks of excitatory neurons. In section 6.3 we state corresponding results for networks composed of neurons with generalized Hodgkin-Huxley dynamics. In order that the theoretical estimates are as accurate as possible, our attention is focused mainly on slowly firing networks whose spiking is in singlets. We explore the relation among single neuron dynamics, network size, and the nature of sustainable spontaneous activity.

6.2 Networks of Generalized Leaky Integrate-and-Fire (LIF) Neurons

Let the electrical depolarization from rest for the kth cell in a network of n neurons be $V_k(t)$ at time t. In general, in the LIF scheme, the dynamical equations for the network can be written

$$dV_k = f(V_k)dt + \sum_{j=1}^{n} a_{jk} dN_j(\lambda_j; t - d_{jk}) + \epsilon_k dN_k^{ext}, k = 1, 2, \ldots, n, \quad (6.1)$$

where a_{jk} (element of the matrix A) is the strength of the connection from neuron j to neuron k, which determines the magnitude of the postsynaptic potential elicited when this connection is activated (it is assumed that $a_{jj} = 0$ for all j). The threshold for action potential generation in the kth cell is set at θ_k, here assumed to be constant. $N_j(\lambda_j; t)$ is the number of spikes emitted by neuron j in the time interval $(0, t]$; the parameter λ_j represents the mean rate of spiking so that as usual as an approximation $1/\lambda_j$ can be interpreted as the mean time between spikes from neuron j. The quantities d_{jk} are the time delays between emission of a spike by cell j and the appearance of a postsynaptic potential in cell k. The process N_k^{ext} is an external Poisson input which produces synaptic potentials of magitude ϵ_k. According to (6.1), the input currents are delta functions so that the detailed temporal structure of postsynaptic potentials is here ignored in a first approximation. This structure can easily be incorporated by adding subsidiary variables, but this would make the theory much more complicated and so is omitted.

6.2.1 A Network of Excitatory and Inhibitory Neurons

Suppose there are n_E excitatory (type E) and n_I inhibitory (type I) cells, whose membrane potentials in the absence of synaptic activation satisfy $\dot{V_E} = f_E(V_E)$ and $\dot{V_I} = f_I(V_I)$, respectively. The postsynaptic potential amplitudes are the nonnegative quantities a_{EE}, a_{IE}, a_{EI}, and a_{II}, where the first subscript refers to the presynaptic cell type and the second to the postsynaptic cell type. The corresponding numbers of connections for individual cells are n_{EE}, n_{IE}, n_{EI}, and n_{II}, respectively. Thus each E-cell receives a total of $n_{EE}+n_{IE}$ synaptic inputs and each I-cell receives $n_{EI}+n_{II}$ such inputs. Since each E-cell has effectively the same input, its membrane potential satisfies

$$dV_E = f_E(V_E)dt + a_{EE}dN_{EE}(n_{EE}\lambda_E; t) - a_{IE}dN_IE(n_{IE}\lambda_I; t), \tag{6.2}$$

whereas for each I-cell,

$$dV_I = f_I(V_I)dt + a_{EI}dN_{EI}(n_{EI}\lambda_E; t) - a_{II}dN_{II}(n_{II}\lambda_I; t), \tag{6.3}$$

where N_{EE}, N_{IE} are the pooled excitatory and inhibitory input point processes for E-cells, and N_{EI}, N_{II} are the corresponding processes for the I-cells. Here we do not include external afferent input (or background noise) which can be taken into account as in subsection 2.2. Letting the thresholds of the two kinds of cell be θ_E and θ_I, we then have the following result, where now refractory periods are included. The proof of this follows by considering the total input frequencies for the component neurons and their respective expected time intervals for their membrane potentials to reach threshold.

Theorem

If $\frac{1}{\lambda_E}+t_{R,E}$ *is the mean time interval between spikes in an E-cell and* $\frac{1}{\lambda_I}+t_{R,I}$ *is the mean time interval between spikes in an I-cell of the network, then these quantities may be estimated implicitly by solving the simultaneous differential-difference equations*

$$f_E(v)\frac{dF_E}{dv} + n_{EE}\Lambda_E F_E(v+a_{EE}) + n_{IE}\Lambda_I F_E(v-a_{IE})$$
$$-(n_{EE}\Lambda_E + n_{IE}\Lambda_I)F_E(v) = -1, \quad v < \theta_E, \tag{6.4}$$
$$f_I(v)\frac{dF_I}{dv} + n_{EI}\Lambda_E F_I(v+a_{EI}) + n_{II}\Lambda_I F_E(v-a_{II})$$
$$-(n_{EI}\Lambda_E + n_{II}\Lambda_I)F_I(v) = -1, \quad v < \theta_I, \tag{6.5}$$

with boundary conditions $F_E(v) = 0, v \geq \theta_E$, $F_I(v) = 0, v \geq \theta_I$, *and* $F_E(0) = \frac{1}{\lambda_E}$, $F_I(0) = \frac{1}{\lambda_I}$, *provided such solutions exist.*

Here $F_E(v)$ and $F_I(v)$ are the mean times for the potentials V_E and V_I to reach thresholds from initial values v, and we have put

$$\Lambda_E = \left(\frac{1}{\lambda_E} + t_{R,E}\right)^{-1} \quad \Lambda_I = \left(\frac{1}{\lambda_I} + t_{R,I}\right)^{-1}$$

where $t_{R,E}$ and $t_{R,I}$ are the refractory periods of the excitatory and inhibitory cells, respectively. In general, in the search for such solutions, one may insist that F_E and F_I vanish for $v < v_E < 0$ and $v < v_I < 0$, respectively, and then let v_E and $v_I \to -\infty$ to ensure that the thresholds for action potentials are attained.

Equations (6.4) and (6.5) are difficult to solve exactly but may be solved numerically or approximately via simulation of the corresponding stochastic differential equations. In the simpler situation of a uniform network where each cell has the same membrane potential dynamics, including thresholds θ for action potentials, and the same numbers of excitatory, n_E, and inhibitory, n_I, inputs, with the same amplitudes a_E and a_I for the synaptic potentials, the same refractory periods t_R, and all cells fire at the same mean rate λ, (6.4) and(6.5) reduce to the single differential-difference equation

$$f(v)\frac{dF}{dv} + \Lambda[n_E F(v+a_E) + n_I F(v-a_I)] - (n_E+n_I)\Lambda F(v) = -1, \quad v < \theta, \quad (6.6)$$

where $\Lambda = (1/\lambda + t_R)^{-1}$; (6.6) is solved with the constraints $F(v) = 0, v \geq \theta$ and $F(0) = \frac{1}{\lambda}$. Numerical solutions of (6.6) have been obtained in [28]. Diffusion approximations may be employed to obtain approximate estimates.

6.2.2 A Network of Only Excitatory Cells

Suppose that all connections are excitatory so that $a_{jk} \geq 0$ and each cell *receives* the same number of inputs, n_E. The network may or may not be fully connected, as long as each neuron receives synaptic input from n_E other cells. The number of inputs is thus $n_E \leq n - 1$, with equality applying in the case of a fully connected network. It is feasible that there may be neurons that are not presynaptic to any other cells, although this is unlikely in the mammalian central nervous system. To simplify, we ignore the transmission time intervals d_{jk} and assume further that all nonzero $a_{jk} = a_E$, which quantities may in fact be random but are here assumed to be deterministic or at their mean values. We assume also that all cells have equal thresholds $\theta_j = \theta, j = 1 \ldots, n$ and the same subthreshold dynamics described by the function f. The network is then described by the stochastic equation

$$dV_k = f(V_k)dt + a_E \sum_{j=1}^{n} dN_j(\lambda_j; t - d_{jk}) + \epsilon_k dN_k^{ext}, k = 1, 2, \ldots, n. \quad (6.7)$$

Now each cell receives excitatory postsynaptic potentials (EPSPs) according to n_E similar pooled point processes with the same mean rates, which we set at λ_E, and their amplitudes are all a_E so the generic depolarization V of each neuron satisfies

$$dV = f(V)dt + a_E dN(n_E\lambda_E;t) + \epsilon dN^{ext}, V < \theta, V(0) = v < \theta. \tag{6.8}$$

In general we have the following result whose proof is immediate from the theory of mean exit times of Markov processes [27] and the requirement that the network input frequency in a network of identical neurons each of which receives n_E inputs is n_E times the output frequency of each cell. Here $F(v)$ is the mean exit time of V from $(0,\theta)$ for an initial value $V(0) = v$.

Theorem

If $\frac{1}{\lambda_E}$ is the mean time interval between spikes in a neuron of the network described by (6.7), then it may be estimated implicitly by solving the differential-difference equation

$$f(v)\frac{dF}{dv} + n_E\lambda_E[F(v+a_E) - F(v)] + \lambda^{ext}[F(v+\epsilon) - F(v)] = -1, \quad v \in (0,\theta), \tag{6.9}$$

with boundary conditions $F(v) = 0, v \geq \theta$ and $F(0^+) = \frac{1}{\lambda_E}$, provided such a solution exists.

However, with the above kind of single-neuron model, the time interval between spikes may have to take into account a refractory period, t_R, on the order of a few msec, which may be significant. In this case (6.9) becomes

$$f(v)\frac{dF}{dv} + \frac{n_E}{\lambda_E^{-1} + t_R}[F(v+a_E) - F(v)] + \lambda^{ext}[F(v+\epsilon) - F(v)] = -1, \quad v < \theta, \tag{6.10}$$

but the second boundary condition is still $F(0^+) = \frac{1}{\lambda_E}$.

6.2.2.1 Graphical and Numerical Methods of Solution

We illustrate graphically that a nontrivial solution of (6.10) may fail to exist or there may be three solutions, one of which may correspond at least approximately to a stable connected network with nonzero firing rates. To do this we consider a network of cells whose common frequency transfer characteristic, including the effect of an external input and a refractory period, is known and is as depicted by the blue curve in Figure 6.1. It is assumed that all postsynaptic potentials are the same size and that background activity, represented by the terms $\epsilon_k dN_k^{ext}$, drives each cell to fire at rate f_b, which may be zero. For a cell to be a member of the given network, if each cell fires at rate f_o, the total input frequency to each cell must be nf_o above background, where n is the number of cells which connect to any given cell. Hence it is required that the output frequency is related to input frequency above background f_i by $f_i = nf_o$ or $f_o = f_i/n$. Such straight lines are drawn in Figure 6.1.

For the straight line with the largest slope and hence the smallest value of n, labeled case A, there is no intersection with the frequency transfer function of the neuron so that theoretically no sustainable firing above background is possible in a network with this number of connections to each cell. For a critical value n_c of the number of connections, $f_o = \frac{1}{n_c} f_i$, labeled B, the line is tangent to the frequency transfer curve. For $n > n_c$, as typified by the line labeled C, there are three solutions, one being at $f_i = 0$ (external input only) and output frequency f_b. Of the other two, P_2 is unstable whereas the asymptotically stable point P_1 corresponds to a possible stable network frequency. Furthermore, if a network has an observed firing rate f_s, the number of neurons in the network or the number of connections to each neuron could be estimated by ascertaining where $f_o = nf_s$ intersects the transfer curve. The latter could be obtained empirically. The reciprocal of the slope of the line from the point (λ^{ext}, f_b) to the stable point of intersection gives an estimate of n.

Exact analytical solutions have not been found for (6.9) for $\theta > 3$, but numerical methods are available [22]. However, the numerical methods are not simple to implement so that simulations of network activity are useful and have been performed for a large range of network sizes and for various parameter values. A problem often encountered is that neurons tend to fire synchronously in groups which makes the analytical methods described in this chapter not applicable because of the underlying assumption of randomness. It is has been found that it is unusual to find sustained network activity in the absence of an external afferent drive, ϵdN^{ext}. One example is shown in Figure 6.2. In this simulation there were 1000 LIF neurons with on average 400 connections per cell. The internal excitatory synaptic potential amplitude was 0.04 mV and the external input to each cell had a rate $\lambda^{ext} = 500$ per sec with amplitude $\epsilon = 1$ mV. The time step used was $\Delta t = 0.00005$ sec. The afferent input was terminated at the 200th time step, and yet activity is seen after a short decline, to rise spontaneously to an apparently steady state. Several checks on the validity of the analytical method were performed. For example, for a fully connected excitatory network of 50 neurons, with $\epsilon = 1$mV, $\lambda^{ext} = 500$ per sec, $a_E = 0.2$ mV, threshold 15mV, and time constant 20 msec, the network individual neuronal frequency obtained by simulation was 7.0 per sec whereas that predicted by the analytical method was 8.4 per sec.

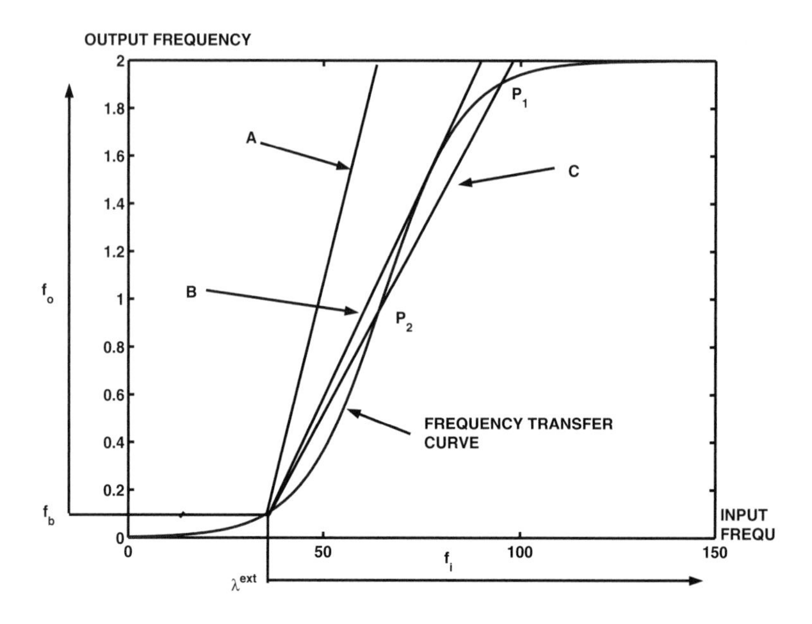

Fig. 6.1. Output frequency is plotted against input excitatory frequency. The frequency transfer characteristic of cells in the network is indicated by the blue curve. f_o is the output frequency above f_b, which corresponds to an external input frequency of λ^{ext}. f_i is input frequency above λ^{ext}. The straight lines A, B and C represent cases where the output frequency is a given fraction of the network input frequency for each cell, so $f_o = f_i/n$ where n is the number of cells which connect to each cell. In case B, $f_o = f_i/n_c$. For more details, see the text.

6.2.3 Diffusion Approximations

An alternative approach is to use smoothed versions of the network voltage processes called diffusion approximations. For such processes it is relatively easy to find the transfer characteristic not only for neurons of purely excitatory networks but also those including both excitation and inhibition. The validity of such an approach depends on parameter values [23]. Further details of such approximations, which have computational advantages, will be reported elsewhere.

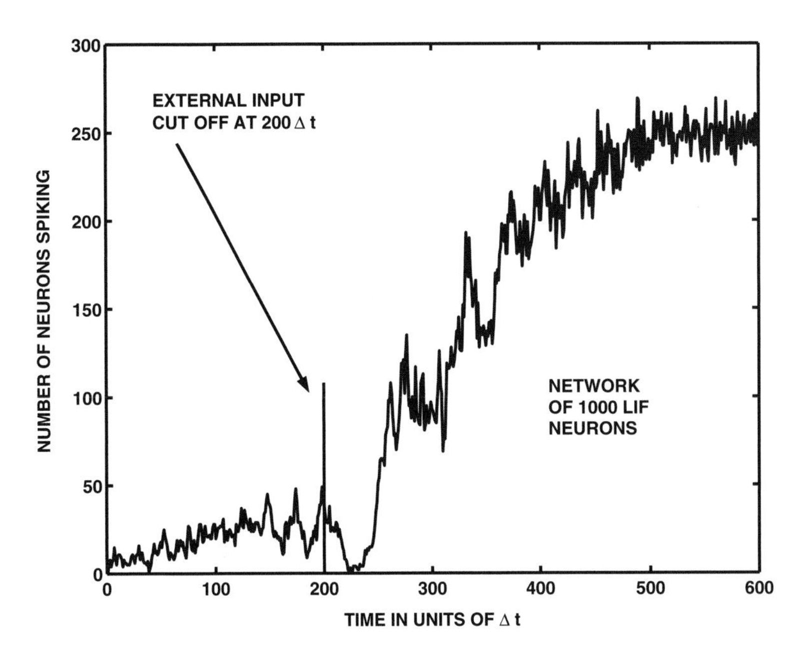

Fig. 6.2. Simulated activity in a network of 1000 LIF model neurons. The number of neurons firing is plotted against time. The external afferent input to each cell was terminated at 200 Δt. Note the sustained activity in the absence of afferent input. The parameter values are given in the text.

6.3 A Network of Neurons with Generalized Hodgkin-Huxley Dynamics

As is well known, the original Hodgkin-Huxley model [24] describes subthreshold activity and action potential generation in squid axon and contains only sodium, potassium, and leak currents. However, many types of ion channels with distinct roles and dynamics [25, 26] have been found in, for example, pyramidal cells, so it is useful to generalize the Hodgkin-Huxley system.

Towards such a generalization we consider $M+1$ neurons described by the dynamical system

$$C_i \frac{dU_i}{dt} = -\sum_{k=1}^{n} \overline{g}_{ki}\Big(U_i - V_{ki}\Big)\prod_{j=1}^{p_k} \gamma_{kji}^{m_{kj}} + I_i,$$

$$\frac{d\gamma_{kji}}{dt} = \alpha_{kj}(U_i)(1-\gamma_{kji}) - \beta_{kj}(U_i)\gamma_{kji},$$

where $i = 1, ..., M+1$, V_{ki}=Nernst potential for the kth ion species for the ith neuron, the maximal conductances are $\overline{g}_{ki}$, there are p_k gating variables

asociated with the kth ion species, and I_i is the total input current for the ith neuron. (Note that $n = 3$ for the Hodgkin-Huxley system.)
In a network of connected such neurons we suppose that

$$C_i \frac{dU_i}{dt} = -\sum_{k=1}^{n} \bar{g}_{ki} \Big(U_i - V_{ki} \Big) \prod_{j=1}^{p_k} \gamma_{kji}^{m_{kj}} + \epsilon_i dN_i^{ext} + \sum_l a_{li} dN_l(\lambda_l; t - \Delta_{li}),$$

where $i, l = 1, ..., M+1$, N_l is the output process describing the spikes of the lth neuron, and where $\{a_{li}, i, l = 1, ..., M+1\}$ is the connection matrix, Δ_{li} is a time delay and $\epsilon_i dN_i^{ext}$ is an external input to cell i.

6.3.1 Network Firing Rate

Now let N be an (approximating) Poisson process representing the input train for each cell. Then for a generic cell, ignoring the time delays of transmission, with each cell firing at rate λ, and assuming a fully connected network,

$$C \frac{dU}{dt} = -\sum_{k=1}^{n} \bar{g}_k \Big(U\_V_k \Big) \prod_{j=1}^{p_k} \gamma_{kj}^{m_{kj}} + \epsilon dN^{ext} + a dN(M\lambda; t),$$

where all nonzero elements of the connection matrix are set at $a > 0$ so that all synapses are excitatory.

Let the infinitesimal (Markov) operator associated with the neuron's dynamics be $\mathcal{L}_{neuron}$. For example, for the usual leaky integrate-and-fire model with

$$\frac{dV}{dt} = \alpha(V) + \quad synaptic \quad input$$

we have

$$\mathcal{L}_{neuron} f(x) = \alpha(x) \frac{df}{dx}.$$

Then let $F(u, \mathbf{v})$ be the first exit time of U to suprathreshold values with an initial value $(u, \mathbf{v})$ where u is the initial voltage and $\mathbf{v}$ contains the initial values of all gating variables (m, n, and h for Hodgkin-Huxley). Then, assuming that the spiking of each cell is not too rapid and occurs in distinct singlets, and that the synaptic potentials are small relative to threshold such that synchronization is not predominant, we have the following result, which states that the network firing rate can then be *estimated*, using the single neuron operator, $\mathcal{L}_{neuron}$, as the solution of a (very complicated) differential-difference equation with the stated boundary conditions. Similar equations hold for the higher moments of the firing time.

Theorem

The network firing rate λ *can be estimated from the solution of the functional differential equation*

$$\mathcal{L}_{neuron}F(u,\mathbf{v})+M\lambda[F(u+a,\mathbf{v})-F(u,\mathbf{v})]+\lambda^{ext}[F(u+\epsilon,\mathbf{v})-F(u,\mathbf{v})]=-1,$$

with the appropriate boundary conditions and the additional constraint

$$F(0,\mathbf{v_0})=\frac{1}{\lambda},$$

where $\mathbf{v_0}$ *contains the chosen initial values of the subsidiary variables.*

Proof

The theorem follows immediately from standard first exit-time theory for discontinuous Markov processes [27], assuming that each neuron fires at rate λ and receives excitation from other cells in the network at rate $M\lambda$.

6.4 Discussion

We have outlined methods for determining the statistics of neuronal spiking activity in networks of neurons with general physiological properties (subthreshold dynamics, threshold and refractory period) in cases where random activity can be sustained. Solutions of functional differential equations yield estimates of the firing rates and higher moments of the interspike interval. These solutions can be found by graphical, numerical or simulation methods. We have concentrated on networks whose neuronal spike trains are roughly Poisson and given results for excitatory networks. We demonstrated graphically that there may be three solutions where $f_o = f_i/n_E$ intersects the frequency transfer curve of each neuronal element and that one of these was stable and gave the frequency of spontaneous activity. With realistic parameter values it is thus possible to estimate the average number of connections to each cell. This approach may provide insight into the factors controlling spontaneous activity of neurons in the mammalian and other nervous systems. Networks with small excitatory postsynaptic potential amplitudes and those with excitatory and inhibitory connections can sometimes be analysed more easily with diffusion approximations. The theory is accurate only for networks in which synchronization is unlikely as this tends to invalidate the assumption of continued randomness in individual neuronal spike trains. This was amply demonstrated in simulations of networks containing up to 100 neurons.

Additional sources of randomness, especially those that can be considered to be of a perturbative nature and tend to decrease the chance of synchronization can easily be included in the analytical framework presented here. Such perturbative noise has been included, and found to play an essential role in maintaining network activity in studies of both a theoretical [6] and more applied nature [17, 18].

References

1. Badoual M, Rudolph M, Piwkowska Z, Destexhe A, Bal T (2005) High discharge variability in neurons driven by current noise. Neurocomputing 65–66: 493–498.
2. Feng J-F, Tuckwell HC (2003) Optimal control of neural activity. Phys Rev Lett 91:018101-5.
3. Jolivet R, Rauch A, Lüscher HR, Gerstner W (2006) Predicting spike timing of neocortical pyramidal cells by simple threshold models. J Comp Neurosci 21:35–49.
4. Heil P (2004) First-spike latency of auditory neurons revisited. Curr Opinion Neurobiol 14:461–467.
5. Mattia M, Del Giudice P (2003) A distribution of spike transmission delays affects the stability of interacting spiking neurons. Scientiae Mathematicae Japonicae 18:335–342.
6. Van Rossum MCW, Turrigiano GG, Nelson SB (2002) Fast propagation of firing rates through layered networks of noisy neurons. J Neurosci 22:1956–1966.
7. Brunel N, Wang XJ (2003) What determines the frequency of fast network oscillations with irregular neural discharges? J Neurophysiol 90:415–430.
8. Shiino M, Yamana M (2004) Statistical mechanics of stochastic neural networks. Phys Rev E 69:011904.
9. Steriade M (2000) Corticothalamic resonance, states of vigilance and mentation. Neuroscience 101:243–276.
10. Fiser J, Chiu C, Weliky M (2004) Small modulation of ongoing cortical dynamics by sensory input during natural vision. Nature 431:573–578.
11. Douglas RJ, Martin KAC (2004) Neuronal circuits of the neocortex. Annu Rev Neurosci 27:419–451.
12. Tuckwell HC, Wan FYM, Rospars J-P (2002) A spatial stochastic neuronal model with Ornstein-Uhlenbeck input current. Biol Cybernet 86:137–145.
13. Burkitt AN (2006) A review of the integrate-and-fire neuron model: I. Homogeneous synaptic input. Biol Cybernetics 95:1–19.
14. Holden AV (1976) Models of the Stochastic Activity of Neurones. Springer-Verlag, Berlin.
15. Gerstner W, Kistler WM (2002) Spiking Neuron Models. Cambridge University Press, Cambridge.
16. Rodriguez R, Tuckwell HC (1996) Statistical properties of stochastic nonlinear dynamical models of single neurons and neural networks. Phys Rev E 54:5585–5590.
17. Durstewitz D, Seamans JK, Sejnowski TJ (2000) Neurocomputational models of working memory. Nature Neurosci 3:1184–1191.

18. Durstewitz D, Seamans JK, Sejnowski TJ (2000) Dopamine-mediated stabilization of delay-period activity in a network model of prefrontal cortex. J Neurophysiol 83:1733–1750.
19. Berns GS, Sejnowski TJ (1998) A computational model of how the basal ganglia produce sequences. J Cogn Neurosci 10:108–121.
20. Durstewitz D, Seamans JK (2002) The computational role of dopamine D1 receptors in working memory. Neural Networks 15:561–572.
21. Amit DJ, Brunel N (1997) Model of global spontaneous activity and local structured activity during delay periods in the cerebral cortex. Cerebral Cortex 7:237–252.
22. Tuckwell HC, Richter W (1978) Neuronal interspike time distributions and the estimation of neurophysiological and neuroanatomical parameters. J Theor Biol 71:167–183.
23. Tuckwell HC, Cope DK (1980) The accuracy of neuronal interspike times calculated from a diffusion approximation. J Theor Biol 83:377–387.
24. Hodgkin AL, Huxley AF (1952) A quantitative description of membrane current and its application to conduction and excitation in nerve. J Physiol 117:500–544.
25. Mainen ZF, Joerges J, Huguenard JR, Sejnowski TJ (1995) A model of spike initiation in neocortical pyramidal neurons. Neuron 15:1427–1439.
26. Migliore M, Hoffman DA, Magee JC, Johnston D (2004) Role of an A-type K+ conductance in the back-propagation of action potentials in the dendrites of hippocampal pyramidal neurons. J Comp Neurosci 7:5–15.
27. Tuckwell HC (1976) On the first-exit time problem for temporally homogeneous Markov processes. J Appl Prob 13:39–48.
28. Cope DK, Tuckwell HC (1979) Firing rates of neurons with random excitation and inhibition. J Theor Biol 80:1–14.

7

Physiology and Related Models of Associative Visual Processing

Reinhard Eckhorn, Alexander Gail, Basim Al-Shaikhli, Andreas Bruns, Andreas Gabriel, and Mirko Saam

Summary. This is a review of our work on multiple microelectrode recordings from the visual cortex of monkeys and subdural recordings from humans, related to the potential underlying neural mechanisms. The former hypothesis of object representation by synchronization in visual cortex (or, more generally, of flexible associative processing) has been supported by our recent experiments in monkeys, which demonstrated local synchrony among rhythmic or stochastic γ-activities (30–90 Hz) and perceptual modulation, according to the rules of figure-ground segregation. However, γ-synchrony in primary visual cortex is restricted to few millimeters, challenging the synchronization hypothesis for larger cortical object representations. We found that the spatial restriction is due to γ-waves, traveling in random directions across the object representations. It will be argued that phase continuity of these waves can support the coding of object continuity. Based on models with spiking neurons, potentially underlying neural mechanisms are proposed: (1) Fast inhibitory feedback loops can generate locally synchronized γ-activities. (2) Hebbian learning of lateral and feed forward connections with distance-dependent delays can explain the stabilization of cortical retinotopy, the limited size of synchronization, the occurrence of γ-waves, and the larger receptive fields at successive levels. (3) Slow inhibitory feedback can support figure-ground segregation. In conclusion, it is proposed that the hypothesis of flexible associative processing by γ-synchronization, including coherent representations of visual objects, has to be extended to more general forms of signal coupling.

7.1 Dynamic Associative Processing by Different Types of Signal Coupling in the Visual Cortex

In the proposed view of the visual system, temporal coding is intimately linked to the neural mechanisms of dynamic cortical cooperativity and flexible associative processing, including the largely unknown mechanisms of perceptual feature binding. How are local features flexibly grouped into actually perceived objects and events, and how do their current representations interact with visual memory and other higher-order processes? It has been proposed that binding of spatially distributed features and inter-areal cooperation are supported by the temporal code of fast synchronization among neurons involved in a common task, for example, the coding of a visual object [1, 2]. This hypothesis attracted attention when synchronized γ-oscillations (30–90 Hz) were found in the primary visual cortex (V1) of anesthetized cats [3, 4, 5] and awake monkeys [6, 7]. Many subsequent experiments were supportive, some challenging with respect to binding of local features by γ-synchronization (reviewed in [8, 9]). For example, synchronization of signals in the γ-range was found to be restricted to few millimeters in primary visual cortex, even with large coherent stimuli [5, 10]. According to a strict interpretation of the original synchronization hypothesis, this should result in locally restricted perceptual feature binding. But this is in contradiction to the capability of perceiving local features of large objects as coherently bound. However, the capability of long-range feature binding across the surface of a large visual object is probably due to continuous binding among overlapping regions of locally restricted feature binding (as demonstrated by the perceptual laws of Gestalt psychology [11]). This view is supported by our observation of γ-waves that propagate across the surface of the representation of visual objects in the primary visual cortex of awake monkeys. Accordingly, we suggest that the phase continuity of such γ-waves (by which we mean a continuum of overlapping near-synchronized patches as opposed to strict long-range synchrony), may be a basis of spatial feature binding across entire objects. Such (locally synchronous) long-range phase coupling has been found to cover larger cortical areas than γ-synchrony as it is measured with spectral coherence [12], and we will argue that it can fill the entire surface representation of visual objects in primary visual cortex.

Such continuity may not be available between separate topographical maps (different visual cortical areas). However, γ-synchronization has been found between neural groups with overlapping receptive fields in adjacent visual cortical areas V1 and V2 in cats [3, 5] and monkeys [7]. It is probable that such synchrony is also present among other visual areas when feed-forward-backward delays are short (e.g., as between V1 and the middle temporal area (MT) [13]). In contrast, when cortical areas are far apart, long conduction delays may cause cooperativity to be reflected in other forms of signal coupling that are less sensitive to any spatiotemporal restriction of synchronization. Taking into account the time-varying amplitude (amplitude envelope) of γ-signals seems to be a particularly promising approach [14].

For our present work different types of neural signals have been recorded, and different forms of temporal coding have been investigated by means of various coupling measures. We will demonstrate dynamic coupling of cortical signals in the form of local intra-areal phase synchrony, and medium-range phase continuity of γ-waves. Our examples show that neural-signal measures correlate with sensory events, and with perceptual and behavioural outputs in monkeys. In essence, we argue that the temporal coding hypothesis of binding-by-synchronization, initially restricted to γ-synchrony of oscillatory signals, has to be extended to more general forms of temporal coding, including non-linear signal coupling across the entire frequency range of cortical activity with phase- and amplitude-coupling among transient and stochastic (non-rhythmic) signals. On the basis of neural models with locally coupled spiking neurons we will discuss most of the physiological results and suggest potential neural mechanisms underlying the presented types of flexible temporal coding.

7.2 Experimental Evidence

7.2.1 γ-Activity in Monkey Primary Visual Cortex is Phase-Coupled within Representations of Scene Segments and Decoupled Across their Contours

The binding-by-synchronization hypothesis suggests coupling among γ-activities representing the same object, or more generally, the same scene segment. Accordingly, neural groups representing different scene segments should decouple their γ-activities. Both predictions have been tested by investigating the effect of a static figure-ground stimulus on local field potentials (LFPs) in primary visual cortex (V1) of awake monkeys, recorded simultaneously from inside and outside a figure's representational area (Fig. 7.1A) [15]. Time-resolved analysis of phase coupling by means of spectral coherence revealed: (1) γ-coherence between neurons representing the same scene segment (figure or ground) is higher than for a homogeneous gray background of the same average luminance (Fig. 7.1B,D); (2) stimulus-specific γ-coherence is strongly reduced across the representation of the figure-ground contour compared to a spatially continuous stimulus (Fig. 7.1B,D); (3) decoupling across the contour emerges with a latency of about 100 ms, and is absent in the earliest neuronal response transients (Fig. 7.1D); (4) coherence of low-frequency components does not show a difference between the figure-ground and the continuous condition (not shown). We propose that the increased γ-coherence between neurons representing the same scene segment and the decoupling of γ-activity at a contour representation are crucial for figure-ground segregation, in agreement with the initial binding-by-synchronization hypothesis.

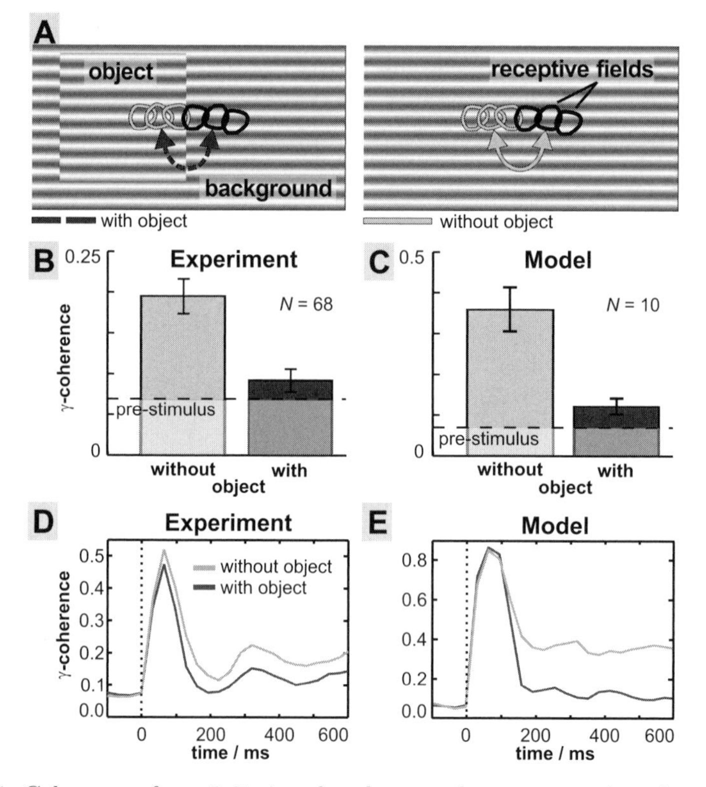

Fig. 7.1. Coherence of γ-activity is reduced across the representation of an object's contour. (**A**) Figure-ground stimulus and schematic positions of receptive fields. Stimuli were arranged in such a way that some of the receptive fields from the simultaneously recorded sites were located on the "object" (only present in the left condition), the others on the "background." (**B**) A grating without object (right condition in A) induced a substantial increase in γ-coherence among local field potentials (LFPs) (light gray) compared to a blank screen condition (pre-stimulus: dashed line). Introduction of the object (left condition in A) reduced LFP γ-coherence between object and background representations almost to pre-stimulus level (dark gray) [15]. Coherence within each segment (object or background) remained high (data not shown). (**C**) A network model (Fig. 7.10) shows equivalent results. (**D,E**) Time courses of coherence in the no-object condition (light gray) and across the object-background contour (dark) in the experiment and the model. Note that decoupling across the contour emerges about 100 ms after stimulus-onset. Data in B are taken from the time intervals with maximal decoupling for each monkey (modified from [15, 16]).

7.2.2 γ-Phase Coupling in Monkey Extrastriate Cortex Correlates with Perceptual Grouping

Are such synchronization effects correlated with perceptual feature grouping and figure-ground segregation? This was tested in a difficult figure-ground task in which a monkey indicated whether he perceived a figure composed of blobs among identical distractor blobs serving as background [17] (Fig. 7.2). This task was sufficiently difficult such that about 25% of responses were incorrect (failed figure detection). Pairs of local populations of figure-activated neurons in visual area V2 showed increased synchronization within the γ-range in correct compared to incorrect responses during a short period before the monkey's behavioural response (Fig. 7.2). Other signal measures were unrelated to perception. These were the first indications that γ-synchronization in V2 not only may represent physical stimulus properties but also supports perceptual grouping.

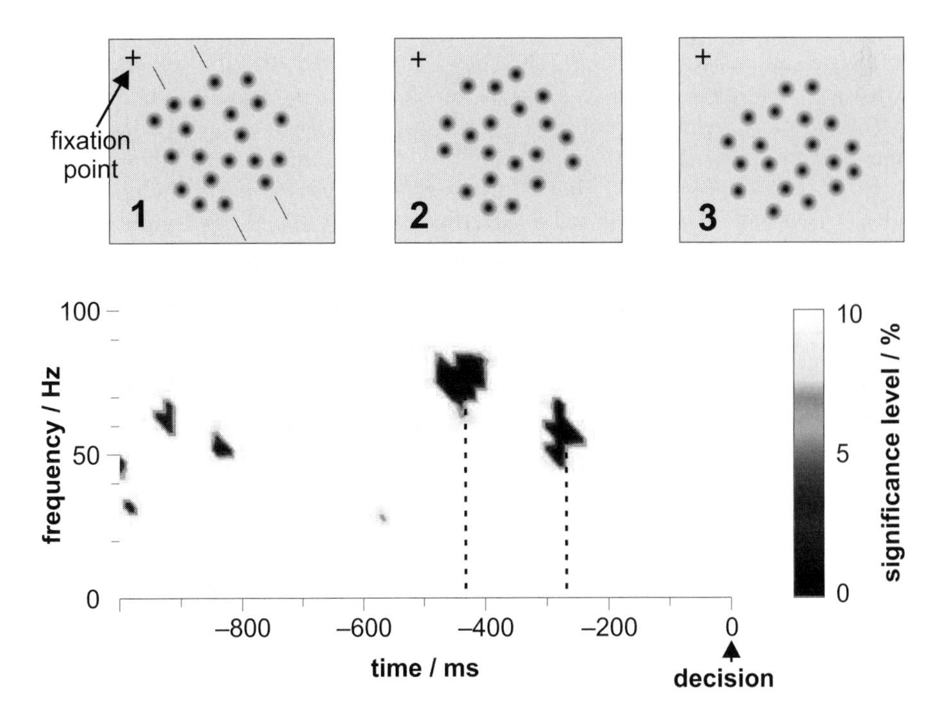

Fig. 7.2. A monkey's correct perception of the orientation of a dual in-line row of dots within a set of distractors (stimuli 1 and 2) caused a short increase in coherence at about 80 Hz and 60 Hz in visual area V2, shortly before the monkey reported his perception (time $t = 0$ ms). The time-frequency map indicates the significance of increase in LFP coherence in trials with correct vs. failed detection of the figure. Three figure-ground stimuli are shown above, with dot rows being left-tilted (left), right-tilted (middle), or absent (right) (modified from [17]).

7.2.3 Phase Continuity but not Synchrony of γ-Waves is Present Across Medium Cortical Distances in Monkey Primary Visual Cortex

Previous work demonstrated that the synchronization range of γ-activity in primary visual cortex is limited to about 5 mm (e.g., [10, 18] and Fig. 7.4A). Hence, objects with larger cortical representations cannot solely be coded by γ-synchrony within their representational area. One explanation for the limited synchronization range lies in the spatiotemporal characteristics of γ-activity. In fact, wave-like phenomena defined by spatially continuous phase-fronts (γ-waves) do extend farther than 5 mm, but phase differences between any two sites change randomly already within 100 ms and also increase with cortical distance (Fig. 7.3A) [15]. Conventional pairwise coupling measures (cross-correlation, coherence) do not capture such nontrivial phase relationships across medium-range cortical distances, which explains the findings of restricted synchronization ranges. To quantify those waves a new method has been developed in our group [12]. It revealed that γ-waves travel at variable velocities and directions. Fig. 7.3C shows the velocity distribution measured with a 4×4 microelectrode array in monkey primary visual cortex during retinally static visual stimulation. Note that this distribution is rather similar to the distribution of spike velocities of horizontal connections in this area (V1) [13]. We suggest that continuity of γ-waves supports the coding of object continuity, in which case their extent over object representations in visual area V1 and the related visual fields should be much larger than that covered by γ-synchronization. We have indeed found that the cortical span of γ-wave fronts is much larger than the span of γ-synchronization (Fig. 7.4A,B) and that y-waves are cut off (damped down) at the cortical representation of object contours.

7.3 Potential Neural Mechanisms of Flexible Signal Coupling

At present it is not possible to identify directly from experimental measurements the neural mechanisms underlying the above mentioned experimental observations of spatiotemporal processing in cortical sensory structures. We therefore use largely reduced model networks with spike-coding neurons to discuss potential mechanisms.

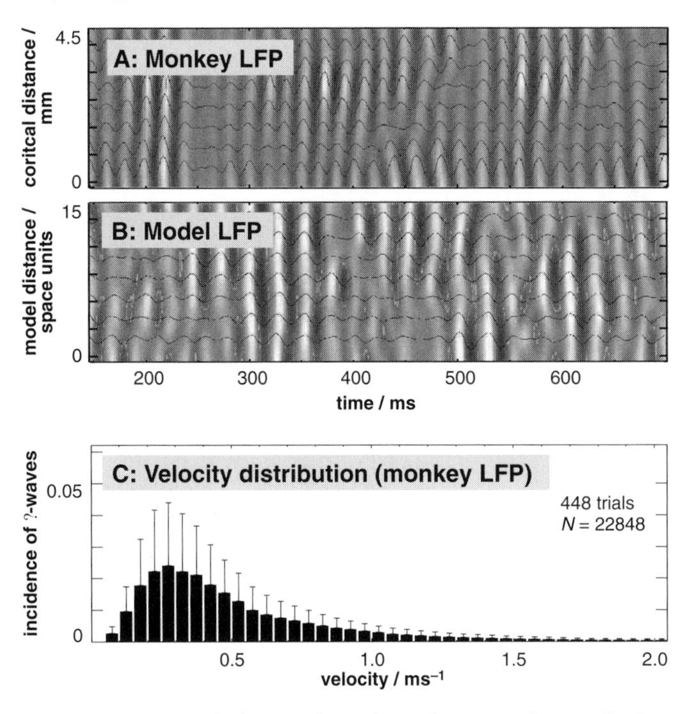

Fig. 7.3. γ-waves occur with fast and random changes of spatial phase relations in monkey primary visual cortex. (**A**) Time-space maps of simultaneously recorded single-trial time courses of local field potentials (LFPs) from a linear array of seven recording positions during sustained static visual stimulation. Gray scales give the instantaneous electrical potentials at the electrodes (in arbitrary units). (**B**) Model-LFPs during presentation of an equivalent stimulus. (Modified from [16]) (**C**) Velocity distribution of traveling waves of γ-activity measured with 4×4 microelectrode arrays in monkey primary visual cortex

7.3.1 γ-Oscillations and Synchrony of Spike Densities in Local Populations, Generated by Feedback Inhibition and Local Lateral Coupling

How can the cortex generate γ-oscillations in local neural groups, as observed in LFP and multiple unit activity (MUA)? We argue (see Fig. 7.5) that membrane potentials of local populations of excitatory neurons are simultaneously modulated by inhibition exerted via a common feedback loop (physiology: [19–21]; models: [22–26]; discussion in [27]). This loop can quickly reduce transient activations, whereas sustained input will lead to repetitive inhibition of the population in the γ-frequency range (Fig. 7.5). In both modes—transient and rhythmic chopping—the common modulation of the neurons' membrane potentials causes their spike trains to become partially synchronized, even if they fire at very different rates. The stronger a neuron

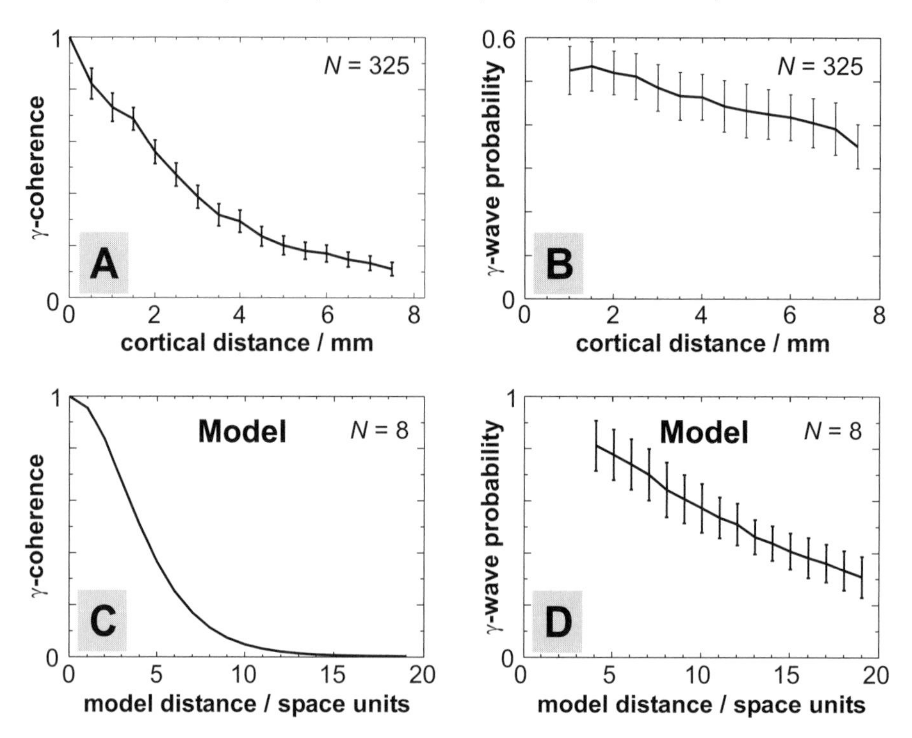

Fig. 7.4. The spatial range of γ-waves is larger than that of γ-coherence. (**A**) Coherence of local field potentials in monkey primary visual cortex is restricted to few millimeters (half-height decline 2.2 mm). (**B**) The probability of occurrence of continuous γ-waves remains high across larger distances (estimated half-height decline: 9.5 mm). (**C,D**) The model shows equivalent dependencies (4.1 vs. 12.8 space units) (modified from [16]).

is activated and depolarized, the earlier it will discharge its first spike during the common repolarization phase, whereby such a population burst will be dominated by the most strongly activated neurons. As local cortical populations generally project to common targets [28], synchronized spike densities (population spike packages) will have stronger impact there than uncorrelated spike densities of equal average amplitudes, because they (1) appear quasi-simultaneously, and (2) mainly comprise spikes of strongly activated neurons, which represent the stimulus at a better signal-to-noise ratio than the neurons that were less activated by the same stimulus. In addition to partial local synchronization by inhibitory feedback, the most relevant mechanism for explaining synchronization in our models are lateral, activity-dependent, facilitatory connections. Local (instead of global) lateral connections are critically important for models of visual feature-binding by synchronization when pattern segmentation (desynchronization) is an important task (e.g., [29, 30, 31]). While Wang used lateral excitatory connections that were modulated in their

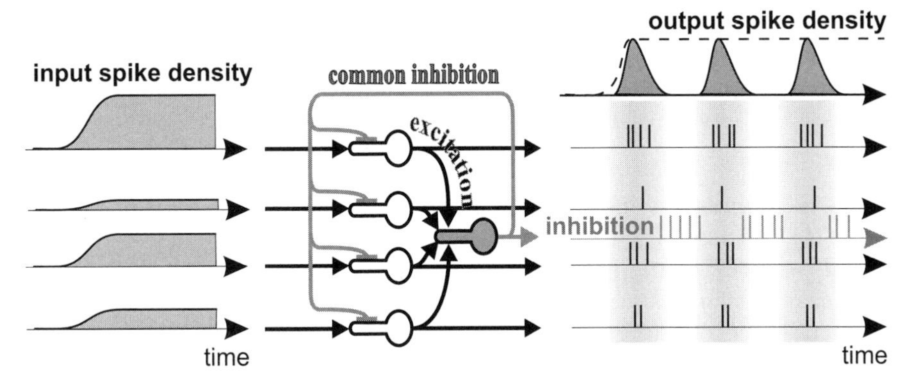

Fig. 7.5. Basic model of common spike density modulation in a local population of excitatory neurons by a common inhibitory feedback neuron. Note that first-spike latencies in each modulation cycle at the outputs (right) are roughly inversely proportional to the input spike densities (profiles at left), whereas the output spike rates are proportional to it (details in text).

efficacy by scene properties (in some respect similar to the facilitatory connections in our models [29]), others used lateral connections from excitatory to inhibitory neurons for synchronization [25, 31]. It is likely that a mixture of these local mechanisms is operative in the generation of rhythmic spiking activities and their partial synchronization. Future experiments have to answer this question.

We can apply the discussed schemes to the primary visual cortex, where local neural clusters represent similar feature values (e.g., receptive field position, contour orientation, etc.). According to the synchronization hypothesis, partial synchronization of spike densities by a common inhibitory feedback means that currently present local combinations of visual feature values are systematically selected by their strength of activation and tagged as belonging together, which is reflected in single or repetitive population discharges.

Other models of visual feature binding use local oscillators, consisting of excitatory and inhibitory units with mutual feedback that generate local oscillations depending on a driving input (e.g., [30, 31, 32]). In these models, the oscillatory signal of a local element stands for the spike density of a local group of partially synchronized spike-coding neurons. Thus, local inhibition in these models implicitly represents the synchrony of local populations of spike-coding neurons with similar receptive field properties, as has been explicitly modeled in our and other simulations (e.g., [23, 25]).

7.3.2 Lateral Conduction Delays Can Limit γ-Synchrony to a Few Millimeters in Cortex, Produce Wave-Like Phenomena, Stabilize Cortical Topography, and Lead to Larger Receptive Fields at Successive Levels of Processing

The synchronization effect of fast orientation-specific inhibitory neurons is probably restricted to an area smaller than a single hypercolumn in primary visual cortex [28]. The most relevant mechanism for explaining flexible synchronization across several millimeters in the cortex in our [29] and Wang's [30] model are the activity-dependent facilitatory connections. They are also highly useful for enabling fast desynchronization as is required for scene segmentation. Their putative physiological substrate in the primary visual cortex are the dense horizontal connections: they cannot directly excite their target neurons, but modulate their activities evoked from their classical receptive fields. The lateral connections project monosynaptically over a range of several hypercolumns [19, 20, 33, 34], and models have shown that this type of connectivity is capable of synchronizing neural populations across larger distances [29, 30]. Another type of local lateral connectivity enabling transient synchronization over larger distances was proposed in the model of König and Schillen [31]. They connected their oscillators by coupling the excitatory units via delay lines to the neighbouring inhibitory units. However, it is difficult to show experimentally which mechanisms are operative in the visual cortex for synchronization across several hypercolumns.

In visual processing for example, one could suppose that neural populations representing the entire surface of a visual object might synchronize their spike packages via horizontal connections. However, γ-synchrony is restricted to about 5 mm of cortical distance in area V1 of awake monkeys (corresponding to five hypercolumns), even if the cortical representation of a visual object is much larger [5, 10]. Hence, feature binding based on γ-synchrony would also be restricted to visual objects being not larger in their cortical representations. In the following we will develop a concept of how distance-dependent spike conduction delays can explain this restricted range of γ-synchrony and the occurrence of wave-like phenomena in a network of spiking neurons. In addition, we will show that Hebbian learning combined with distance-dependent spike conduction delays leads to spatially restricted lateral connectivity within the same layer and restricted feed-forward divergence between different layers. Therefore, such a mechanism is also suitable to explain the emergence of larger receptive fields at successive levels of processing while preserving a topographical mapping. Note that these conditions are also present in topographically organized cortical areas of other sensory modalities, including auditory and somatosensory.

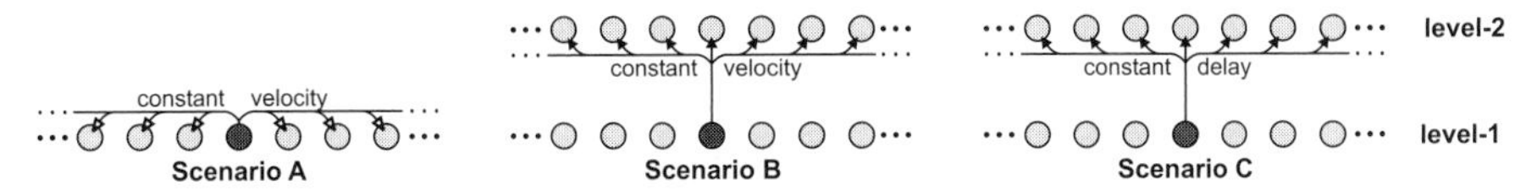

Fig. 7.6. One-dimensional sketch of the initial connectivity in the Hebbian learning model with distance-dependent lateral conduction delays. For a given level-1 neuron (dark), the scheme shows lateral modulatory (facilitatory) connections (scenario A), and feed-forward connections with either distance-dependent (scenario B) or constant (scenario C) conduction delays (modified from [35]).

7.3.2.1 Hebbian-Learning Model with Finite Conduction Velocities

The local generation of γ-oscillations and their spatial synchronization are two separate mechanisms. For the sake of simplicity, the following model solely investigates synchronization effects, thereby neglecting any inhibition and the generation of oscillations. The model [35] consists of spike-coding neurons (as in Fig. 7.10B) at two successive, two-dimensional retinotopic visual processing stages named level 1 (representing visual cortical area V1) and level 2 (V2) (Fig. 7.6). Learning of lateral weights and level-1 to level-2 weights is implemented using a Hebbian spike correlation rule [36]. Feed-forward connections are additive excitatory and determine the properties of the classical receptive fields. Lateral connections are multiplicatory (with a positive offset of one), which means they cannot directly evoke spikes in a target neuron (as exitatory synapses can do), but require quasi-simultaneous feed-forward input to that neuron (model: [29]; physiology: [37]). Spikes evoked by quasi-simultaneous feeding input to neighbouring neurons can synchronize via their mutual lateral facilitatory connections because these spikes will often occur within the so-called capture range of the spike encoder's dynamic threshold [29, 38, 39]. The lateral connections have constant conduction velocities, that is, conduction delays become proportionally larger with distance. This reduces the probability of neurons becoming quasi-synchronized because constructive superposition of locally evoked and laterally conducted activities gets less probable for increasing delay. Hence, synchrony is laterally restricted to a spatial range that is proportional to the conduction velocity of the lateral connections.

7.3.2.2 Spatiotemporal Structuring of Lateral Connectivity with Hebbian Learning

The relation between conduction velocity and synchronization range suggests an influence of temporal neighborhood (defined by the distance-dependent delays) on the ontogenetic, possibly prenatal formation of functionally relevant structures from an initially unstructured system [40, 41, 42]). This effect can be simulated with our model. In the beginning, neurons are fully interconnected within level 1 (Fig. 7.6, scenario A). Feed-forward input spike trains have spatially homogeneous random patterns and are given a temporally confined, weak co-modulation, mimicking activity before visual experience. This type of spike pattern appears, slightly modified by the connections, at the output of the level-1 neurons (Fig. 7.7) and hence, is used for Hebbian learning. The only topography in the network is given by the distance-dependent time delays of the lateral connections. During a first learning period, the homogeneous coupling within level 1 shrinks to a spatially limited coupling profile for each neuron, with a steep decline of coupling strength with increasing distance (Fig. 7.8). The diameter of the resulting coupling profile for each neuron is near the lateral synchronization range, and hence directly proportional to the lateral conduction velocity (Fig. 7.9A).

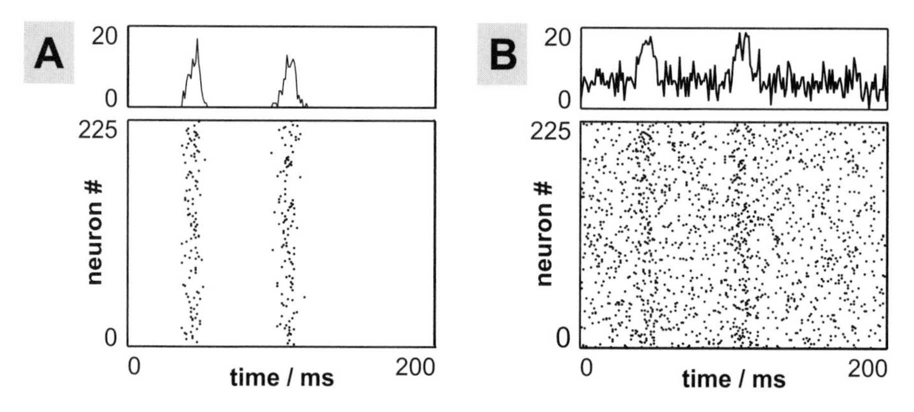

Fig. 7.7. Spatiotemporal properties of level-1 output activity in the Hebbian learning model. (**A**) Two events of spatially homogeneous, transient spike-rate enhancement (upper panel: total population spike density; lower panel: single spike traces). (**B**) As in A, but with additional independent Gaussian white noise at the inputs. Note that the activity is spatially homogeneous in the sense that any two spike trains have the same weakly correlated temporal statistics (modified from [35]).

7.3.2.3 Spatiotemporal Structuring of Interlevel Connectivity

In a second learning period following the learning period within level 1, the excitatory level-1 to level-2 connections are adapted, also starting from full connectivity (Fig. 7.6, scenario B). Again, as a result of Hebbian correlation learning [36], the feed-forward divergence retracts to a limited spatial range that is given by the size of the level-1 synchronization fields, that is, excitatory forward connections from neurons within a level-1 synchronization field (sending near-synchronized spike packages) converge onto one level-2

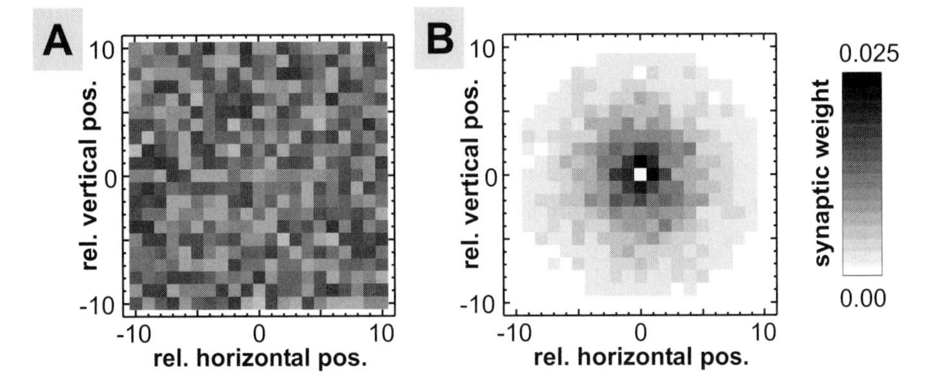

Fig. 7.8. Spatial coupling profile between a single level-1 neuron (center position) and its neighbours. (**A**) Before learning weights are randomly initialized. (**B**) After learning synaptic strength decays with increasing distance (modified from [35]).

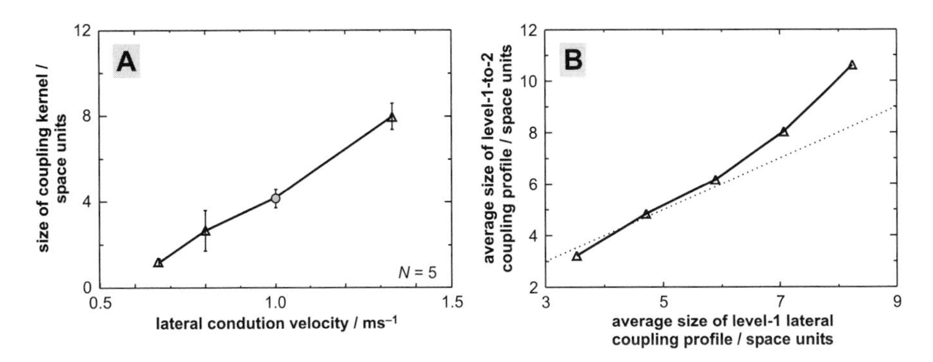

Fig. 7.9. (**A**) As a result of Hebbian learning, the size of the synaptic weight profile (coupling kernel) of lateral linking connections within level 1 becomes directly proportional to the lateral conduction velocity. (**B**) The size of level-1 synchronization fields determines the size of receptive fields of level-2 neurons. Synaptic weight profiles of level-1-to-2 feeding connections evolve correspondingly to synaptic weight profiles of level-1 lateral linking connections (modified from [35]).

neuron (Fig. 7.9B). This convergent projection pattern even emerges if the feed-forward connections and the level-2 lateral connections are modeled with distance-independent constant delays (Fig. 7.6, scenario C). The physiological interpretation of this result is that the size of level-1 synchronization fields (in visual area V1) can determine the size of level-2 receptive fields (in area V2). Indeed, synchronization fields in V1 and classical receptive fields in V2 of the monkey do have similar sizes. Since equivalent considerations should hold for projections from the retina to V1, the model accounts for the emergence not only of a spatially regular, but also of a retinotopically organized connectivity.

7.3.2.4 Traveling γ-Waves with Altering Phase Relations

After the learning period, we wanted to investigate the network dynamics. To compare the results with experimental data, we added local inhibitory feedback neurons and provided a sustained spatially homogeneous input to level-1 neurons. This inhibition did not invalidate the previous results, because its dynamics rarely overlap with the learning process. This model reproduces the phenomenon of waves with jittering phase relations, traveling in random directions, just as it was observed in the primary visual cortex (Fig. 7.3) [12, 16]. Traveling waves of γ-activity, though concentrically expanding, have already been described in different cortical areas of different animal species (e.g., [43]). The varying phase relations in our model as well as the more rapid spatial decline of γ-coherence (compared to γ-wave probability) are consistent with the experimental data (Figs. 7.4A, B and 7.5). Formation of γ-waves in the model results from the locally restricted inhibition, the lateral conduction velocity, and the steep spatial decline of coupling strength [35]. It seems probable that cortical γ-waves are also strongly depending on the lateral conduction velocities, because the distribution of γ-wave velocities (Fig. 7.3C) is similar to the distribution of spike conduction velocities of lateral connections in primary visual cortex. These velocities have been estimated in different preparations, including rat slices and in vivo recordings from cats and monkeys, to range between 0.1 and $1.0\,\mathrm{m \cdot s^{-1}}$ (review: [13]).

In conclusion, the lateral conduction velocities in primary visual cortex, combined with Hebbian correlation learning, can explain the restricted spatial range of γ-synchrony and the occurrence of traveling γ-waves with random phase relations. They can also account for the larger receptive fields at higher processing levels and for the emergence and stability of topographic visual (and other sensory) representations without the need for visual (sensory) experience. During visual experience, of course, similar influences on synchronization-field and receptive-field size and on topographic stability are probably operative at successive levels of processing, including other parts of the visual system. As the traveling waves can cover the entire representation of an object's surface in the primary visual area, we propose that phase continuity of γ-waves may constitute a mechanism that supports the coding of object continuity in visual cortex [12].

7.3.3 Model Explaining Figure-Ground Segregation and Induced Modulations at Lower Frequencies by Slower and More Globally Acting Feedback Circuit

In a further approach the above model of the visual cortex has been expanded by using orientation-selective excitatory neurons and two types of inhibitory neurons with different spatiotemporal properties. However, the lateral connections among the excitatory neurons were modeled without delay and learning has been excluded in order to keep the complexity of the network within limits in order to understand its processing. Fig. 7.10A shows a simplified wiring diagram of this model. The spiking neurons (Fig. 7.10B) have linearly and nonlinearly acting synapses and are retinotopically arranged in two primary layers with receptive fields at perpendicular orientation preferences. Additionally to the fast inhibitory feedback loop, serving neurons with similar orientation preference and generating local γ-rhythms (see above and Fig. 7.5), a slow shunting inhibition is added in this model that forms a feedback circuit among neurons with overlapping receptive fields and receive input from, as well as feed output to, excitatory neurons of all orientation preferences.

7.3.3.1 Decoupling of γ-Signals Across Figure-Ground Contour

In the figure-ground experiment, representations of different scene segments in primary visual cortex (area V1) were decoupled in their γ-activities (Fig. 7.1 B,D), while the same sites showed substantial γ-coupling when representing one coherent scene segment. Analogous results are obtained with the model (Fig. 7.1C,E). It explains the reduced γ-coherence as a blockade of lateral coupling at the position of the contour representation due to several effects: First, neurons responding preferentially to the horizontal grating are only weakly activated by the vertical contour. Second, their activity is even more reduced by the orientation-independent slow shunting inhibition that is evoked by the strongly activated vertically tuned neurons at the contour. As a consequence, neurons activated by the horizontal grating near both sides of the contour cannot mutually interact, because the orientation-selective lateral coupling is interrupted by the inhibited horizontally tuned neurons at the contour representation. The resulting decoupling of inside and outside representations is not present during the first neural response transient after stimulus onset (Fig. 7.1D,E) since the sharp, simultaneous response onset common to both orientation layers denotes a highly coherent broad-band signal that dominates internal dynamics. Note that orientation-selectivity was implemented for the sake of comparability with the experimental data. This does not limit the general validity of this model, since any object border constitutes a discontinuity in at least one visual feature dimension, and therefore an analogous argumentation always holds for other local visual features.

7.3.4 Spatial and Temporal Aspects of Object Representations

We have seen that within an object's representation in primary visual cortex (V1), locally synchronized γ-activations emerge that overlap in space and

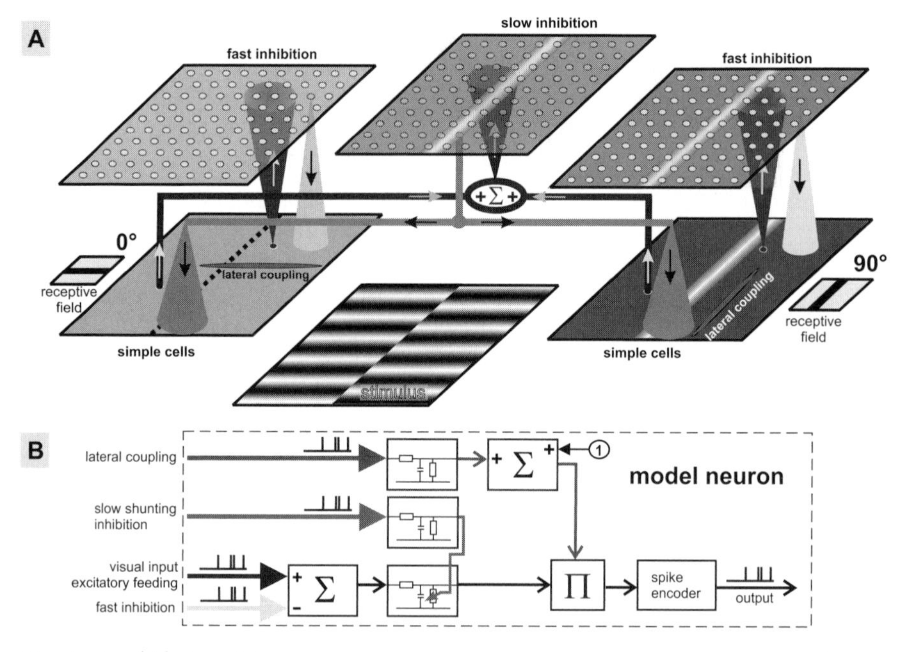

Fig. 7.10. (**A**) Simplified model of primary visual cortex with two retinotopic layers (left and right) of orthogonal orientation detectors. Each layer consists of an excitatory (simple cells at 0 and 90) and an inhibitory sublayer forming fast, local feedback loops preferring the γ-frequency range (Fig. 7.7). The cones between layers indicate direction and width of divergent projections of each single neuron. Modulatory (facilitatory) lateral coupling is confined to excitatory neurons of the same orientation preference with coaxially aligned receptive fields. Both orientation maps share an additional sublayer (middle) mediating slow, local shunting inhibition. To account for stochastic input from brain regions excluded from the model, all neurons receive independent broad-band noise. (**B**) The spike-coding model neuron with dynamic threshold [27] is extended by inputs exerting shunting inhibition on the feeding pathway. Synapses are modeled by leaky RC-filters, lateral modulatory input is offset by a value of +1 and then coupled to the feeding pathway by multiplication.

time and thereby support the formation of γ-waves traveling across the object's surface representation with random phase relations. When waves of γ-activity travel across V1, this is paralleled by quasi-synchronous activity of those neurons in area V2 having corresponding receptive field positions [3, 5, 7], that is, those receiving input from, and sending γ-feedback to, corresponding V1 neurons. Thus, adjacent V2 neurons, driven simultaneously by adjacent parts of a traveling wave, will also form a traveling wave of γ-activity (with similar extent, velocity, and direction if projected to visual space). We expect such an argumention to hold for subsequent stages of processing, provided that they are retinotopically arranged, are activated by bottom-up input, and have fast inter-areal feedback (compared to a half-cycle of a γ-wave).

Accordingly, quasi-synchrony should generally be present among neurons with overlapping receptive field positions across cortical levels connected via fast feed-forward-backward loops (e.g., as among V1-V2 and V1-MT [13, 44]). As the traveling waves are γ-activities and we observed γ-decoupling across the cortical representation of a figure-ground contour (explained in our model by the slow inhibition of neurons at the contour), we assume that the waves do not pass object contour representations with any figure-ground feature contrast. *Object continuity* across the entire surface may thus be coded by *phase continuity* of traveling γ-waves.

7.4 Conclusion

We investigated neural mechanisms of associative processing by considering a variety of flexible forms of signal coupling. In particular, we were interested in associations of local visual features into perceptually coherent visual objects by transient synchronization. In our recent experiments in monkeys, we have shown that local synchrony among γ-activities correlates with perceptual modulation, which supports the hypothesis of object representation by synchronization in the visual cortex. The synchronization hypothesis for larger cortical object representations, however, has been challenged by our experimental finding that γ-synchrony in primary visual cortex (V1) already drops to noise levels across few (4–6) millimeters of cortex. We can explain this restriction by the randomly changing phase relations among locally synchronized patches, which, however, form continuous waves of γ-activity, traveling across object representations. Extending the initial synchronization hypothesis, we propose that phase continuity of these waves may support the coding of object continuity across intermediate and longer ranges within V1.

We have discussed the different types and ranges of experimentally observed signal coupling on the basis of visual cortex models with locally coupled, spike-coding neurons. In these models, the lateral, activity-dependent facilitatory connections with distance-dependent delays are the most important feature for explaining synchronous activity. They can account for local and medium-range γ-synchronization, the occurrence of γ-waves and the limited extent of γ-synchrony. Hebbian learning of these connections can explain the stabilization of cortical retinotopy and the larger receptive fields at successive levels of visual cortical processing. Fast local feedback inhibition in our models can generate local γ-oscillations and support local synchrony, while slow shunting-inhibitory feedback supports figure-ground segregation by decoupling activities within neighbouring cortical representations of figure and background. We propose that the hypothesis of associative processing by γ-synchronization be extended to more general forms of signal coupling.

References

1. Reitboeck HJ (1983) A multi-electrode matrix for studies of temporal signal correlations within neural assemblies. In: Basar E et al, (eds). Synergetics of the Brain Springer, Berlin 174–182.
2. von der Malsburg C, Schneider W (1986) A neural cocktail-party processor. Biol Cybern 54:29–40.
3. Eckhorn R, Bauer R, Jordan W, Brosch M, Kruse W, Munk M, et al (1988) Coherent oscillations: a mechanism of feature linking in the visual cortex? Multiple electrode and correlation analyses in the cat. Biol Cybern 60:121–130.
4. Gray CM, König P, Engel AK, Singer W (1989) Oscillatory responses in cat visual cortex exhibit inter-columnar synchronization which reflects global stimulus properties. Nature 338:334–337.
5. Eckhorn R (1994) Oscillatory and non-oscillatory synchronizations in the visual cortex of cat and monkey. In: Pantev C, Elbert T, Lütkenhöner B (eds). Oscillatory Event-Related Brain Dynamics. Plenum Press, New York, 115–134.
6. Kreiter A, Singer W (1992) Oscillatory neuronal responses in the visual cortex of the awake macaque monkey. Eur J Neurosci 4:369–375.
7. Frien A, Eckhorn R, Bauer R, Woelbern T, Kehr H (1994) Stimulus-specific fast oscillations at zero phase between visual areas V1 and V2 of awake monkey. Neuroreport 5:2273–2277.
8. Gray CM (1999) The temporal correlation hypothesis of visual feature integration: still alive and well. Neuron 24:31–47.
9. Eckhorn R (1999) Neural mechanisms of visual feature binding investigated with microelectrodes and models. Vis Cogn 6:231–265.
10. Frien A, Eckhorn R (2000) Functional coupling shows stronger stimulus dependency for fast oscillations than for low-frequency components in striate cortex of awake monkey. Eur J Neurosci 12:1466–1478.
11. Wertheimer M (1923) Untersuchungen zur Lehre von der Gestalt: II. Psychologische Forschung 4:301–350.
12. Gabriel A, Eckhorn R (2003) Multi-channel correlation method detects traveling gamma-waves in monkey visual cortex. J Neurosci Methods 131:171–184.
13. Nowak LG, Bullier J (1997) The timing of information transfer in the visual system in extrastriate cortex in primates. In: Rockland et al, (eds). Cerebral Cortex. Plenum Press, New York, 205–240.
14. Bruns A, Eckhorn R (2004) Task-related coupling from high- to low-frequency signals among visual cortical areas in human subdural recordings. Int J Psychophysiol 51:97–116.
15. Gail A, Brinksmeyer HJ, Eckhorn R (2000) Contour decouples gamma activity across texture representation in monkey striate cortex. Cereb Cortex 10:840–850.
16. Eckhorn R, Bruns A, Saam M, Gail A, Gabriel A, Brinksmeyer HJ (2001) Flexible cortical gamma-band correlations suggest neural principles of visual processing. Vis Cogn 8:519–530.
17. Woelbern T, Eckhorn R, Frien A, Bauer R (2002) Perceptual grouping correlates with short synchronization in monkey prestriate cortex. Neuroreport 13:1881–1886.
18. Steriade M, Amzica F, Contreras D (1996) Synchronization of fast (30–40 Hz) spontaneous cortical rhythms during brain activation. J Neurosci 16:392–417.

19. McGuire BA, Gilbert CD, Rivlin PK, Wiesel TN (1991) Targets of horizontal connections in macaque primary visual cortex. J Comp Neurol 305:370–392.
20. Bosking WH, Crowley JC, Fitzpatrick D (2002) Spatial coding of position and orientation in primary visual cortex. Nature Neurosci 5:874–882.
21. Jefferys JGR, Traub RD, Whittington MA (1996) Neuronal networks for induced 40 Hz rhythms. Trends Neuosci 19:202–208.
22. Chang HJ, Freeman WJ (1996) Parameter optimization in models of the olfactory neural system. Neural Networks 9:1–14.
23. van Vreeswijk C, Abbott LF, Ermentrout GB (1994) When inhibition not excitation synchronizes neural firing. J Comput Neurosci 1:313–321.
24. Freeman WJ (1996) Feedback models of gamma rhythms. Trends Neurosci 19:468–470.
25. Bush P, Sejnowski T (1996) Inhibition synchronizes sparsely connected cortical neurons within and between columns in realistic network models. J Comput Neurosci 3:91–110.
26. Wennekers T, Palm G (2000) Cell assemblies, associative memory and temporal structure in brain signals. In: Miller R (ed). Time and the Brain. Conceptual Advances in Brain Research, vol. 3. Harwood Academic Publishers, 251–273.
27. Eckhorn R (2000) Cortical processing by fast synchronization: high frequency rhythmic and non-rhythmic signals in the visual cortex point to general principles of spatiotemporal coding. In: Miller R (ed). Time and the Brain. Gordon & Breach, Lausanne, 169–201.
28. Braitenberg V, Schüz A (1991) Anatomy of the Cortex. Springer, Berlin.
29. Eckhorn R, Reiboeck HJ, Arndt M, Dicke P (1990) Feature linking via synchronization among distributed assemblies: simulations of results from cat visual cortex. Neural Comput 2:293–307.
30. Wang DL (1995) Emergent synchrony in locally coupled neural oscillators. IEEE Trans Neural Netw 6:941–948.
31. König P, Schillen TB (1991) Stimulus-dependent assembly formation of oscillatory responses: I. Synchronization. Neural Comput 3:155–166.
32. Li Z (1998) A neural model of contour integration in the primary visual cortex. Neural Comput 10:903–940.
33. Bosking WH, Zhang Y, Schofield B, Fitzpatrick D (1997) Orientation selectivity and the arrangement of horizontal connections in tree shrew striate cortex. J Neurosci 17:2112–2127.
34. Gilbert C (1993) Circuitry, architecture, and functional dynamics of visual cortex. Cereb Cortex 3:373–386.
35. Saam M, Eckhorn R (2000) Lateral spike conduction velocity in visual cortex affects spatial range of synchronization and receptive field size without visual experience: a learning model with spiking neurons. Biol Cybern 83:L1–L9.
36. Kempter R, Gerstner W, van Hemmen JL (1999) Hebbian learning and spiking neurons. Phys Rev E 59:4498–4514.
37. Fox K, Daw N (1992) A model of the action of NMDA conductances in the visual cortex. Neural Comput 4:59–83.
38. Johnson JL (1993) Waves in pulse-coupled neural networks. In: Proceedings of the World Congress on Neural Networks, vol. 4, 299–302.
39. Johnson JL (1994) Pulse-coupled neural networks. In: Proceedings of the SPIE Critical Rev, 55:47–76.
40. Trachtenberg JT, Stryker MP (2001) Rapid anatomical plasticity of horizontal connections in the developing visual cortex. J Neurosci 21:3476–3482.

41. Ruthazer ES, Stryker MP (1996) The role of activity in the development of long-range horizontal connections in area 17 of the ferret. J Neurosci 16:7253–7269.
42. Crair MC, Gillespie DC, Stryker MP (1998) The role of visual experience in the development of columns in cat visual cortex. Science 279:566–570.
43. Freeman WJ, Barrie JM (2000) Analysis of spatial patterns of phase in neocortical gamma EEGs in rabbit. J Neurophysiol 84:1266–1278.
44. Girard P, Hupé JM, Bullier J (2001) Feedforward and feedback connections between areas V1 and V2 of the monkey have similar rapid conduction velocities. J Neurophysiol 85:1328–1331.

8

Gestalt Formation in a Competitive Layered Neural Architecture

Helge Ritter, Sebastian Weng, Jörg Ontrup, and Jochen Steil

Summary: The dynamical organization or "binding" of elementary constituents into larger structures forms an essential operation for most information processing systems. Synchronization of temporal oscillations has been proposed as a major mechanism to achieve such organization in neural networks. We present an alternative approach, based on the competitive dynamics of tonic neurons in a layered network architecture. We discuss some properties of the resulting "Competitive Layer Model (CLM)" system and show that the proposed dynamics can give rise to Gestalt-like grouping operations for visual patterns and can model some characteristics of human visual perception. Finally, we report on an approach how the necessary, task-dependent interactions can be formed by learning from labeled grouping examples and sketch as an application of the system segmentation of biomedical images.

8.1 Introduction

Visual input is extremely high-dimensional: the brightness of each pixel in an image represents one degree of freedom (actually, three degrees of freedom, when we allow for color), and a typical picture offers on the order of several millions of them. Remarkably, our visual system has found highly efficient ways to extract from this extremely high-dimensional input rapidly and reliably a much smaller number of constituents that appear to us as conspicuous "patterns" or "gestalts." This amazing capability has already puzzled the psychologists of the early 20th century, and the Gestaltists have attempted to characterize good gestalts by a number of rules to explain general features to which our perception is tuned when organizing its inputs into "meaningful" constitutents [10].

Some of these classical gestalt rules are shown schematically in Fig. 8.1, where each row illustrates the gestalt rules of proximity, similarity, topological features such as closure or good continuation, or symmetry properties. As another major grouping principle the Gestaltists observed the visual system's tendency to organize many inputs into figure and ground (bottom row). Also dynamical features, such as shared movement direction ("common fate") can induce the perception of a gestalt, for example, when we watch fireworks or perceive a directed movement in a chain of merely alternatingly flashing light bulbs.

These "laws" show that our perception is strongly biased: we prefer particular organizations of the visual input over other ones, which on purely logical grounds would be defendable as well. However, the particular gestalt biases of the visual system are well in line with the computational task of decomposing the visual input into constituents that have a high chance of "good correspondence" with physical counterparts in our surroundings [12].

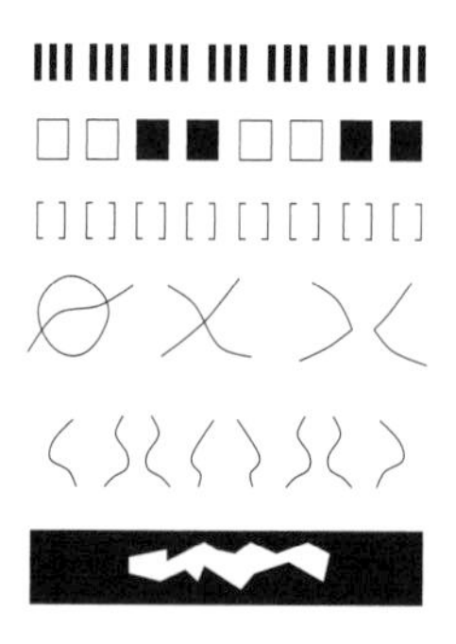

Fig. 8.1. Gestalt laws: proximity, similarity, closure, good continuation, symmetry and figure-ground.

Fig. 8.2. An example where prior knowledge is required.

However, not all of these biases are "hardwired" into our visual system. There are many patterns whose recognition relies on higher-level world knowledge that we only can have acquired by learning. Obvious examples are highly trained patterns such as digits or letters. Another example is depicted in Fig. 8.2. Here, we first discern only irregular white patches on a black background. Only when looking longer, we discern a pattern,[1] which, once we see it, forcefully masks the perception of the former patches as individual units (turning the page by 180 degrees restores the perception of the patches, illustrating that higher level grouping processes need not be rotationally invariant).

[1] A hidden face of a man in three-quarter profile in the middle of the picture.

8.2 Gestalt Grouping and Feature Binding

Forming a group or gestalt seems inevitably to require some operation of binding to implement the property that the gestalt is more than the sum of its parts (for instance, a suitable subset of the patches in Fig. 8.2 has to be "bound" into the unity of the face, separating it from the background of the remaining patches).

The need for such binding operations arises also in numerous other contexts that require the formation of units from simpler constituents, for example, elementary features. Fig. 8.3 illustrates the issue for the simple example of representing for two objects the conjunction of their shape and their color: to represent the simultaneous presence of a red round ball and a green triangle requires a binding within the four-element feature set {red, green, round, triangular} such that "red" goes with "round" and "green" with "triangular" and not vice versa.

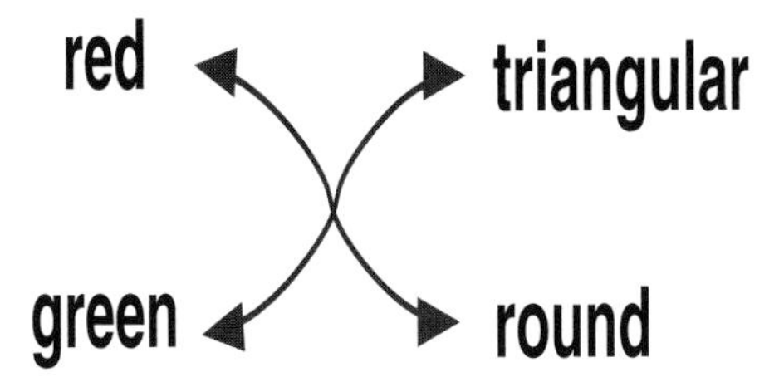

Fig. 8.3. Binding between color and shape attributes.

The problem that such a binding cannot be represented by just the copresence of neural activity for the individual constituents was first emphasized by Christoph von der Malsburg almost three decades ago [13], together with an elegant proposal for a solution: to use neural synchrony as a means to represent togetherness.

Subsequent research attempting to corroborate or refute this hypothesis has a long and controversial history (for a recent discussion, see [20] and the accompanying special issue of *Neuron*, as well as chapter 7 by Reinhard Eckhorn et al in this volume). While there exist numerous experimental results that have been interpreted in favor of the brain's use of neural synchrony to represent togetherness of visual stimuli [4, 23], opponents of this hypothesis have pointed out that these results are largely indirect and not necessarily conclusive [22].

In computer science, the binding problem usually is solved by the use of *pointers*. In fact, pointers are a major means to construct the complex data structures required in almost all computer vision or graphics programs.[2]

[2] Since the use of pointers has proven to be very error-prone for human programmers, the development of modern programming languages, such as Java, has

Therefore, considering neural computation as a serious alternative to conventional programming confronts us with the question of how to implement the functionality of pointers within these systems [7]. However, from a computational perspective, we can explore different possibilities more freely than in the context of careful biological modeling. This has motivated us to explore a computational approach that is complementary in that it uses *space* instead of *time* (such as, e.g., [21, 24, 25]) to express binding between features. This is achieved by using a stack of competitively interacting neural layers, each layer competing for a group (and, thereby, roughly playing the role of a "time slice" in synchrony models [21, 24, 25] of binding). We have dubbed this the competitive layer model (CLM) [19, 29].

8.3 The Competitive Layer Model (CLM)

An intuitive explanation of the idea underlying the CLM is most easily given in the context of the simple grouping situation of Fig. 8.3. For each potential group we have to introduce one layer, which has to implement a map of all features that might be present in a group. So, for the present example, two layers, each consisting of four feature units implementing a feature map for the set {red, green, round, triangular}, would be sufficient (Fig. 8.4).

The structuring of a collection of active features into groups [such as (red, round), (green, triangular)] is then expressed by a corresponding sharing of layers by the associated feature cell activities (i.e., a "red" and "round" feature cell active in one layer, and a "green" and "triangular" feature cell active in a second layer).

Of course, we wish this representation to emerge as the result of suitable interaction dynamics among the units in the layers. To this end we split the required interactions into a lateral (within-layer) and a vertical (inter-layer) component. The within-layer component should be excitatory for features that are compatible with a group (e.g., shape-color) and inhibitory between incompatible features (e.g., shape-shape or color-color).

The inter-layer component only connects corresponding locations in all layers (e.g., runs perpendicularly through the layer stack). It is pairwise inhibitory in order to enforce for each feature a unique assignment to one layer (i.e., one group). An additional, excitatory interaction to an additional input layer (not shown in Fig. 8.4) imposes the constraint that the superposition of all groups matches a prescribed, to-be-decomposed, input pattern, which is represented by the joint activity of all feature cells of all groups.

invested great efforts to hide pointers from the programmer. Still, pointers are "working under the hood" even in these languages.

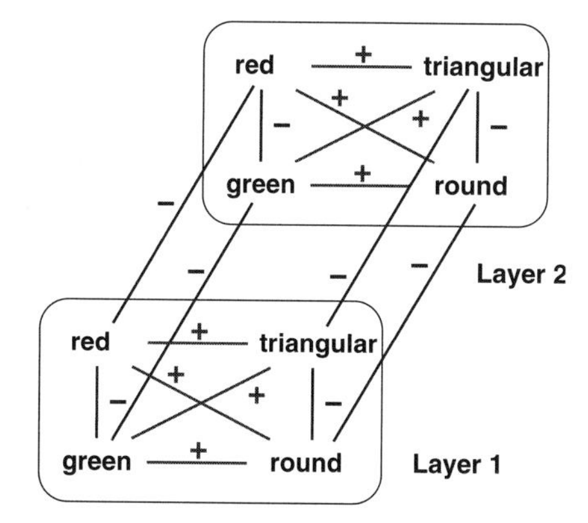

Fig. 8.4. CLM scheme for the simple toy problem of Fig. 8.3.

More formally, this model can be described as follows. The neurons are organized in layers $\alpha = 1, \ldots, L$ and columns $r = 1, \ldots, N^3$, where each layer contains all neurons belonging to the same label α and each column contains the neurons belonging to the same image location r. The neurons of each column have access to a local feature vector $\mathbf{m}_r$, which may augment the spatial position r by additional local image properties, such as color $\mathbf{c}_r$, texture features $\mathbf{t}_r$, or orientation vectors $\mathbf{o}_r$ of a local edge: $\mathbf{m}_r = (r, \mathbf{c}_r, \mathbf{t}_r, \mathbf{o}_r, \ldots)^T$.

The lateral interactions between neurons at locations r, r' of a layer are assumed to be symmetric functions $f(\mathbf{m}_r, \mathbf{m}_{r'})$ of the feature vectors of the associated columns (we will use the shorthand notation $f_{rr'}$ in many places below). To achieve the intended assignment of input features to layers, weights f_{rr} will be excitatory (> 0), if features $\mathbf{m_r}$,$\mathbf{m_{r'}}$ are compatible in the same group, inhibitory (< 0) otherwise. For instance, choosing $f_{rr'}$ as the simple "on-center-off-surround" function

$$f_{rr'} = \begin{cases} 1 & : \quad \| r - r' \| < R \\ -1 & : \quad \| r - r' \| \geq R \end{cases}, \tag{8.1}$$

leads to a grouping according to spatial proximity, where R sets the spatial scale. For more complex tasks, like texture or colour segmentation, the lateral connections have to be specified by more complex functions as the later application examples will demonstrate.

The unique assignment of each feature to a single layer is realized by a columnar "winner-takes-all" (WTA) circuit, which uses mutually symmetric inhibitory connections with absolute strength $J > 0$ between the neural activities $x_{r\alpha}$ and $x_{r\beta}$ that share a common column r. Due to the WTA coupling, for a stable state of the CLM only a single neuron from one layer can be active within each column [28].

[3] To simplify notation, we "overload" symbol r by the (discrete) index $\in \{1 \ldots N\}$ identifying a column and the two-dimensional coordinate vector indicating the associated location in the image.

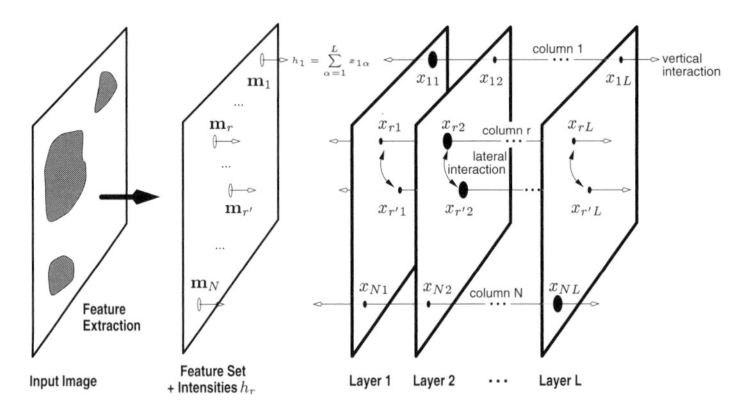

Fig. 8.5. CLM architecture: Feature extraction from an input image (left) gives rise to a set of local feature vectors $\mathbf{m}_r$ with activities h_r represented in an input layer (second left). Subsequent layers arranged in a stack decompose each h_r into a sum of layer-wise responses $x_{r,\alpha}$ as a result of lateral (within-layer) and vertical (across-layer) interactions. After convergence, each location r elicits a nonzero activity $x_{r,\alpha} > 0$ in maximally one layer α, defining a grouping of the features $\mathbf{m}_r$ into layers $\alpha(r) = \arg\max_\alpha x_{r,\alpha}$.

$$\forall r: \quad \forall \beta \neq \hat{\alpha}(r): \quad x_{r\hat{\alpha}(r)} > 0, \quad x_{r\beta} = 0, \tag{8.2}$$

where $\hat{\alpha}(r)$ is the index of the maximally supporting layer characterized by

$$\forall r: \quad \forall \beta \neq \hat{\alpha}(r): \quad \sum_{r'} f_{rr'} x_{r'\hat{\alpha}(r)} > \sum_{r'} f_{rr'} x_{r'\beta}. \tag{8.3}$$

A binding between two feature, associated with the columns r and r', is expressed by simultaneous activities $x_{\hat{\alpha}(r)}$ and $x_{\hat{\alpha}(r')}$, $\hat{\alpha}(r) = \hat{\alpha}(r') = \hat{\alpha}$ that share a common layer $\hat{\alpha}$. Each neuron receives a constant input Jh_r, where h_r specifies the basic amount of activity that is available in each column r. The activity of the single active neuron per column is given by

$$\forall r: \quad x_{r\hat{\alpha}(r)} = h_r + \frac{1}{J} \sum_{r'} f_{rr'} x_{r'\hat{\alpha}(r)}. \tag{8.4}$$

Thus, it suffices to consider only the subset of columns for which $h_r > 0$, that is, where input is present.

The combination of afferent inputs and lateral and competitive interactions can be formulated into the standard additive activity dynamics

$$\dot{x}_{r\alpha} = -x_{r\alpha} + \sigma\Big(J(h_r - \sum_{\beta} x_{r\beta}) + \sum_{r'} f_{rr'} x_{r'\alpha} + x_{r\alpha}\Big). \tag{8.5}$$

An essential role for the functionality of the CLM plays the threshold function $\sigma(\cdot)$, which is defined by a unsaturated linear threshold function $\sigma(x) =$

$\max(0, x)$ that cuts off negative interactions. As a result, the CLM dynamics corresponds to gradient descent in the energy function

$$E = -J \sum_{r\alpha} h_r x_{r\alpha} + \frac{1}{2} J \sum_{r} \sum_{\alpha\beta} x_{r\alpha} x_{r\beta} - \frac{1}{2} \sum_{\alpha} \sum_{rr'} f_{rr'} x_{r\alpha} x_{r'\alpha}, \tag{8.6}$$

under the constraint $x_{r\alpha} > 0$, since E is nonincreasing under the dynamics (8.5) [28]. Dynamical systems of this type have recently attracted interest as candidates for "cortex-inspired" analog silicon circuits [6].

8.4 Some Properties of the Binding Process

The special form of the linear threshold function makes it possible to formulate concrete statements about the behaviour of the CLM dynamics. The main statements can be summarized in two theorems about the convergence and assignment properties of the CLM [29].

Convergence Theorem: If $J > \lambda_{\max}\{f_{rr'}\}$, where $\lambda_{\max}\{f_{rr'}\}$ is the largest eigenvalue of the lateral interaction matrix, or $J > \max_r(\sum_{r'} \max(0, f_{rr'}))$, then the CLM dynamics is bounded and convergent.

Assignment Theorem: If the lateral interaction is self-excitatory, $f_{rr} > 0$ for all r, then an attractor of the CLM has in each column r either
(i) at most one positive activity $x_{r\hat{\alpha}(r)}$ with

$$x_{r\hat{\alpha}(r)} = h_r + \frac{1}{J} \sum_{r'} f_{rr'} x_{r'\hat{\alpha}(r)}, \qquad x_{r\beta} = 0 \quad \text{for all } \beta \neq \hat{\alpha}(r), \tag{8.7}$$

where $\hat{\alpha}(r)$ is the index of the maximally supporting layer characterized by

$$\sum_{r'} f_{rr'} x_{r'\hat{\alpha}(r)} > \sum_{r'} f_{rr'} x_{r'\beta}, \quad \text{for all } r, \beta \neq \hat{\alpha}(r) \tag{8.8}$$

or,
(ii) all activities $x_{r\alpha}$, $\alpha = 1, \ldots, L$ in a column r vanish and $\sum_{r'} f_{rr'} x_{r'\alpha} \leq -J_r h_r$ for all $\alpha = 1, \ldots, L$.

The dynamics (8.5) have an energy function of the form

$$E = -J \sum_{r\alpha} h_r x_{r\alpha} + \frac{1}{2} J \sum_{r} \sum_{\alpha\beta} x_{r\alpha} x_{r\beta} - \frac{1}{2} \sum_{\alpha} \sum_{rr'} f_{rr'} x_{r\alpha} x_{r'\alpha}. \tag{8.9}$$

The energy is nonincreasing under the dynamics (8.5) [28] :

$$d/dt\ E = -\sum_{r\alpha} E_{r\alpha} \dot{x}_{r\alpha} = -\sum_{r\alpha} E_{r\alpha}(-x_{r\alpha} + \sigma(E_{r\alpha} + x_{r\alpha})) \leq 0, \tag{8.10}$$

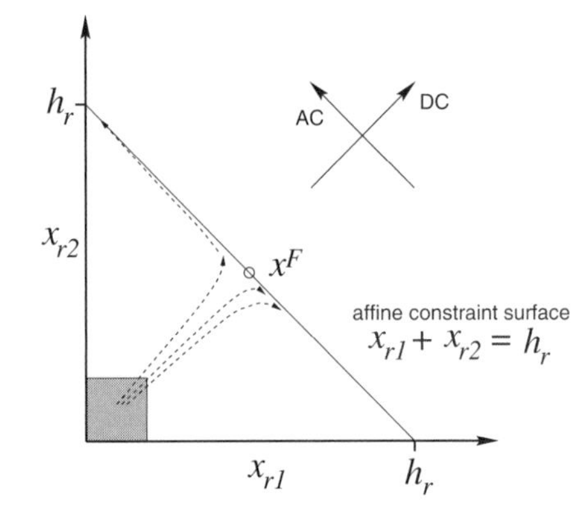

Fig. 8.6. Sketch of the AC and DC eigenmodes of the CLM dynamics for two neuron activities in the same column. The AC modes occur in the hyperplane $\sum_{\alpha} x_{r,\alpha} = h_r = \text{const}$, and the DC mode is responsible for enforcing this constraint.

where

$$E_{r\alpha} = -\partial E/\partial x_{r\alpha} = Jh_r - J\sum_{\beta} x_{r\beta} + \sum_{r'} f_{rr'} x_{r'\alpha}. \tag{8.11}$$

Thus the attractors of the dynamics (8.5) are the local minima of (8.9) under constraints $x_{r\alpha} \geq 0$. Additionally a kind of annealing process can be included in the dynamics, by extending the energy function with

$$E' = E + T\sum_{r\alpha} x_{r\alpha}^2, \tag{8.12}$$

which adds a convex term that biases the local minima toward graded assignments and thus "softens" the WTA dynamics. Within the dynamics this introduces a new self-inhibitory term

$$\dot{x}_{r\alpha} = -x_{r\alpha} + \sigma\Big(J(h_r - \sum_{\beta} x_{r\beta}) + \sum_{r'} f_{rr'} x_{r'\alpha} + (1-T)x_{r\alpha}\Big). \tag{8.13}$$

Through gradually lowering the self-inhibition T, (8.12) becomes (8.9) and (8.13) becomes (8.5).

A detailed analysis of the annealing process can be found in [28]. The result of this linear analysis is that the CLM-dynamics is driven by two kinds of eigenmodes called the AC- and DC-eigenmodes (sketched in Fig. 8.6), whose eigenvalues and eigenvectors mainly depend on the matrix F of lateral weights $f_{rr'}$. The DC-eigenmodes are driving the equilibration of the layer activities, whereas the AC-eigenmodes are responsible for the winner-takes-all behaviour. The stability of both sets of modes can be controlled with the "temperature" parameter T.

8.5 Implementation Issues

The CLM dynamics can be simulated efficiently with a Gauß-Seidel approach of solving iteratively the fixed point equations of (8.5) for a randomly chosen activity $x_{r\alpha}$ while all other activities are held constant [5, 28]. The algorithm can be implemented in the following way:

1. Initialize all $x_{r\alpha}$ with small random values around $x_{r\alpha}(t=0) \in [h_r/L - \epsilon, h_r/L + \epsilon]$.
   Initialize T with greatest eigenvalue of matrix $\{f_{rr'}\}$.
2. Do $N \cdot L$ times: choose (r, α) randomly and update $x_{r\alpha} = \max(0, \xi)$, where $\xi := \frac{J(h_r - \sum_{\beta \neq \alpha} x_{r\beta}) + \sum_{r' \neq r} f_{rr'} x_{r'\alpha}}{J - f_{rr} + T}$
3. Decrease T by $T := \eta T$, with $0 < \eta < 1$. Go to step 2 until convergence.

8.6 Image Preprocessing

Before we can apply the CLM to images, we require some preprocessing in order to achieve a dimension reduction from the extremely high-dimensional pixel input to a more manageable dimensionality. In our visual system, such dimensionality reduction starts already in the retina, where the activity values of the about 10^8 rods and cones are "compressed" into an activity pattern of the about 1.5 million fibres of the optical nerve connecting each eye (via the geniculate bodies, where further recoding occurs) to the visual cortex.

The response properties of many cells in the visual cortex have been found to coarsely resemble local Gabor filters [8]. A Gabor filter can be represented by a basis function parameterized by a position vector $r \in \mathrm{R}^2$, a wave vector $k \in \mathrm{R}^2$ and a range σ:

$$g_{r,k,\sigma}(x) = \cos(k \cdot (x - r)) \exp(-(x - r)^2 / 2\sigma^2) \tag{8.14}$$

The "overlap" or scalar product

$$c_{r,k,\sigma} = \sum_x g_{r,k,\sigma}(x) I(x) \tag{8.15}$$

of a Gabor filter with an image intensity function $I(x)$ can be interpreted as a measure of the contribution of a spatial frequency k to the local image structure within a radius σ around location r. A neuron implementing a response described by $c_{r,k,\sigma}$ thus represents spatial and frequency information for a local image patch simultaneously (and it can be shown that the choice of Gabor profiles leads to a compromise for representing the two conflicting observables location and spatial frequency that in a certain sense is optimal [2]).

From the set of Gabor filter responses $c_{r,k,\sigma}$ one can reconstruct an approximation $\hat{I}(x)$ of the original image as

$$\hat{I}(x) = \sum_{r,k,\sigma} c_{r,k,\sigma} \tilde{g}_{r,k,\sigma}(x) \tag{8.16}$$

where $\tilde{g}_{r,k,\sigma}(x)$ denotes the dual basis functions for the $g_{r,k\sigma}$.

Data from psychophysical measurements have led to sampling grids (r, k, σ) in position-frequency-scale space that permit rather accurate modeling of the measured position-frequency-scale variation of our ability to discriminate local contrast patterns [26].

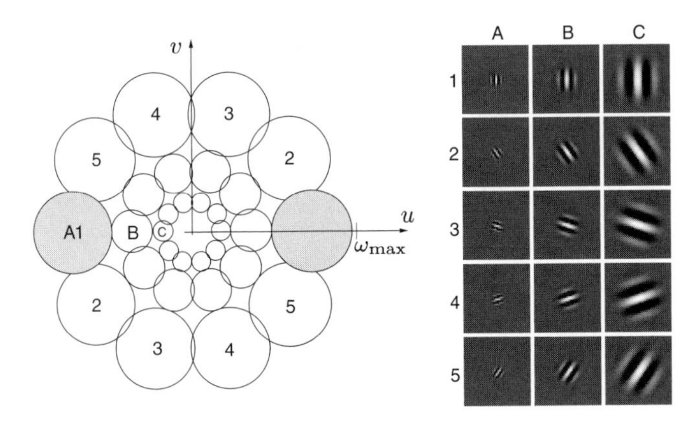

Fig. 8.7. Artificial "fovea" for image preprocessing.

For the following experiments, we have used a simplified filter arrangement, consisting of an artificial fovea comprising three octaves of spatial frequency, with five directions uniformly separated by 72 degrees. The range σ is chosen proportional to the wave length of the associated wave number k (Fig. 8.7).

In addition to the sampling of r, k, σ the visual cortex also performs some sort of pooling process where input stimuli from neighbouring points are pooled together. Following a proposal by [15], we compute the following pair of local feature coefficients:

$$M_k(r) = c_k(r) * G_k(r) \tag{8.17}$$

$$S_k(r) = \sqrt{(c_k(r) - M_k(r))^2 + G_k(r)} \tag{8.18}$$

where $*$ denotes the convolution operation, $c_k(r)$ is the response in channel k after a nonlinear scaling with a contrast transfer function, and $G_k(r)$ is the corresponding Gaussian filter kernel given by

$$G_k(r) = e^{-\frac{k^2}{2\rho_k^2}},$$

where the parameter ρ_k is a critical choice reflecting the size of the pooling area. A reliable description of the local feature statistics around r calls for large sizes. On the other hand, an accurate localization of feature positions demands smaller sizes. In our experiments we found a heuristical value of three times the size of the Gabor range σ as a good compromise. As a result, for each position r in the input image, we get a 30-dimensional feature vector $\mathbf{m}(r)$ describing the local texture at that point:

$$\mathbf{m}_r = (M_1(r), \ldots, M_{15}(r), S_1(r), \ldots, S_{15}(r))^T \tag{8.19}$$

This leads to a 30-dimensional feature vector at each image pixel $r = (x, y)$. To further reduce processing burden, we use subsampling on 8×8 pixel blocks and project at each (subsampled) location the 30-dimensional feature distribution on the four principal components of maximal variance. For more detail, see [18].

8.7 Interaction Function

As already indicated, it is the interaction function $f_{rr'}$ that determines which features are "compatible" (such as color and shape in the introductory toy example) to be bound together as a group, and which features cannot be bound for lack of compatibility. For the perceptual grouping experiments reported below, a suitable interaction was chosen as

$$f_{rr'} = \exp\left[-\frac{d_{text}^2(r, r')}{R_1^2}\right] + \gamma \cdot \exp\left[-\frac{d_{pos}^2(r, r')}{R_2^2}\right] - I \tag{8.20}$$

Here, $d_{pos}(r, r')$ is the position distance of locations r and r', and $d_{text}(r, r')$ is a measure of the difference of the local textures around r and r'. Interpreted in Gestalt psychology terms, the first part can be identified with the Gestalt law of similarity, which states that similar features are bound together; the second part corresponds to the Gestalt law of spatial proximity, with the constant γ controlling the weighting between these two principles. The parameter $I > 0$ causes inhibition of features that are both dissimilar and far apart.

Although d_{pos} is straightforward, our choice for d_{text} was motivated by a proposal of Ma and Manjunath [14] and chosen as the L_1 norm of the difference $\hat{p}_r - \hat{p}_{r'}$ between the (normalized[4]) vectors $\hat{p}_r$, where the unnormalized

[4] Normalization consisted of dividing each component of p_r by the standard deviation of all its values over the input grid

p_r is the feature vector $\mathbf{m}(r)$ of (8.19), projected into the eigenspace spanned by the four leading principal components (for details, cf. [18]). Parameters were chosen as $\gamma = 0.6$, $I = 0.5$, $R_1 = 6.6$, $R_2 = 0.63$ and all held constant for the examples shown in the next section.

8.8 Perceptual Experiments

With the described preprocessing, we can perform artificial perceptual experiments to compare the grouping behaviour of the CLM for various images with that of human subjects. Fig. 8.8 shows examples of arrangements of dot patterns (upper row) that exemplify the laws of proximity and similarity.

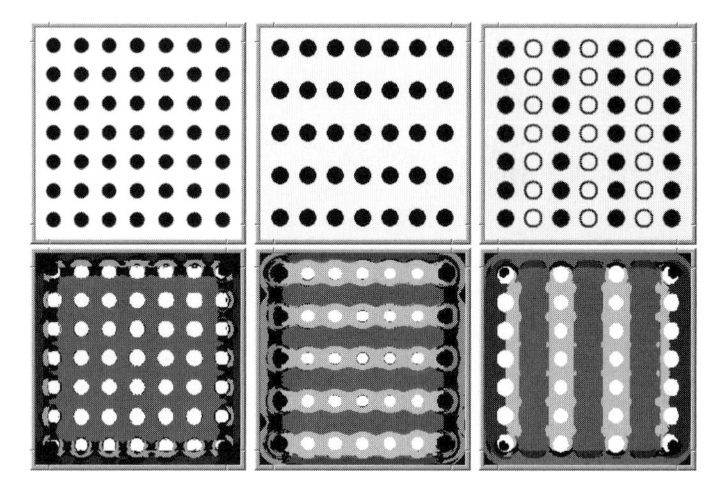

Fig. 8.8. Grouping of dot patterns. Top: input patterns. Bottom: CLM layer activities.

The lower row shows the superimposed responses of the CLM layers. Since in the converged state each location r can at most have a single layer $\alpha(r)$ with nonvanishing activity $x_{r,\alpha(r)} > 0$, it suffices to directly display the layer indicator function $\alpha(r)$ (using colors to distinguish different layers and black for the absence of activity in *all* layers).

As can be seen, in the case of equidistant dots (left), the system just performs (except for edge effects) figure-ground separation. Closer proximity, for example, along the horizontal direction (middle), causes in addition the activation of horizontal bars, indicating a grouping by proximity, without losing the identity of the dots themselves. Returning to an equidistant arrangement, but instead introducing different types of dots again creates a corresponding grouping—this time according to similarity (right).

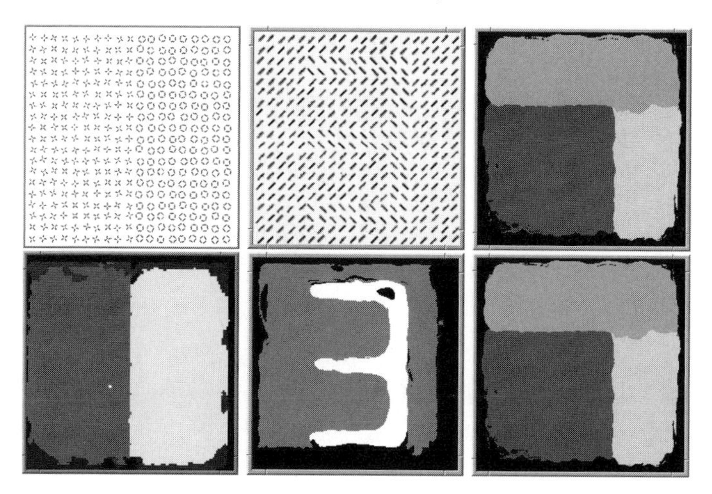

Fig. 8.9. Top row: types of test images used for the benchmark. The rightmost image is an example in which preattentive human perception misses the segregation of the lower left square area, in line with the CLM result.

Figure 8.9 shows a different set of examples, this time with regions demarcated by similarity in local texture features. Remarkably, in the rightmost column the CLM only "sees" three areas. However, if we take a closer (serial) look, we can distinguish one further border, which is not perceived preattentively. Thus, the model can mimic some effects that have been explained in terms of interactions of elementary texture elements ("textons") [9].

Finally, Fig. 8.11 depicts the last set of examples, demonstrating the grouping of line terminations into an apparent contour (left) and its disappearance when the lines are getting too sparse (middle). The right column finally presents an example for natural textures from the Brodatz [1] album.

To give a quantitative comparison of our model's performance with human data, we have constructed a set of seven texture pairs according to [11] (one example was already depicted in Figure 8.9 on the left). For each pair, we created 10 texture images of 512×512 pixels with random orientation of the elements plus an additional positional random jitter of 4 pixels. Table 8.1 shows that the rank order of human discriminability of these patterns according to [11] is almost identical to the rank order of the classification rates of the CLM, indicating a significant similarity of the CLM grouping to human texture perception (for details, see [18]).

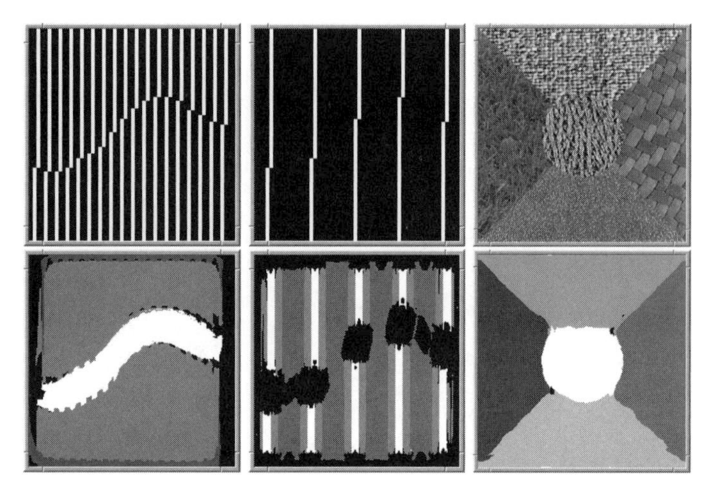

Fig. 8.10. In the top left a pattern of broken lines is shown. Because the points marked by the discontinuities lie on a smooth line, we perceive a wavelike illusory contour. The grouping result obtained with the CLM also shows this wavelike contour in the bottom row.

Table 8.1. Comparison of psychophysical data from [11] with CLM Model's performance. The *rank order* of the classification rate of the CLM matches the rank order of the psychophysical data remarkably well.

Texture Pair	Psychophysical Data (Discriminability)	CLM Model (Classification rate)
(a)	100	89.1
(b)	88.1	89.3
(c)	68.6	82.1
(d)	52.3	79.7
(e)	37.6	74.7
(f)	30.6	69.7
(g)	30.3	69.0

8.9 Learning the Interaction Function

While we have shown that the heuristic specification of the interaction functions $f_{rr'}$ given in section 8.7 can also be adapted to other exacting applications, such as the identification of cell contours in microscope images [16, 17], the required hand tuning can be cumbersome and certainly is undesirable for a broader use of the CLM in practical computer vision tasks.

This has motivated the search for learning approaches that permit the data-driven construction of a good interaction function from a set of example images with specified groupings in the form of target labels $\hat{\alpha}(r)$. These groupings are transferred into target states of the CLM $\mathbf{y} = (y_{11}, \ldots, y_{N1}, \ldots, y_{NL})^T$.

Some useful guidance is given by the stability conditions that must be met for any activity pattern in the converged state:

$$\forall r: \quad : \forall \beta \neq \hat{\alpha}(r): \quad \sum_{r'} f_{rr'} y_{r'\hat{\alpha}(r)} > \sum_{r'} f_{rr'} y_{r'\beta}. \tag{8.21}$$

If we only require the stability of a single pattern, we can write the following analytic solution from group contrasts [27]:

$$\hat{f}_{rr'} = \sum_{\mu\nu} (\mathbf{y}_\mu - \mathbf{y}_\nu)(\mathbf{y}_\nu - \mathbf{y}_\mu)^T, \tag{8.22}$$

where $\mathbf{y}_\nu$ are the layer-wise subvectors $\mathbf{y}_\nu = (y_{1\nu}, \ldots, y_{N\nu})$ of the target state $\mathbf{y}$ and

$$y_{\nu,r} = \begin{cases} 1 & : \quad \hat{\alpha}(r) = \nu \text{ i.e., } m_r \text{ belongs to group } \nu \\ 0 & : \quad \hat{\alpha}(r) \neq \nu \text{ else} \end{cases} \tag{8.23}$$

However, this is only an expression in (r, r') space and does not include any spatial variation of the features $\mathbf{m}_r$. A straightforward way to generalize this to arbitrary $(\mathbf{m}_r, \mathbf{m}_{r'})$ pairs can be implemented by the following steps:

1. Sample $(\mathbf{m}_r, \mathbf{m}_{r'})$ pairs from training inputs.
2. Use associated $f_{r,r'}$ values from the above one-pattern-solution as target values.
3. Apply a standard machine learning algorithm [e.g. SVM classifier or multilayer perceptron (MLP)] to create a mapping $\phi : (\mathbf{m}_r, \mathbf{m}_{r'}) \rightarrow f_{r,r'}$ that yields the desired interpolation in $(\mathbf{m}_r, \mathbf{m}_{r'})$-space.

A major drawback of this approach is the need for one evaluation of the trained network *for each* interaction pair in the CLM; this makes the method very time-consuming and practically rules out the application of the SVM. The MLP is considerably faster, but we found it rather difficult to achieve good training results for the required interaction function: the MLP strongly suffers from the danger of overfitting to the feature positions r if only one pattern is presented for learning.

Therefore, in [27] we developed an alternative approach based on an expansion of $f_{rr'}(m_r, m_{r'})$

$$f_{rr'}(m_r, m_{r'}) = \sum_k c_k \cdot b^k_{rr'}(m_r, m_{r'}) \tag{8.24}$$

into a suitable set of basis functions b_k, thereby reducing the learning task to the determination of the required expansion coefficients c_k. Since in most grouping tasks the important information mainly lies in the similarity or dissimilarity of the features $(\mathbf{m}_r, \mathbf{m}_{r'})$ it has proven beneficial to first transform the $(\mathbf{m}_r, \mathbf{m}_{r'})$ space into a dissimilarity space so that "relevant" dissimilarities can be made more explicit. A rather flexible way to achieve this is to choose a number of (application specific) dissimilarity functions $d_1(\cdot,\cdot), \ldots, d_P(\cdot,\cdot)$ for mapping each pair $(\mathbf{m}_r, \mathbf{m}_{r'})$ into a P-dimensional vector $\mathbf{D}(\mathbf{m}_r, \mathbf{m}_{r'})$:

$$\mathbf{D}(\mathbf{m}_r, \mathbf{m}_{r'}) = (a_1 d_1(\mathbf{m}_r, \mathbf{m}_{r'}), \ldots, a_P d_P(\mathbf{m}_r, \mathbf{m}_{r'}))^T. \tag{8.25}$$

Examples of dissimilarity functions are the local distance $\| \mathbf{r} - \mathbf{r}' \|$ or distances in color $\| \mathbf{c}_r - \mathbf{c}_{r'} \|$ of texture space $\| \mathbf{t}_r - \mathbf{t}_{r'} \|$ as also component-wise distances in these feature vectors. The dissimilarity functions do not necessarily need to fulfill properties of a distance metric, and choices such as, for example, the scalar product $\mathbf{o}_r \cdot \mathbf{o}_{r'}^T$ of two orientation vectors $\mathbf{o}_r$ and $\mathbf{o}_{r'}$ are admissible as well. The only necessary condition is that the distance functions are symmetric under feature exchange $d_p(\mathbf{m}_r, \mathbf{m}_{r'}) = d_p(\mathbf{m}_{r'}, \mathbf{m}_r)$ to guarantee the symmetry of the lateral interaction weights. The scaling factors a_p permit implementing a normalization of the different distance functions $d_p(\cdot,\cdot)$ to the same range of values.

Finally, we have to make a choice of a set of basis functions in the transformed space. A computationally very attractive scheme results if we use as basis functions the characteristic functions of a Voronoi tessellation generated from a number of prototype vectors adjusted by means of a training data set: this allows reducing the computation of the expansion coefficients c_k to a simple counting of hits in Voronoi cells. The necessary computations can be compactly summarized in the following two steps:

1. **Adaptive tessellation in the transformed space:** using a suitable vector quantization algorithm, position a number of prototype vectors

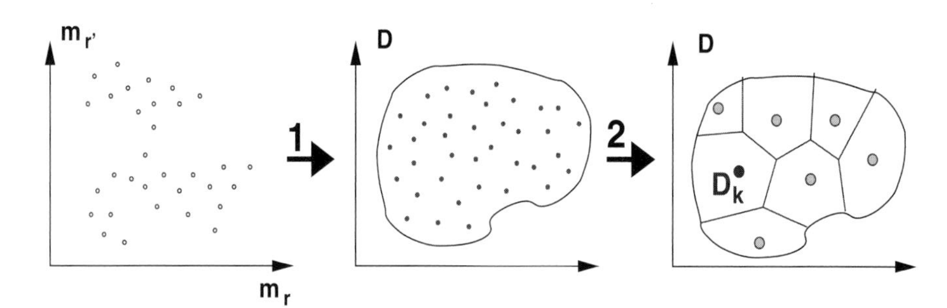

Fig. 8.11. Interaction quantization for learning: A transformation of feature pairs from a training set (left) into a "dissimilarity space" (center) and a subsequent Voronoi tessellation (generated by prototype vectors $\mathbf{D}_k$) provides the basis to approximate the desired interaction function as a stepwise constant function on the Voronoi cells (right).

$\mathbf{D}_1 \ldots \mathbf{D}_K$ to define a Voronoi tessellation of the training data in the transformed space. The main parameter to be chosen here is the number K of prototype vectors to use.

2. **Computing the interaction function:** Using the characteristic functions on the Voronoi cells V_k as basis functions, the interaction function can be cast into form

$$\hat{f}_{rr'}(\mathbf{m}_r, \mathbf{m}_{r'}) = c_k^+ - \Lambda c_k^-, \tag{8.26}$$

with

$$k = arg \min_k \parallel \mathbf{D}(\mathbf{m}_r, \mathbf{m}_{r'}) - \mathbf{D}_k \parallel, \tag{8.27}$$

where Λ is a weighting parameter and the expansion coefficients c_k can be obtained by simple counting:

$$\begin{aligned} c_k^+ &= \text{No. of within-group } \mathbf{D}(\mathbf{m}_r, \mathbf{m}_{r'}) \text{ samples in } V_k \\ c_k^- &= \text{No. of across-group } \mathbf{D}(\mathbf{m}_r, \mathbf{m}_{r'}) \text{ samples in } V_k \end{aligned} \tag{8.28}$$

The choice of Λ allows controlling the degree of segmentation: small values of Λ favour few and large groups, while large values of Λ favour many small groups. For more details, the reader is referred to [27].

8.10 Experiments with Artificial Data

In this section, we illustrate how the above learning algorithm can generate interaction fields for the grouping of line segments into contours. Here, the line segments $\mathbf{m}_r = (\mathbf{r}, \mathbf{o}_r)$ are represented by 2D position $\mathbf{r}$ and an undirected orientation vector $\mathbf{o}_r$ of the lines, while the distance space between features is given by the local distance and the inner angle between two lines segments.

Training data consisted always of a single image, containing several hundred line segments forming the contours of five randomly positioned contours of circles (Fig. 8.12 top left), squares (Fig. 8.12 top center) and triangles (Fig. 8.12 top right), respectively (since a single image usually already contains a very large number of line segment pairs, a single image can usually offer enough training data to estimate a good interaction function). The learned interaction function is depicted in the corresponding diagrams below: each figure depicts for a horizontally oriented "reference" line segment (positioned at the center of the figure) the distribution of excitatorily connected line orientations within a 11×11 neighbourhood around the reference line segment. Longer line segments indicate stronger excitation, whereas only weakly excitory connections are indicated with correspondingly shorter lines. From this visualization it becomes apparent that the loci of excitatory connectivity are

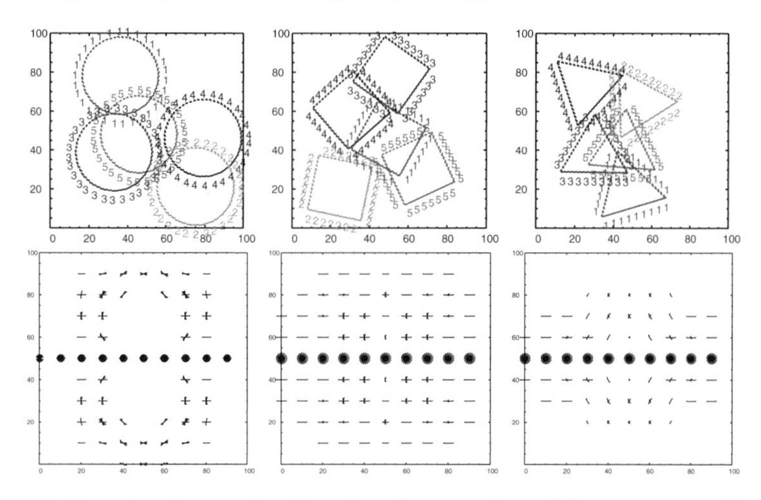

Fig. 8.12. Learning interaction functions for grouping of line segments. Top row: groupings used for training. Bottom row: resulting interaction functions (for explanation, see text).

distributed such as to favour the continuation of the reference line segment in a way that follows the geometrical shape that was used during training.

The second set of pictures (Fig. 8.13) shows some grouping results obtained with a CLM using the trained interaction functions. The first two pictures show grouping results for almost "perfect" input shapes, while the last two pictures illustrate the effect of the grouping in the presence of strong positional or orientational jitter in the input arrangements.

8.11 Application Example: Cell Image Segmentation

In this section we illustrate the learning approach for a more complex domain from medical image processing. The goal is to determine in microscope images the number, positions, and shape of lymphocyte cells that have been stained with a fluorescent dye. An automated system for this has been presented in [17]. In the context of that work, we have found the CLM as a very suitable method for an accurate determination of cell boundaries; however, good results required a careful "hand tuning" of the underlying interaction function.

To apply the present learning method for an automated generation of the interaction function, we use as features directed edge vectors $\mathbf{m}_r = (\mathbf{r}, \mathbf{o}_r)$ whose orientation is estimated from the responses of a pair of 3×3 Sobel filter masks $\mathbf{o}_r = (S^x_r, S^y_r)^T$. The interaction function $f_{rr'}(\mathbf{m}_r, \mathbf{m}_{r'})$ can then be parameterized by the local distance $\| \mathbf{r} - \mathbf{r}' \|$ and the two relative angles between the connection vector $(\mathbf{r} - \mathbf{r}')$ and the orientation vectors of the two edges $\mathbf{o}_r$ and $\mathbf{o}_{r'}$ (for details, see [27]). Images consist of 45×45 pixel squares,

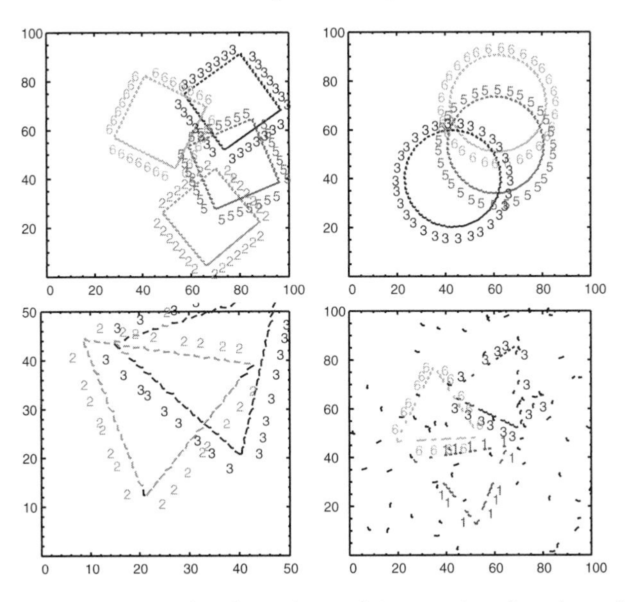

Fig. 8.13. Grouping examples from learned interaction functions for noiseless arrangements of line segments (left) and for noise in line segment parameters or in background (right).

leading to a total number of 2025 input features. For the grouping, we use a 10-layer CLM.

For the interaction vector quantization, we use 100 prototype vectors [leading to the same number of basis functions in the expansion (8.24)]. We then use (8.26) to define interactions for three different values of Λ. With these interactions, we obtain the segmentation results depicted in Fig. 8.14. While the value of $\Lambda = 1$ yields the best segmentation, we also see that the system is rather robust to the setting of this single main parameter and there is a rather good segmentation also for the other values of Λ. For comparison, the last line depicts the (much worse) segmentation results obtained with direct k-means vector quantization when applied directly to the feature vectors $\mathbf{m}_r$ (using seven prototype vectors, expecting seven cells or less).

8.12 Discussion

Pattern recognition is frequently equated with the process of classification and often mathematically idealized as a mapping from some feature space into a (usually discrete) set of output values ("categories"). Although this view has been very useful and led to the development of powerful classifiers, such as feed-forward neural networks or support vector machines, it tends to

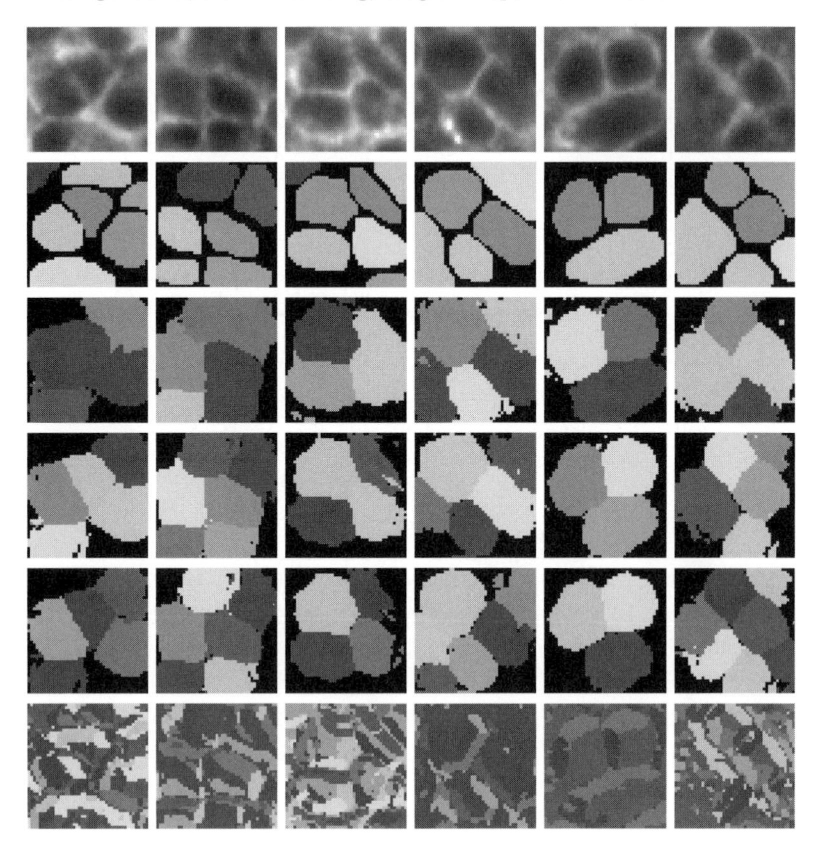

Fig. 8.14. CLM cell image segmentation (from top row to bottom row): input images, target regions, CLM segmentation results after learning for $\Lambda = 1$, $\Lambda = 2$ and $\Lambda = 3$, "naive" k-NN-segmentation results (adapted from [27]).

de-emphasize an aspect already brought to our attention by the early work of the Gestaltists: in many cases, the desired result of a pattern recognition system is not just an "opaque" category, but a richer structure that represents the "pattern" in a more explicit and invariant way than the original data (e.g., the image pixels). Perceptual grouping appears to be one of the fundamental operations that produce such representations by establishing "bindings" between suitable, simpler constituents, such as predefined features.

Mappings that can create such richer representations are much harder to construct, motivating an approach that generates the required structures as the response of a suitable dynamical system to constraints representing the input. For the task of feature grouping or binding, we have presented the competitive layer model (CLM), consisting of a "stack" of layers of feature-selective units ("neurons") coupled in such a way that an input pattern "projected" onto the stack as an activity constraint for the activity sums of its vertical columns can become partitioned into groups of constituents by dividing the activity among the different layers.

A critical element of this system is the choice of interactions between cells sharing a layer; together with the chosen basis features, they determine the grouping characteristics of the resulting dynamics. We have shown that we can choose the interactions in such a way that the resulting grouping dynamics resemble human perceptual grouping for quite a wide range of input patterns. In addition, we have described a learning procedure that can construct suitable interactions from a small set of training patterns specifying examples of desired target groupings.

The emphasis of this work is toward the development of computational systems that implement nontrivial computations in the form of distributed dynamical systems that can be constructed efficiently from data examples. While there is a large number of approaches focusing on the learning of temporal trajectories, for example, for sensorimotor control, we think that dynamical systems *for performing binding operations* are another very important class of building blocks for the following reason: real environments do not provide "carefully formatted" input and output spaces with stable and obvious sets of low-dimensional input and output variables that we then could correlate by mappings constructable from a training data sets of realistic size. Currently, it is largely the work of the engineer to prestructure the "binding" of its devices to the relevant degrees of freedom in their environment, sacrificing in this way a large part of their flexibility. Only when we succeed in treating the selection of relevant degrees of freedom on the input (and output) side as an integral part of the task that a perceptual (and actor) system has to solve, can we approach the flexibility that we admire in living animals. We think that perceptual grouping can provide us with an interesting starting point to tackle this task, and that the next step will be an extension of perceptual grouping to situations, where "good gestalts" have to be formed *at the level of coordination patterns between an agent and its environment.* We are confident that the CLM can offer guidance for the design of systems with such capabilities, which are closely related with the important issues of attention and dynamic dimension reduction.

References

1. Brodatz P (1966) Texture: A Photographic Album for Artists and Designers. Dover, NewYork.
2. Daugman JG (1985) Uncertainty relation for resolution in space, spatial frequency, and orientation optimized by two-dimensional visual cortical filters. J Opt Soc Am A 2(7):1160–1169.
3. Dunn D, Higgins WE, Wakeley J (1994) Texture segmentation using 2D Gabor elementary functions. IEEE Trans Pattern Anal Machine Intell 16(2):130–149.

4. Eckhorn R (1999) Neural mechanisms of visual feature binding investigated with microelectrodes and models. Vis Cogn 6:231–265.
5. Feng J (1997) Lyapunov functions for neural nets with nondifferentiable input-output characteristics. Neural Comput 9:43–49.
6. Hahnloser RHR, Sarpeshkar R, Mahowald MA, Douglas RJ, Seung HS (2000) Digital selection and analogue amplification coexist in a cortex-inspired silicon circuit. Nature 405:947–951.
7. Hinton GE (1990) Mapping part-whole hierarchies into connectionist networks. Artific Intell 46 47–75.
8. Jones J, Palmer L (1987) An evaluation of the two-dimensional Gabor filter model of simple receptive fields in cat striate cortex. J Neurophysiol 58:1233–1258.
9. Julesz B (1981) Textons, the elements of texture perception and their interaction. Nature 290:91–97.
10. Köhler W (1929) Gestalt Psychology. Liveright, New York.
11. Kröse BJ (1987) Local structure analyzers as determinants of preattentive pattern discrimination. Biol Cybernet 55:289–298.
12. Lowe DG (1985) Perceptual Organization and Visual Recognition, Kluwer Academic Publishers, Boston.
13. Von der Malsburg C (1981) The Correlation Theory of Brain Function. Internal report 81-2, Max Planck Institute for Biophysical Chemistry, Göttingen, Germany.
14. Ma WY, Manjunath BS (1996) Texture features and learning similarity. In Proceedings of IEEE International Conference on Computer Vision and Pattern Recognition.
15. Manjunath BS, Ma WY (1996) Texture features for browsing and retrieval of image data. IEEE Trans Pattern Anal Machine Intell 18(8):837–842.
16. Nattkemper TW, Wersing H, Schubert W, Ritter H (2000) Fluorescence micrograph segmentation by gestalt-based feature binding. In Proceedings of the International Joint Conference on Neural Networks (IJCNN), 248–254.
17. Nattkemper TW, Ritter H, Schubert W (2001) A neural classificator enabling high-throughput topological analysis of lymphocytes in tissue sections. IEEE Trans Inf Techn in Biomed 5 (2):138–149.
18. Ontrup J, Wersing H, Ritter H (2004) A computational feature binding model of human texture perception. Cogn Process 5 (1):32–44.
19. Ritter H (1990) A spatial approach to feature linking. In Proceedings of the International Neural Network Conference, vol 2, 898–901, Paris.
20. Roskies AL (1999) The binding problem. Neuron 24:7–9.
21. Schillen TB, König P (1994) Binding by temporal structure in multiple feature domains of an oscillatory network. Biol Cybernet 70:397–405.
22. Shadlen MN, Movshon JA (1999) Synchrony unbound: a critical evaluation of the temporal binding hypothesis. Neuron 24:67–77.
23. Singer W (1999) Neuronal synchrony: a versatile code for the definition of relations? Neuron, 24:49–65.
24. Terman D, Wang DL(1995) Global competition and local cooperation in a network of neural oscillators. Physica D 81:148–176.
25. von der Malsburg C, Buhmann J (1992) Sensory segmentation with coupled oscillators. Biol Cybernet 54:29–40.
26. Watson AB (1987) Efficiency of a model human image code. J Optical Soc Am A 4(12):2401–2417.

27. Weng S, Wersing H, Steil JJ, Ritter H (2006) Lateral interactions for feature binding and sensory segmentation from prototypic basis interactions. IEEE Trans Neural Networks 17 (3): (to appear).
28. Wersing H (2000) Spatial Feature Binding and Learning in Competitive Neural Layer Architectures. Cuvillier, Goettingen.
29. Wersing H, Steil JJ, Ritter H (2001) A competitive layer model for feature binding and sensory segmentation. Neural Comput 13(2):357–387.

Part III

Applications in Bioinformatics

9

Regulatory Signals in Genomic Sequences

Sonja J. Prohaska, Axel Mosig, and Peter F. Stadler

Summary. Gene expression is a complex multiple-step process involving multiple levels of regulation, from transcription, nuclear processing, export, post-transcriptional modifications, translation, to degradation. Over evolutionary timescales, many of the interactions determining the fate of a gene have left traces in the genomic DNA. Comparative genomics, therefore, promises a rich source of data on the functional interplay of cellular mechanisms. In this chapter we review a few aspects of such a research agenda.

9.1 Introduction

Gene expression is the process by which a gene's information is converted into a structural or functional gene product. This product is typically a protein, but might also be an RNA molecule or a complex of RNA and protein. A specific spatial and temporal distribution of these units within a cell is crucial for their function. The process of gene expression involves multiple steps, starting with the DNA in a state that makes the information accessible, transcription (DNA $\rightarrow$ RNA) and perhaps translation (RNA $\rightarrow$ protein), which is then followed by protein folding, posttranslational modification, and targeting. Once started, gene expression does not run through unaffected. Every step in the process is under tight control and actively regulates, or at least modulates, the flow through each checkpoint. Trapping intermediates at any step of the process may halt or even abort gene expression. Together, all regulatory effects from the gene to the functional gene product determine whether a gene product exceeds its threshold of expression to be effective. Therefore, the state of expression is not simply on or off.

In recent years it has become apparent that gene expression is a complex network comprising different, often-intertwined, regulatory layers. A few of these mechanisms, such as the binding of transcription factors to the DNA, leave direct traces in the genomic sequences that can be detected and deciphered by comparative approaches. In other cases, gene regulation is afforded by *trans*-acting RNAs, first and foremost micro-RNAs. In this situation, one first has to identify the transacting regulator before it becomes possible to search for the target that it regulates.

From a comparative genomics perspective, on the other hand, we can identify a plethora of evolutionary conserved DNA sequences that apparently do not code for proteins. Among these signals are also sizable regions with very high levels of sequence conservation and no reported function [29, 82, 127, 128]. The question that we at least begin to address in this chapter is how we can identify evolutionary conserved DNA, and how we can determine the mechanisms that they are involved in. While it is clear that comparative genomics cannot by itself elucidate the complete complexity of cellular regulation, it has also become clear in recent year that over evolutionary time scales, this regulatory network has left trace evidence at the DNA level. This requires, however, an understanding of the many distinct mechanisms.

Nucleic acid sequence motifs are, with a few examples such as self-splicing introns and some noncoding RNAs (ncRNAs), not catalytically active. Additional proteins or (sometimes) ncRNAs are therefore involved that exert their regulatory function by binding either to the DNA or to the transcribed RNA at regulatory elements. Regulatory mechanisms that alter the components binding directly or indirectly to the sequence motifs are beyond the scope of this chapter. The generic mechanisms involving the motifs and their accessibility are listed in Table 9.1.

Changes at the nucleotide sequence level occur either by recombination/repair processes or covalent modification of single bases. Direct DNA modifications yield stable or even irreversible gene expression patterns in descending cells. Genomic imprinting reduces gene expression to one parental allele through DNA methylation. Once established, the methylation pattern is rather constant and hardly reversible.

Eukaryotic DNA is packed into a compact structure, the chromatin. Every 150 base pairs (bp), the linear DNA molecule is wrapped around a protein core in 1.65 turns, forming the nucleosome. Regulation at the epigenetic level concerns modifications of the histone protein core. The pattern of acetylation, (mono-, di-, and tri-)methylation, phosphorylation, and other covalent modifications at about 40 different amino acids of the five different histone proteins (H4, H3, H2A, H2B, and H1) is also referred to as the histone code. Histone modification patterns are able to recruit specific protein complexes just like binding sites on the DNA, or set the chromatin state. Heterochromatin or condensed chromatin is in the "silent" state, while euchromatin or open chromatin is "transcribable" mainly due to histone acetylation.

Insulators describe a phenotype rather than a single kind of element with a fixed mechanism of action. They have the ability to protect genes they surround from the influence either of outside enhancers or inactivating chromatin structures. An important part of the underlying mechanism might be

Table 9.1. Overview of major regulatory modes in eukaryotic cells

Regulatory mechanism	Effect on gene expession	Example (organism)	Reference
DNA rearrangements	Selective/irreversible	V(D)J-joining (human)	[105]
Site-specific recombination	Selective/reversible	Mating-type switching (yeast)	[26]
DNA amplification	Enhancing	Chorion genes (*Drosophila*)	[22]
DNA methylation of CpG dinucleotides	Silencing/imprinting	Parent-of-origin-specific silencing (human)	[120]
DNA demethylation by DNA repair	Enhancing	Glycosylase at polycomb genes (*Arabidopsis*)	[96]
Histone code	Silencing/enhancing	Everywhere	[81]
Heterochromatin barrier (fixed/flexible)	Insulator/silencing/enhancing	USF binding at HS4 (chicken)	[146]
Enhancer blocker	Insulator/silencing	su(Hw) (*Drosophila*), CTCF (human)	[146]
Enhancer promoter contact	Insulator/silencing/enhancing	Trithorax at Ubx promoter (*Drosophila*)	[146]
Nuclear matrix attachment region (MAR)	Insulator/silencing/enhancing	Lysozyme locus (chicken), tyrosinase locus (mouse)	[146]
Subnuclear compartment	Silencing/enhancing	LCR at β-globin gene (human)	[145]
Gene competition for enhancers	Silencing/enhancing	β-globin gene (human)	[131]
Chromatin remodeling	Silencing/enhancing	SWI/SNF at PHO5 or PHO8 genes (yeast)	[15]
RNA-directed transcriptional gene silencing	Silencing	X chromosome inactivation (human)	[7]
Promoters (TF binding sites)	Basal	Everywhere	[14]
Enhancer (TF binding sites)	Enhancing	Everywhere	[148]
Silencer/repressor (TF binding sites)	Silencing	Ume6 at URS1-containing promoters (yeast)	[148]
Alternative transcription start sites	Silencing/enhancing	IGF-1 (human)	[77]
Antisense transcripts	Silencing/enhancing	Frq gene (*Neurospora*)	[18]
Regulation of elongation phase	Silencing/enhancing	Fkh at CLB2 locus (yeast)	[98]
Pre-mRNA processing (nucleus)	Silencing/enhancing	Everywhere	[68]
Alternative splicing (nucleus)	Selective/silencing/enhancing	Sex lethal (*Drosophila*)	[73]
Trans-splicing (nucleus)	Selective/alteration	SL RNA at all genes (*Trypanosoma*	[78]
mRNA editing (nucleus)	Alteration	ADAR at GluR mRNA (human)	[6]
Sequestration of mRNA (nucleus)	Silencing	Rrp6 at exosomes (yeast)	[117]
Nonsense-mediated mRNA decay (nucleus)	Silencing	Pseudogenes	[144]
mRNA export	Silencing/enhancing	EJC at spliced mRNAs (*Drosophila*)	[117]
RNA-directed mRNA degradation	Silencing	DCL1 mRNA (*Arabidopsis*)	[80]
Degradation and stability of RNAs (cytoplasm)	Silencing/enhancing	HuR at ARE-containing RNAs (human)	[147]
mRNA localization (cytoplasm)	Silencing/enhancing	ASH1 (yeast)	[41]
Alternative translation start sites	Silencing/enhancing	IRES	[69]
Scanning for translation start sites	Silencing/enhancing	uORF at GCN4 (yeast)	[38]
Translation initiation regulators	Silencing/enhancing	CPEB at CPE-containing mRNA (human)	[38]
RNA-directed translational gene silencing	Silencing	lin-4 miRNA at lin-14 mRNA (*C. elegans*)	[38]

USF = upstream stimulatory factor, HS4 = hypersensitive site at the LCR of the β-globin gene, su(Hw) = suppressor of hairy wing protein, CTCF = CCCTC-binding factor, MAR = nuclear matrix attachment region, LCR = locus control region, SWI/SNF = remodeling complex, PH05 = repressible acid phosphatase precursor, PHO8 = repressible alkaline phosphatase precursor, TF = transcription factor, Ume6 = transcriptional regulator, URS1 = Ume6-binding site, frq = frequenin, Fkh = fork head protein, CLB2 = G2/mitotic-specific cyclin 2, ADAR = double-stranded RNA-specific adenosine deaminase, GluR = glutamate receptor, mRNA = messenger RNA, Rrp6 = exosome complex exonuclease, EJC = exon-junction complex, DCL1 = Dicer-like 1, HuR = Human-antigen R (stability factor), ARE = AU-rich element, ASH1 = daughter cells HO repressor protein, IRES = internal ribosomal entry site, uORF = upstream open reading frame, GCN4 = general control protein, CPEB = cytoplasmic polyadenylation element binding protein, CPE = cytoplasmic polyadenylation element, miRNA = micro RNA.

the formation of insulator bodies. Insulator binding proteins form complexes that divide the chromatin into looped domains that are functionally isolated from one another. This could be a step toward regulation by discrete subnuclear compartment. For example, actively expressed genes migrate to nuclear compartments enriched in RNAPol II, so-called transcription factories, while inactive genes loop out. Such agglomerations may serve to raise the local concentration of associated components, favouring interactions that might not otherwise occur.

The promoter is the assembly site for the basal transcription machinery right next to the transcription start site. Transcription factors (TFs) binding to enhancers facilitate recruitment of RNA polymerase to the promoter if physical contact can be established by cofactors. Silencers, on the other hand, circumvent such an interaction and therefore initiation of transcription. In general, binding sites for TFs are short (4–12 bp) and occur clustered upstream of the promoter sequence. While there are numerous examples where the context (i.e., order, orientation, distance, presence of certain TFs) of TF binding sites is functionally relevant, there is an equally large number of examples where the context is not relevant.

The following elongation of transcription and all regulatory steps at the RNA level that take place in the nucleus are coupled to a large extent. For example, nuclear export of RNAs is linked to the subnuclear compartment of transcription, transcription elongation, mRNA processing (splicing), and mRNA stability. Once the mRNA is exported to the cytoplasm, it is either degraded or translated, but it might also be stored for later use.

Translation of mRNA is the final step in gene expression. It involves nucleic acid sequence elements in control. Not only upstream regulatory elements like secondary structures or upstream ORF may effect scanning of the small ribosomal subunit for the initiation codon. Close proximity of the $5'$ and $3'$ end of the mRNA allows protein binding sites located in the $3'$-UTR to control translation initiation. In fact, most known regulatory sequences, and mRNA binding sites are found within the $3'$ UTR.

The regulatory mechanisms and phenomena described above leave more or less visible traces in their genome sequences. For some regulatory elements, the corresponding traces on DNA level are very well understood and have been studied in much detail. This chapter reviews the known sequence characteristics of regulatory elements. Whenever one is available, we will give an overview over the corresponding computational methods for unveiling those traces in genomic sequences.

9.2 Gene Finding

The most conspicuous traces found in a genome sequence arise from protein coding regions. Since proteins are key players in the gene regulatory network, identifying the positions of the protein coding genes in whole genome sequences is an elementary step. Beside identifying the protein coding sequences, genome annotations serve a second purpose, namely to obtain those regions in the vicinity of the annotated genes that contain *cis*-regulatory elements, such as transcription factor binding sites. Furthermore, gene annotations give an estimate of the number of players involved in regulatory networks.

Once a complete genome sequence is available, a first step typically is to identify protein coding genes in the sequence by computational means, a task commonly referred to as *gene prediction* or *gene finding*. Due to statistically noticeable features such as being grouped in coding triplets or protruding traits such as start or stop codons, protein coding genes typically show comparatively strong signals in genome sequences. Consequently, a number of well-established methods have contributed to detecting protein coding genes in genomes. The first type of gene prediction methods, so-called *ab initio* methods, are based on considering a single genome sequence in combination with a probabilistic model involving multiple characteristic traits of transcriptional, translational, or splicing sites that are typically visible on sequence level. Approaches such as `GENSCAN` [13] or `Genie` [72] incorporate this information into a hidden Markov model for unveiling genomic regions that have a striking probability of being protein coding.

While the accuracy of *ab initio* gene prediction methods turned out to be principally limited [43, 118], more reliable results can be obtained by *comparative gene prediction* approaches, which incorporate pairwise alignments of the underlying genomes produced by programs such as `Blastx`. Due to a very specific selectional pressure on the coding triplets of protein coding regions, predictions produced by programs such as `Twinscan` [67] or `Procrustes` [40] yield much more reliable results than *ab initio* methods.

As has been demonstrated by several studies, incorporating issues such as protein similarity or expressed sequence tags may enhance the reliability of gene prediction methods [54, 113, 150]. For surveys on gene prediction methods, we refer to [12, 34, 39, 71]. Gene prediction methods yield estimates of the number of (protein coding) genes, some of which are shown in Table 9.2.

Table 9.2. Estimated number of protein coding genes

Species	**Estimated No. of Genes**	Ref.
Homo sapiens	20,000–25,000	[132]
Drosophila melanogaster	12,000	[133]
Caenorhabditis elegans	19,000	[135]

While detecting protein coding genes appears to be a largely resolved problem, finding noncoding RNA genes is much more involved. For details on RNA gene prediction, see section 9.4.

9.3 Identifying *Cis*-Regulatory Elements

Once the protein coding genes and noncoding RNAs (see section 9.4), as the key players in the regulatory network and their coding regions, are known, one is naturally interested in their *cis*-regulatory elements, that is, sequence elements associated with the gene to be regulated that serve as sequence-based "addresses" for their regulators. On the level of transcription regulation, the most striking sequence signals of a gene are given by the basal promoter and the proximal promoter. In mammals, 60% of all promoters colocalize with regions of high C+G content, known as CpG islands. A feature that can also be used to find unknown genes. In the immediate upstream region of the basal an proximal promoter, auxiliary binding sites can be located for further transcription factors, which are often observed to be organized in regulatory modules.

9.3.1 Polymerases and Associated Promoters

Transcription of DNA into RNA is performed by the three different types of RNA polymerases. For the modes of transcription associated with the different RNA polymerases, see Table 9.3. Each of the polymerases requires certain *cis*-acting elements in order to initiate transcription; due to its crucial relevance in transcribing mRNA necessary for protein coding genes, much effort has been spent on studying the polymerase II core promoter as the minimal stretch of contiguous DNA sufficient for initiating transcription. In most (yet not all) polymerase II transcribed genes, the core promoter contains the *TATA box*, which is located 25 bases upstream of the transcription start site. The TATA box is a usually 6-nucleotide-long sequence motif characterizing the binding site for the tata-binding-protein (TBP). TBP usually interacts with other transcription factors, whose binding sites are typically found within 40 nucleotides (nt) upstream to the transcription start site. For details on the RNA polymerase II core promoter, we refer to the survey by Butler and Kadonaga [14]. The action of the polymerase II promoter is often enhanced by several distal promoters organized in *cis*-regulatory modules (see section 9.3.3).

Polymerase I transcripts are also regulated by a core promoter, which is separated by about 70 bp from a second elementary promoter element, the

Table 9.3. Major modes of transcription.

RNA polymerase	Promoter	Location relative to start site	Transcript	Function
Pol I	Core element UCE (upstream control element)	−45 to +20 −180 to −107	pre-rRNA (28S, 18S, 5.8S)	Components of the ribosome; translation
Pol II	TATA-Box Initiator CpG islands no	−25 to −35 −100	mRNA snRNA (U1-4) LINEs	Protein coding genes Components of the spliceosome; mRNA splicing Retrotransposon
Pol III	Type 1: A-box, C-box Type 2: A-box, B-box Type 3: TATA-Box Internal	+50 to +80 +10 to +60 −30 to −70	5S rRNA tRNA snRNA (U6) 7SL RNA SINEs	Component of large ribosomal subunit translation Components of the spliceosome; mRNA splicing Component of the SRP (signal recognition particle); protein transport to ER (endoplasmatic reticulum) Retrotransposon

so-called *upstream control element* (UCE). In place of the TBF for polymerase II, pol I requires two core transcription factors, namely UBF1 (*upstream binding factor*) binding to a GC-rich region and SL1 (*selectivity factor*).

Promoters for RNA polymerase III occur in several variants. First, they may consist of bipartite sequences downstream of the start point, with promoter element *boxA* separated from either of the promoter elements *boxC* or *boxB*. Second, some U snRNA genes are regulated by upstream type promoters involving an octamer binding site, a so-called *proximal sequence element* and a TATA box.

9.3.2 Identification of Transcription Factor Binding Sites (TFBSs)

Transcription factors are known to bind to short, specific sequences of DNA. Experimental evidence obtained by techniques such as DNase footprinting [36] and gel-shift assays [37] suggests that protein-DNA binding of transcription factors involves a relatively short, contiguous DNA segment, whose length

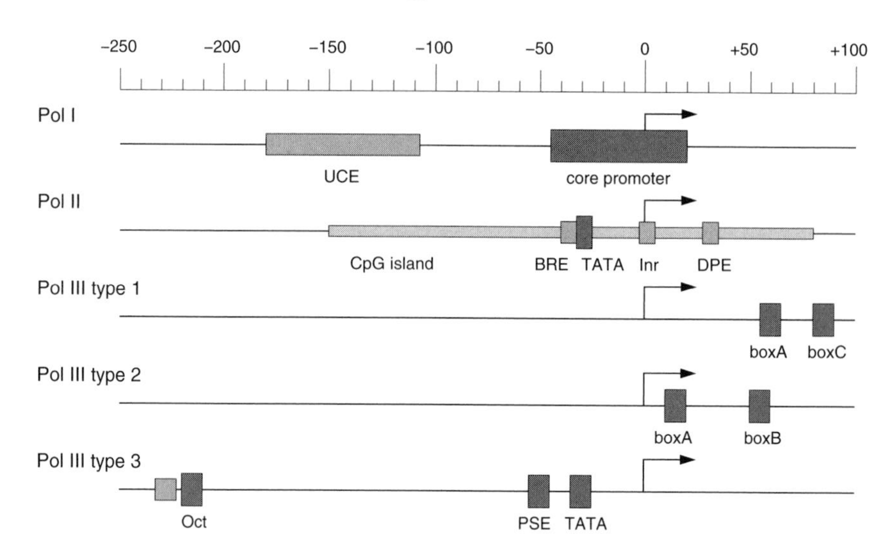

Fig. 9.1. Core motifs of the different promoter types. Motifs in dark gray are less dispensable than motifs in light gray. Any specific promoter may contain just a subset or, in the worst case, none of these motifs. UCE = upstream control element, BRE = TFIIB recognition element, Inr = initiator element, DPE = downstream core promoter element, Oct = octamer binding site, PSE = proximal sequence element. The arrow indicates the transcription start site at +1.

usually ranges between 8 and 15 nucleotides. Repositories of known transcription factors and their experimentally derived binding site motifs can be found in databases such as TRANSFAC [45] or JASPAR [123]. However, experimental determination of binding sites and their relevance *in vivo* takes a significant effort, so that numerous computational approaches have been proposed for determining candidates for TFBSs *in silico.*

While the length of the binding sites corresponding to one specific TF is observed to be essentially constant, the individual positions of the binding site sequences may vary up to a certain degree. Hence, to derive a suitable model of the sequences that a TF binds to, different notions of describing TFBS sequence variability have been proposed. Such models are also important in the context of computationally determining TFBSs based on comparative genomics approaches.

Different observed binding sites corresponding to a given transcription typically show a high degree of sequence similarity; moreover, the observed binding site motifs have the same length ℓ. To capture the observed binding sites in one unique structure, we define a *binding site model of length* ℓ as a mapping $M\colon \Sigma^{\ell} \to \mathbb{R}_{\geq} 0$ assigning a weight to each sequence of length ℓ over the DNA alphabet Σ. While $M(s)$ ideally should be related to the physical binding affinity of sequence s binding to the transcription factor modeled by M, $M(s)$ usually is obtained on the basis of the frequency of observed or putative binding sites in a given set of genomic sequences.

The concept of binding site models introduced above is too general in many situations: first, there are usually not enough data to derive reasonable weights for each DNA sequence of length ℓ, and second, storing and retrieving the complete mapping M would be too expensive. Hence, several simplified models of TFBSs have been established. The most simple model of a binding site model is to derive a *consensus sequence*. In this model, each of the ℓ positions is associated with a subset of the DNA nucleotide alphabet. A given sequence fragment s of length ℓ is assigned score $M(s) = 1$ if at each position, the nucleotide of the fragment is contained in the corresponding nucleotide set of the model; otherwise, we have $M(s) = 0$. Alternatively, one can define $M(s)$ as the number of positions in s where the nucleotide in s is contained in the corresponding set in M.

Consensus sequences disregard major information contained in the sequences used for deriving the consensus model; namely, they do not take into account frequencies of occurrence. This is overcome in the most established way of TFBS modeling, namely *position weight matrices* (PWMs). In the PWM model (sometimes also referred to as a *position specific weight matrix*), each of the ℓ positions of the binding site is assumed to be distributed independently: for each position i, we are given a probability density function $pi\colon \Sigma \to [0,1]$ over the four nucleotides. Given $s := s1 \dots s\ell \in \Sigma^\ell$, this allows us to define

$$M(s) := p1(s1) + \cdots + p\ell(s\ell).$$

PWMs can be derived canonically from a collection of sequences $S1, \dots, SM \in \Sigma^\ell$: for $x \in \Sigma$ and $i \in \{1, \dots, \ell\}$, let $\nu(x, i)$ denote the number of sequences in which letter x occurs at position i. By setting $pi(x) := \nu(x, i)/M$, we indeed obtain a PWM model. In practice, the sequences $S1, \dots, SM$ are typically obtained either from a set of experimentally determined binding sites or from motif discovery methods.

PWMs, however, disregard any information about the correlation between sites that may be contained in the sequences that a matrix was derived from. As a remedy, Pudimat et al [112] have developed a more sophisticated way of modeling TFBSs based on parameter estimation in a Bayesian belief network. As opposed to most other approaches of TFBS modeling, their approach allows us to model correlations between the individual sequence positions. Another approach for modeling dependencies between positions in PWMs based on χ^2 statistics has been investigated in [33].

While obtaining models for TFBSs from experimental data is relatively easy, deriving them computationally from genomic sequences is a complex problem. Essentially all approaches are based on comparative genomics in the sense that they seek for motifs contained in each, or at least most, of K promoter regions belonging to K co-regulated (or orthologous) genes. The returned motifs usually result from optimizing a scoring function that measures how well a candidate motif statistically differs from global properties of the promoter sequences.

Among the earliest nontrivial approaches to extracting overrepresented short motifs as potential TFBSs, Hertz and Stormo [48] proposed a greedy algorithm. Their `CONSENSUS` approach starts with a position weight matrix derived from a single sequence of a fixed length, which is extended to a pairwise alignment of the same width by considering a best-matching subsequence of the second sequence. The algorithm proceeds by successively adding one subsequence of each remaining input sequence to obtain the final PWM, along with a p value that allows us to assess the statistical significance of the result.

A different approach based on the expectation maximization (EM) algorithm is investigated in `MEME` [4], improving a previous approach by Lawrence and Reilly [75]. The EM-based approach starts with an a priori guess for a position weight matrix representing a binding site of fixed length ℓ, which is then improved according to the input sequences in each of the subsequent iteration steps. A single iteration step works as follows: for each subsequence ℓ of the input sequences, the score of the current matrix is computed. After normalization, the matrix entries are updated by summing up the individual position contributions of each of the length ℓ subsequences, weighted by its corresponding normalized probability computed before. The resulting new matrix is then used as input for the next iteration step, until convergence of the process is observed.

`AlignACE` developed by Roth and Hughes's group [56, 119] is yet another approach to obtain PWMs from genomic regulatory sequences. `AlignACE` is based on Gibbs sampling, enhancing approaches previously used for locally aligning motifs in protein sequences such as [74] in a way such that both strands of the input sequences are considered. Furthermore, single motifs that were found are masked iteratively to allow for the extraction of more than one binding site motif.

9.3.3 Discovering Regulatory Modules

As numerous studies demonstrate, transcription factors exhibit their function synergistically through complexes of several transcription factors activating or deactivating gene expression by binding to their corresponding binding sites [27, 152], which thus form the building blocks of regulatory modules. On the genome level, regulatory modules are characterized by binding sites being located close to each other, usually within a segment whose length does not exceed a few hundred nucleotides.

In recent years, a number of approaches have been developed in the context of discovering *cis*-regulatory modules. Kel-Margoulis et al [65] propose a method based on identifying clusters with the property that pairwise distances between occurrences of TFBSs range within certain bounds; sets of binding sites that maximize a certain cluster score are searched by the means of a genetic algorithm. Other methods are based on probabilistic methods [109] or require (only sparsely available) knowledge about interactions between transcription factors such as the algorithm presented in [130].

Among the most established methods, Sharan et al proposed an approach implemented in the program `CREME` [129], which is conceptually somewhat related to our approach. Given a set of candidate binding sites, `CREME` seeks to identify motif clusters of limited length that occur more than once in a set of regulatory sequences. However, the capabilities of the `CREME` approach is limited to discovering repetitive occurrences of modules that contain *precisely* the same set of binding sites. While biological data indeed indicate that functionally related modules share a significant number of binding sites, modules observed in a given regulatory region might as well contain occurrences of known binding site motifs, which are not functional in the given context. If this number of additional, nonshared binding sites is nonzero, the method underlying `CREME` does not allow us to discover such functional modules reliably.

To overcome this shortcoming, the hypothesis underlying the `bbq` approach [99] is that CRMs are characterized by sharing a significant number of common binding sites, but do not necessarily contain precisely the same set of binding sites. More formally, we are given a set of candidate binding sites $s1, \ldots, sm$ together with a set of genomic sequences $T1, \ldots, TK$. The role of the genomic sequences Tj is taken by the regulatory regions of genes that are suspected to share a regulatory module (due to being orthologous or having a similar expression scheme), while the binding site motifs can be derived from databases such as `TRANSFAC` or `JASPAR`. Alternatively, these motifs can be derived from $T1, \ldots, TK$ using the motif discovery approaches discussed in section 9.3.2. Finally, an upper bound for the length L (specified as a number of nucleotides) of the regulatory module is given as an input parameter.

The `bbq` approach starts with determining the occurrences of each motif si in each Tj and associating a color i with binding site si. For each occurrence of si, an interval of length $(L - |si|)$ ending at the position of the occurrence is introduced, so that one finally obtains K arrangements of colored intervals. By "stabbing" into this arrangement, one obtains a cell in this arrangement. Such a cell is associated with a set of colors, which corresponds to a set of binding sites occurring within a genomic subsequence whose length is at most L nucleotides. Finally, attempting to stab a maximum number of common colors in each of the K arrangements leads to the so-called *best-barbecue problem.* This problem leads to a natural combinatorial and geometric optimization problem that is NP-complete in general.

9.3.4 Phylogenetic Footprinting

Just as genomic regions that code for proteins or functional RNAs, regulatory elements are also subject to stabilizing selection. They evolve much more slowly than adjacent nonfunctional DNA, so that one can observe conserved

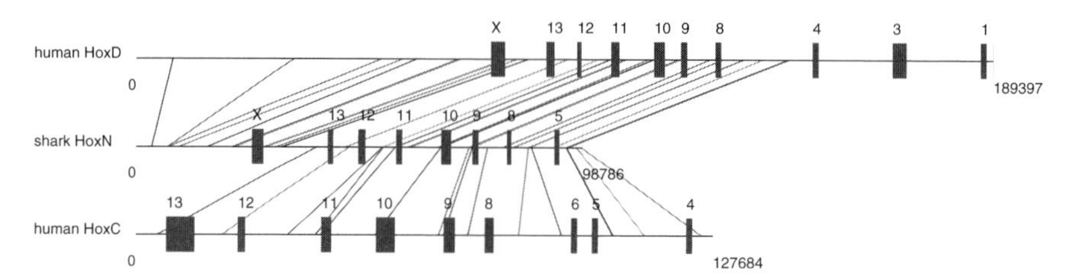

Fig. 9.2. Distribution of phylogenetic footprints in regulatory regions between the horn shark HoxN sequence and the human HoxC and HoxD sequences obtained by the `tracker` tool. Using this information for phylogenetic inference, this supports the hypothesis that the shark HoxN sequence is orthologous to the mammalian HoxD sequence [110]. Boxes indicate the location of the coding regions for the paralog groups 1 to 13, X denotes the *evx* gene. Lines conecting sequences represent phylogenetic footprints shared with the shark HoxN sequence.

islands of regulatory regions within intergenic or intronic regions. These conserved islands are commonly referred to as *phylogenetic footprints*, which can be detected by comparison of the sequences surrounding orthologous genes in different species. The loss of phylogenetic footprints as well as the acquisition of conserved noncoding sequences in some lineages, but not others, can provide evidence for the evolutionary modification of *cis*-regulatory elements.

While the motifs detected by the discovery methods discussed in section 9.3.2 can be seen as one particular type of footprints, one can often observe conserved islands that are much longer, up to several hundred nucleotides, than individual binding sites. Therefore, phylogenetic footprinting tools are usually based on pairwise or multiple local alignment algorithms such as `blastz` [125] or `Dialign` [97]. The tools `PipMaker` and `MultiPipMaker` [125] (among others) process these alignments in order to provide information on significantly conserved regions. The `Tracker` tool [111] assembles individual pairwise `blastz` alignments into cliques of overlapping alignments. This results in the possiblity of listing alternative multiple local alignments if the pairwise matches are not consistent with one multiple alignment.

As demonstrated in [111], the analysis of sequence conservation of non-protein-coding DNA can be used to unveil the evolutionary origin of phenomena such as the duplication of Hox clusters in shark, human, and the duplicated zebrafish and Takifugu (Fig. 9.2). In this context, information contained in the regulatory regions yields insights that are not visible on the level of the corresponding protein coding regions.

9.4 Regulatory ncRNAs and RNA Motifs

9.4.1 Diversity of the RNA Inventory

Noncoding RNAs form a diverse group of transcripts with often poorly understood function. In contrast to protein-coding mRNAs there is little that they all have in common. One group, which itself is composed of a heterogeneous set of RNA families including tRNAs, the U6 snoRNA, the RNA component of the signal recognition particle, and a small number of less well known ncRNAs including 7SK RNA and Y RNAs, is transcribed by RNA polymerase-III. Ribosomal RNAs, transcribed by pol-I, form a group by themselves. Almost all of these ncRNAs are evolutionarily very well conserved, and most of them are evolutionarily ancient.

In contrast, the majority of the known ncRNAs are transcribed by pol-II. These can be subdivided into messenger-RNA–like transcripts, such as *Xist*, which are typically spliced and polyadenylated, "structural ncRNAs" such as spliceosomal RNAs (snRNAs) and many microRNAs, which are neither spliced nor polyadenylated, and a class of functional RNAs that is processed from introns (in particular snoRNAs). Informally, it is useful to distinguish a restricted group of "classical" ncRNAs containing the rRNAs, the pol-III transcripts listed above, spliceosomal RNAs, box-C/D and hox-H/ACA small nucleolar RNAs (snoRNAs), microRNAs, as well as telomerase RNA. As far as we know, these RNAs are evolutionarily old, they have distinctive RNA secondary structure, and most of them are reasonably well conserved at sequence level.

Recently, a number of small, non-mRNA-like ncRNAs have been found, for example in the nematode *Caenorhabditis elegans*, which does not appear to belong to one of the classical families, although at least some of them share the distinctive promoter features of tRNAs or pol-II transcribed snRNAs [28]. Bacterial genomes also contain a large and diverse set of small RNAs (sRNAs) in addition to the classical ncRNAs. A recent survey discusses 55 known *E. coli* sRNAs [47] and their conservation patterns within Enterobacteria. For a review of functional aspects of various bacterial sRNAs see [42]. An additional class of small anti-sense transcripts derived from UTRs is discussed in [64]. For a recent survey focusing on the regulatory effects of ncRNAs in eucaryotes, see [25].

The function of almost all mRNA-like ncRNAs remains unknown. The few well-studied examples, such as *Xist* or *H19*, have functions in imprinting [103].

Regulation by means of RNA can follow at least three distinct principles: `RNA switches` sense changes in temperature or chemical environment and react by conformational changes. *Cis*-acting RNA signals, often located in untranslated regions of mRNAs, are bound by proteins. *Trans*-acting RNAs, such as microRNAs, perform their function by binding to complementary nucleic acid sequence motifs.

9.4.2 RNA Secondary Structure Prediction and Comparison

From a theoretical perspective, computational RNomics draws much of its appeal from the fact that most quantities of interest can be computed exactly within the secondary structure model. In contrast to proteins, nucleic structures are dominated by a single, very specific type of interaction: the formation of Watson-Crick and wobble (G-U) base pairs. The resulting contact structures, which are predominantly stabilized by the stacking interactions of adjacent base pairs, are not only a convenient and routinely used representation [70, 104, 106, 124], they also quantitatively describe the energetics of RNA structure formation, and they form intermediate steps in the folding process itself.

Formally, a *secondary structure* is a set Ω of base pairs such that (1) each nucleotide position i is paired with at most one other nucleotide (i.e., Ω is a matching), and (2) base pairs do not cross, i.e., $(i,j),(k,l) \in \Omega$ with $i < j, k < l$ implies $j < k$ or $l < j$. The second condition ensures that two base pairs are either separated along the sequence or nested within each other. A secondary structure therefore can be seen as a circular matching. Drawing the bases along a circle, the base pairs form chords that do not cross. It follows that RNA secondary structures can be dealt with by means of exact dynamic programming algorithms (Fig. 9.3).

A plethora of careful thermodynamic measurement confirmed that the energetics of RNA structures can be understood in terms of additive contributions of "loops" (Fig. 9.3) see [83, 84] and the references therein. Exact dynamic programming algorithms can be used to compute, for example, the minimum energy structure given any RNA sequence s [136, 153, 155]. The most frequently used implementations of these algorithms are `mfold` [153, 155] and the `Vienna RNA Package` [50, 53].

An RNA molecule, however, does not only form a single (ground state) structure; rather, there is an ensemble $\Sigma(s)$ of different structures Ψ that depend on the sequence s, which are populated in proportion to their Boltzmann factors $F(\Psi)/RT$. The partition function

$$Z = \sum \Psi \in \Sigma(s) \exp\left(-\frac{F(\Psi)}{RT}\right) , \qquad (9.1)$$

from which all thermodynamics quantities of interest can be readily derived, can be computed by the same type of dynamic programming approach [87].

9.4.3 Suboptimal Structures and RNA Switches

Some RNA molecules exhibit two competing conformations, whose equilibrium can be shifted easily by molecular events such as the binding of another

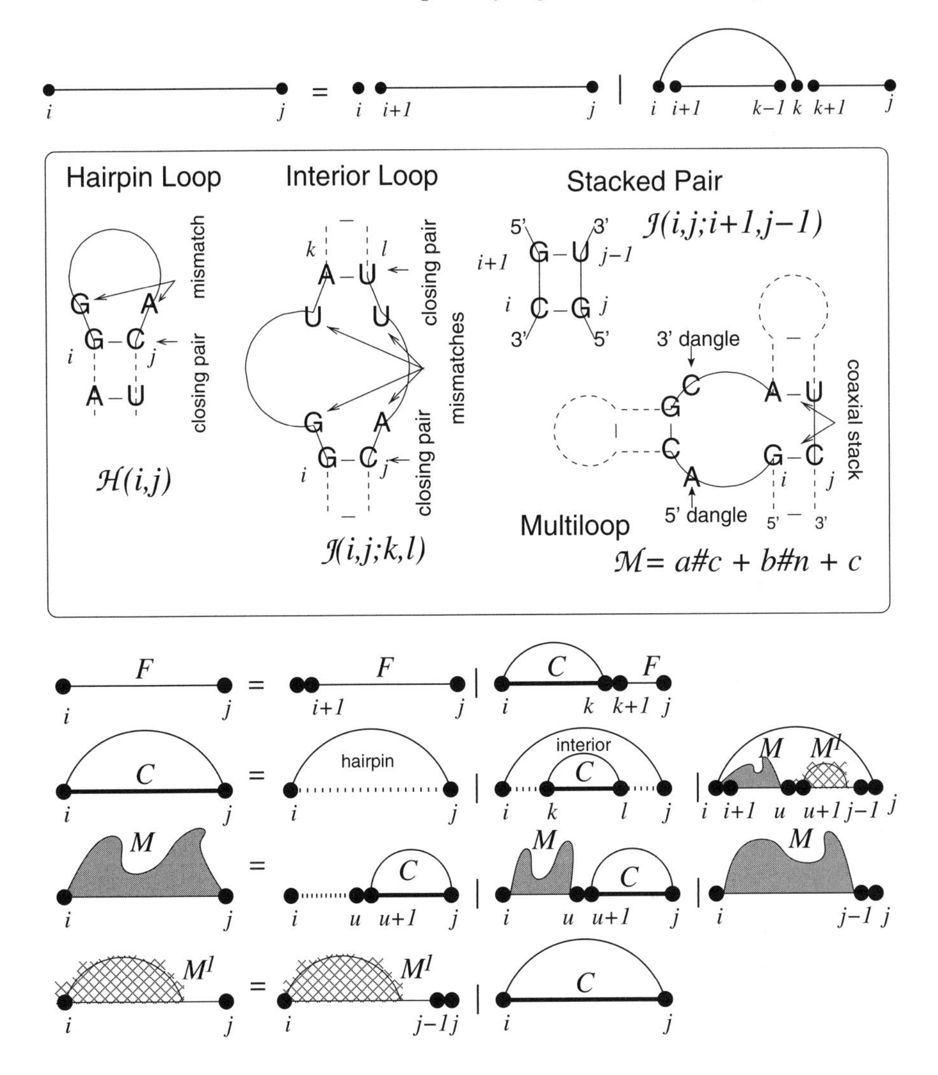

$$
\begin{aligned}
Fij &= \min\{Fi+1,j,\ \min i<k\leq jCik + Fk+1,j\}\\
Cij &= \min\{\mathcal{H}(i,j),\ \min i<k<l<jCkl + \mathcal{I}(i,j;k,l),\\
&\qquad \min i<u<jMi+1,u + M^1u+1,j-1+a\}\\
Mij &= \min\{\min i<u<j(u-i+1)c + Cu+1,j+b,\\
&\qquad \min i<u<jMi,u + Cu+1,j+b,\ Mi,j-1+c\}\\
M^1ij &= \min\{M^1i,j-1+c,\ Cij+b\},
\end{aligned}
$$

Fig. 9.3. RNA folding in a nutshell. *Caption continued overleaf...*

molecule. This can be used to regulate of gene expression, when the two mutually exclusive alternatives correspond to an active and inactive conformation of the transcript [46, 90]. While most riboswitches were found in bacteria, where they regulate several key metabolic pathways [11, 101], metabolite-binding RNA domains are also present in some eukaryotic genes [134]. An early computational study concluded that RNA switches are readily accessible in evolution and are therefore probably not exceptional instances of unusual RNA behaviour [35]. These findings, and the fact that riboswitches bind their effectors directly without the need of additional factors, suggest that riboswitches represent one of the oldest regulatory systems [139].

9.4.4 Detection of Functional RNAs in Genomic DNA

Large-scale efforts to uncover the human and mouse transcriptomes, using very different experimental techniques including tiling arrays [8, 21, 61, 63], cDNA sequencing [58, 102], and unbiased mapping of transcription factor binding sites [17], agree that a substantial fraction of these genomes is transcribed and that the majority of these transcripts do not code for proteins. It is still unclear at present, however, which fraction represents functional noncoding RNAs (ncRNAs), and which constitutes "transcriptional noise" [57].

Genome-wide computational surveys of ncRNAs, on the other hand, have been impossible until recently, because ncRNAs do not share common signals that could be detected at the sequence level. An exception are bacterial

---

Fig. 9.3, continued.
Top: The basic recursion for RNA folding is based on the observation that each structure either terminates in an unpaired base or in a base pair that then separates the structure into two *independent* parts: the one enclosed by the base pair, and the one outside.
Box: The standard energy model distinguishes three types of loops: hairpin loops with a single closing pair, interior loops (including bulges and the stabilizing stacking pairs) that are delimited by two base pairs, and multiloops at which the structure branches. For the latter the energy model assume additive contributions depending on the number of branches $\#c$ and the number of unpaired bases $\#n$ in the loop.
Middle: Using the loop-based energy model complicated the recursion since one now has to distinguish the different types of loops because of their distinct energy contributions. Instead of a single array storing the optimal energies Fij for substructure on the subsequence $x[i..j]$, one now need a few auxiliary arrays that correspond to restricted classes of structures. For instance, Cij is the optimal energy subject to the constraint that i and j form a base pair.
Bottom: We give the complete recursion for energy minimization in the loop-based energy model. Replacing minima by sums, and sums by products leads leads to the recursions for the partition function Z.

genomes, where a purely sequence-based machine learning approach was fairly successful [122].

Most of the "classical" ncRNAs mentioned above, however, have characteristic (secondary) structures that are functional and hence are well conserved over evolutionary time scales. The stabilizing selection acting on the secondary structure causes characteristic substitution patterns in the underlying sequences: Consistent and compensatory mutations replace one type of base pair by another one in the paired regions (helices) of the molecule. In addition, loop regions are more variable than helices. These patterns not only have a significant impact on phylogenetic inference based on ribosomal RNA sequences (see, e.g., [62] and the references therein), but it also can be exploited for ncRNA detection in comparative computational approaches. Examples are the `alidot` [51] and `qrna` [116] programs. Related approaches predict consensus secondary structures for a set of aligned sequences [52, 107].

A second effect of stabilizing selection for RNA secondary structure is even easier to measure. It was first suggested by Maizel's group that functional RNA elements should have a more stable secondary structure than comparable random sequences [19, 76].

As demonstrated in [137], selection for structure implies that in the long run sequences evolve that are more robust against mutations, that is, for which a larger fraction of mutations does not lead to a change in the ground state structure. This effect can be detected, for example, in viral RNA structures [140]. Mutational robustness, however, is in turn strongly correlated with the thermodynamic stability of the ground state structure [2, 149]. Thus we expect that the ground states of functional RNA structures should be thermodynamically more stable than expected by chance, independently of whether there is a direct selection pressure for thermodynamic stability or not. While this effect can indeed be demonstrated [23], it is not statistically significant enough for reliable ncRNA detection [115]. It can be quite large for specific classes of ncRNAs, in particular microRNAs, however [10, 141].

Combinations of thermodynamic stability and information on gene structure such as positions of rho-independent terminators were quite successful for ncRNA prediction in intergenic regions of prokaryotic genomes [16, 79]. Such methods cannot be employed in eukaryots because of their much larger genome size and the much more complex gene structures.

Sufficient statistical power for ncRNA detection in eukaryotic genomes can be obtained, however, by combining measures for both thermodynamics stability and structural conservation. An implementation of such a combined approach is the `RNAz` program [143]: A structure conservation index (SCI) is computed by comparing the predicted minimum free energies of the sequences in an alignment with a consensus energy, which is computed by incorporating covariation terms into a free energy minimization computation [52]. Thermodynamic stability is quantified by means of a z-score that measures the folding energy relative to shuffled sequences (a regression approach replaces time-consuming shuffling methods). A support vector machine then classifies an alignment as "structured RNA" or "other" based on z-score and SCI. The significance of the classification is quantified as "RNA-class probability" p.

Various computational screens [1, 16, 20, 79, 116, 122] predict several hundred ncRNA candidates. These predictions, however, show relatively little mutual overlap in general. Indeed, the majority of bacterial sRNAs was discovered based on computational predictions and subsequent experimental verification.

A `RNAz` survey based on the most conserved parts of the vertebrate genomes estimates that the ncRNA content of mammalian genomes is comparable to their protein-coding genes [142], and hence at least an order magnitude larger than in nematodes. In contrast, only a few thousand structured RNAs in the urochordate *Ciona intestinalis* [93] and in the nematode *C. elegans* [28, 94]. Only a few hundred ncRNAs appear to be present in the yeast *Saccharomyzes cerevisiae* [88]. This indicates that higher vertebrates have dramatically expanded their ncRNA inventory relative to their complement of protein coding genes. This is consistent with the assumption that the function of the ncRNAs is primarily regulatory [85, 86].

9.4.5 RNA-RNA Interaction

Algorithmically, the "co-folding" of two RNAs can be dealt with in the same way as folding a single molecule by concatenating the two sequences and using different energy parameters for the loop that contains the cut-point between the two sequences. A corresponding `RNAcofold` program is described in [53]; the `pairfold` program [3] also computes suboptimal structures in the spirit of `RNAsubopt` [149]. A restricted variant of this approach is implemented in the program `RNAhybrid` [114] as well as `RNAduplex` from the Vienna RNA package, see also [30, 154]: here secondary structures within both monomers are neglected so that only intermolecular base pairs are taken into account. The program `bindigo` uses a variation of the Smith-Waterman sequence alignment algorithm for the same purpose [49].

The most prominent application of RNA co-folding algorithms is the prediction of microRNA target genes [9, 70, 95, 104, 124, 151]. The biological activity of siRNAs and miRNAs is influenced by local structural characteristics of the target mRNA. In particular, the binding site at the target sequence must be accessible for hybridization in order to achieve efficient translational repression. Recent contributions [106, 124] suggest two significant parameters: the stability difference between the $5'$ and $3'$ end of the siRNA, which determines which strand is included into the RISC complex [66, 126] and the local secondary structure of the target site [9, 70, 95, 104, 124, 151].

The energetics of RNA-RNA interactions can be understood in terms of two contributions: the free energy of binding consists of the contribution ΔGu that is necessary to expose the binding site in the appropriate conformation, and the contribution ΔGh that describes the energy gain due to hybridization at the binding site. The first term can be computed from a partition function computation as described above, and the second term is obtained through a version of the co-folding algorithm. Comparison with the partition function of the isolated systems and standard statistical thermodynamics can be used to explicitly compute the concentration dependence of RNA-RNA binding [30].

9.4.6 RNA-Protein Interaction

In recent years an increasing number of functional features has been reported in the untranslated regions of eukaryotic mRNA [60, 92, 108]. Well-known motifs include internal ribosomal entry sites (IRESs) in viral as well as cellular mRNAs [55, 108, 121], and the AU-rich elements (AREs) [5, 89]. In many cases, secondary structure motifs are recognized by regulatory proteins with only highly degenerate, or no sequence constraints at all [91, 138]. In such cases, the thermodynamics of RNA folding can influence binding specificities.

Consider a (protein) ligand that can bind to certain set RNA$*$ of structural conformations a given RNA molecules:

$$\text{Ligand} + \text{RNA}* \rightleftharpoons \text{Ligand} \cdot \text{RNA}$$

The law of mass action implies that the concentrations [RNA$*$], [Ligand], and [Ligand $\cdot$ RNA] of free accessible RNA, free protein, and complex are related through the dissociation constant

$$Kd = \frac{[\text{RNA}*]\,[\text{Ligand}]}{[\text{Ligand} \cdot \text{RNA}]} \tag{9.2}$$

Writing $A(s) \subseteq \Sigma(s)$ for the accessible structures of our RNA molecule s we obtain

$$[\text{RNA}*] = p*\,[\text{RNA}] \tag{9.3}$$

where $p*$ is the fraction of accessible secondary structures, which can be computed as a ratio of two partition functions

$$p* = \sum \Psi \in A(s) p(\Psi) = \frac{1}{Z} \sum \Psi \in A(s) \exp\left(-\frac{F(\Psi)}{RT}\right) = \frac{Z*}{Z}. \tag{9.4}$$

$Z*$, the partition function of all RNAs with suitable structure can be computed by dynamic programming [87, 100] or by means of stochastic backtracking and sampling [31, 32].

Using conventional methods to measure RNA protein interactions, only the total concentration of unbound RNA, [RNA], can be measured. Hence, only the apparent dissociation constant $Kd^{\text{app}} = Kd/p*$ can be determined

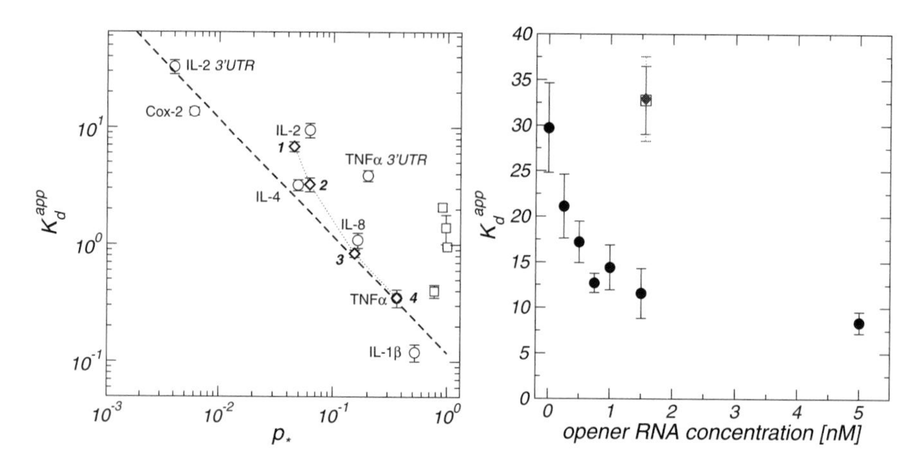

Fig. 9.4. Left: Apparent dissociation constants for *HuR*-mRNA complexes at 23.5°C for natural ARE and UTR sequences (○), artificial molecules (□), and designed mutants of the tumor necrosis factor-α (TNF-α) 3′UTR (◇) [44]. Right: Effect of a complementary opener of length $N0 = 20$ on *in vitro HuR*/RNA affinities. The apparent affinity of recombinant HuR to IL-2 3′UTR was determined in the presence and absence of the opener Op3 (black circles) and of the negative controls with 1D-FIDA detection. Data redrawn from [89].

experimentally. The theory therefore predicts structure dependence of the measured values of Kd^{app}. Under the assumption that the true value of Kd depends only on the ligand and the sequence-structure motif that binds the ligand, we can predict sequence-dependent variations in RNA-ligand binding affinity by means of a computational analysis of the ensemble of RNA structures. In [44, 89] it has been shown that the interaction of the HuR protein with ARE-carrying mRNAs indeed follows this scheme.

An immediate consequence of this mechanism is the possibility of using small RNA "modifiers" to modulate the binding affinities of RNAs and ligands by binding to their target RNA in such a way that it alters the local structure at the ligand binding site. The *HuR*-mRNA interaction again serves as a well-studied *in vitro* example for such a technique [44, 89] (Fig. 9.4). The regulation of *HuR*-ARE-mediated export and RNA stability *in vivo*, however, remains enigmatic. There is only the single ubiquitously expressed protein *HuR* (and a handful of tissue specific relatives such as the neuronal specific homologue *HuD*) that upregulates the export and stability of potentially thousands of ARE-carrying mRNAs. It is tempting to speculate that modifying RNA "openers" could be involved in target gene–specific regulation of *HuR* activity.

9.5 Conclusion

In this chapter we have discussed at least some of the regulatory mechanism that leave traces at the DNA level. A significant fraction of the non-repetitive DNA of higher eukaryotes is subject to stabilizing selection. It has been estimated, for example, that about 5% of the human genome is under stabilizing selective pressure [24, 59], while less than 2% are protein-coding genes. It is a major challenge for bioinformatics to elucidate the meaning of the remaining conserved DNA.

The information about at least a large part of the regulatory circuitry of a species is accessible by means of comparative genomics. Without a large body of independent experiments, however, we have little chance to decode this information. The first, and maybe crucial step, beyond identifying the DNA footprints themselves is to discriminate between regulatory elements that exert their function at the DNA level, *cis*-acting elements that function at the mRNA level, and noncoding RNAs.

We have reviewed here some of the currently available computational approaches that can be used to detect and analyze such elements. Few general tools are available. A subclass of noncoding RNAs and *cis*-acting mRNA elements, for example, can be recognized because of its conserved secondary structure. On the other hand, at present there is not even a way to distinguish protein binding sites on the genomic DNA from those on the mRNA, unless specific knowledge about a particular sequence motifs is available from experiments.

References

1. Agarman L, Hershberg R, Vogel J et al (2001) Novel small RNA-encoding genes in the intergenic regions of *Escherichia coli*. Curr Biol 11:941–950.
2. Meyers LA Lee JF, Cowperthwaite M et al (2004) The robustness of naturally and artificially selected nucleic acid secondary structures. J Mol Evol 58:681–691.
3. Andronescu M, Zhang Z, Condon A (2005) Secondary structure prediction of interacting RNA molecules. J Mol Biol 345:987–1001.
4. Bailey T, Elkan C (1995) The value of prior knowledge in discovering motifs with MEME. Proc Int Conf Intell Syst Mol Biol 3:21–29.
5. Bakheet T, Frevel M, Williams BR et al (2001) ARED: human AU-rich element-containing mRNA database reveals an unexpectedly diverse functional repertoire of encoded proteins. Nucl Acids Res 29:246–254.
6. Barlati S, Barbon A (2005) RNA editing: a molecular mechanism for the fine modulation of neuronal transmission. Acta Neurochir Suppl 93:53–57.
7. Bayne E, Allshire R (2005) RNA-directed transcriptional gene silencing in mammals. Trends Genet 21:370–373.

8. Bertone P, Stoc V, Royce TE et al (2004) Global identification of human transcribed sequences with genome tiling arrays. Science 306:2242–2246.
9. Bohula EA, Salisbury AJ, Sohail M et al (2003) The efficacy of small interfering RNAs targeted to the type 1 insulin-like growth factor receptor (IGF1R) is influenced by secondary structure in the IGF1R transcript. J Biol Chem 278:15991–15997.
10. Bonnet E, Wuyts J, Rouzé P et al (2004) Evidence that microRNA precursors, unlike other non-coding RNAs, have lower folding free energies than random sequences. Bioinformatics 20:2911–2917.
11. Brantl S (2004) Bacterial gene regulation: from transcription attenuation to riboswitches and ribozymes. Trends Microbiol 12:473–475.
12. Brent M, Guigó R (2004) Recent advances in gene structure prediction. Curr Opin Struct Biol 14:264–272.
13. Burge C, Karlin S (1997) Prediction of complete gene structures in human genomic DNA. J Mol Biol 268:78–94.
14. Butler J, Kadonaga J (2002) The RNA polymerase II core promoter: a key component in the regulation of gene expression. Genes Dev 16:2583–2592.
15. Cairns B (2005) Chromatin remodeling complexes: strength in diversity, precision through specialization. Curr Opin Genet Dev 15:185–190.
16. Carter RJ, Dubchak I, Holbrook SR (2001) A computational approach to identify genes for functional RNAs in genomic sequences. Nucl Acids Res 29:3928–3938.
17. Cawley S, Bekiranov S, Ng HH et al (2004) Unbiased mapping of transcription factor binding sites along human chromosomes 21 and 22 points to widespread regulation of noncoding RNAs. Cell 116:499–509.
18. Chen J, Sun M, Hurst L et al (2005) Genome-wide analysis of coordinate expression and evolution of human cis-encoded sense-antisense transcripts. Trends Genet 21:326–329.
19. Chen JH, Le SY, Shapiro B et al (1990) A computational procedure for assessing the significance of RNA secondary structure. Comput Appl Biosci 6:7–18.
20. Chen S, Lesnik EA, Hall TA et al (2002) A bioinformatics based approach to discover small RNA genes in the *Escherichia coli* genome. Biosystems 65:157–177.
21. Cheng J, Kapranov P, Drenkow J et al (2005) Transcriptional maps of 10 human chromosomes at 5-nucleotide resolution. Science 308:1149–1154.
22. Claycomb J, Orr-Weaver T (2005) Developmental gene amplification: insights into DNA replication and gene expression. Trends Genet 21:149–162.
23. Clote P, Ferré F, Kranakis E et al (2005) Structural RNA has lower folding energy than random RNA of the same dinucleotide frequency. RNA 11:578–591.
24. Cooper GM, Brudno M, Stone EA et al (2004) Characterization of evolutionary rates and constraints in three mammalian genomes. Genome Res 14:539–48.
25. Costa FF (2005) Non-coding RNAs: New players in eucaryotic biology. Gene, in press.
26. Dalgaard J, Vengrova S (2004) Selective gene expression in multigene families from yeast to mammals. Sci STKE 256:re17.
27. Davidson E (2001) Genomic Regulatory Systems. Academic Press, San Diego.
28. Deng W, Zhu X, Skogerbo G et al (2006) Organisation of the *Caenorhabditis elegans* small noncoding transcriptome: genomic features, biogenesis and expression. Genome Res, submitted.

29. Dermitzakis ET, Reymond A, Scamuffa N et al (2003) Evolutionary discrimination of mammalian conserved non-genic sequences (CNGs). Science 302:1033–1035.
30. Dimitrov RA, Zuker M (2004) Prediction of hybridization and melting for double-stranded nucleic acids. Biophys J 87:215–226.
31. Ding Y, Chan CY, Lawrence CE (2004) Sfold web server for statistical folding and rational design of nucleic acids. Nucl Acids Res 32:W135–141.
32. Ding Y, Lawrence CE (2003) A statistical sampling algorithm for RNA secondary structure prediction. Nucl Acids Res 31:7280–7301.
33. Feng X, Lin W, Minghua D et al (2005) An efficient algorithm for deciphering regulatory motifs. In These Proceedings.
34. Fickett J. (1996) Finding genes by computer: the state of the art. Trends Genet 12:316–320.
35. Flamm C, Hofacker IL, Maurer-Stroh S et al (2000) Design of multi-stable RNA molecules. RNA 7:254–265.
36. Galas D, Schmitz A (1978) DNAse footprinting: a simple method for the detection of protein-DNA binding specificity. Nucl Acids Res 5:3157–3170.
37. Garner M, Revzin A (1981) A gel electrophoresis method for quantifying the binding of proteins to specific DNA regions: application to components of the Escherichia coli lactose operon regulatory system. Nucl Acids Res 9:3047–3060.
38. Gebauer F, Hentze M (2004) Molecular mechanisms of translational control. Nat Rev Mol Cell Biol 5:827–835.
39. Gelfand M (1995) Prediction of function in DNA sequence analysis. J Comput Biol 2:87–115.
40. Gelfand M, Mironov A, Pevzner P (1996) Gene recognition via spliced sequence alignment. Proc Natl Acad Sci U S A 93:9061–9066.
41. Gonsalvez G, Urbinati C, Long R (2005) RNA localization in yeast: moving towards a mechanism. Biol Cell 97:75–86.
42. Gottesman S (2004) The small RNA regulators of *Escherichia coli*: roles and mechanisms. Annu Rev Microbiol 58:303–328.
43. Guigó R, Dermitzakis E, Agarwal P et al (2003) Comparison of mouse and human genomes followed by experimental verification yields an estimated 1,019 additional genes. Proc Natl Acad Sci U S A 100:1140–1145.
44. Hackermüller J, Meisner NC, Auer M et al (2005) The effect of RNA secondary structures on RNA-ligand binding and the modifier RNA mechanism: a quantitative model. Gene 345:3–12.
45. Heinemeyer T, Wingender E, Reuter I et al (1998) Databases on transcriptional regulation: TRANSFAC, TRRD, and COMPEL. Nucl Acids Res 26:364–370.
46. Henkin TM, Yanofsky C (2002) Regulation by transcription attenuation in bacteria: how RNA provides instructions for transcription termination/antitermination decision. BioEssays 24:700–707.
47. Hershberg R, Altuvia S, Margalit H (2003) A survey of small RNA-encoding genes in *Escherichia coli*. Nucl Acids Res 31:1813–1820.
48. Hertz G, Stormo G (1999) Identifying DNA and protein patterns with statistically significant alignments of multiple sequences. Bioinformatics 15:563–577.
49. Hodas N, Aalberts D (2004) Efficient computation of optimal oligo-RNA binding. Nucl Acids Res 32:6636–6642.
50. Hofacker IL (2003) Vienna RNA secondary structure server. Nucl Acids Res 31:3429–3431.

51. Hofacker IL, Fekete M, Flamm C et al (1998) Automatic detection of conserved RNA structure elements in complete RNA virus genomes. Nucl Acids Res 26:3825–3836.
52. Hofacker IL, Fekete M, Stadler PF (2002) Secondary structure prediction for aligned RNA sequences. J Mol Biol 319:1059–1066.
53. Hofacker IL, Fontana W, Stadler PF et al (1994) Fast folding and comparison of RNA secondary structures. Monatsh Chem 125:167–188.
54. Hooper P, Zhang H, Wishart D (2000) Prediction of genetic structure in eukaryotic DNA using reference point logistic regression and sequence alignment. Bioinformatics 16:425–438.
55. Huez I, Créancier L, Audigier S et al (1998) Two independent internal ribosome entry sites are involved in translation initiation of vascular endothelial growth factor mRNA. Mol Cell Biol 18:6178–6190.
56. Hughes J, Estep P, Tavazoie S et al (2000) Computational identification of cis-regulatory elements associated with groups of functionally related genes in Saccharomyces cerevisiae. J Mol Biol 296:1205–1214.
57. Hüttenhofer A, Schattner P, Polacek N (2005) Non-coding RNAs: hope or hype? Trends Genet, in press.
58. Imanishi T et al (2004) Integrative annotation of 21,037 human genes validated by full-length cDNA clones. PLoS Biology 2:0856–0875.
59. International Mouse Genome Sequencing Consortium (2002) Initial sequencing and comparative analysis of the mouse genome. Nature 420:520–562.
60. Jacobs GH, Rackham O, Stockwell PA et al (2002) Transterm: a database of mRNAs and translational control elements. Nucl Acids Res 30:310–311.
61. Johnson JM, Edwards S, Shoemaker D et al (2005) Dark matter in the genome: evidence of widespread transcription detected by microarray tiling experiments. Trends Genet 21:93–102.
62. Jow H, Hudelot C, Rattray M et al (2002) Bayesian phylogenetics using an RNA substitution model applied to early mammalian evolution. Mol Biol Evol 19:1591–1601.
63. Kampa D, Cheng J, Kapranov P et al (2004) Novel RNAs identified from an in-depth analysis of the transcriptome of human chromosomes 21 and 22. Genome Res 14:331–342.
64. Kawano M, Reynolds AA, Miranda-Rios J et al (2005) Detection of 5′- and 3′-UTR-derived small RNAs and *cis*-encoded antisense RNAs in *Escherichia coli*. Nucl Acids Res 33:1040–1050.
65. Kel-Margoulis O, Ivanova T, Wingender E et al (2002) Automatic annotation of genomic regulatory sequences by searching for composite clusters. In Proc Pac Symp Biocomput 7:187–198.
66. Khvorova A, Reynolds A, Jayasena SD (2003) Functional siRNAs and miRNAs exhibit strand bias. Cell 115:209–216.
67. Korf I, Flicek P, Duan D et al (2001) Integrating genomic homology into gene structure prediction. Bioinformatics 17 Suppl 1:140–148.
68. Kornblihtt A, de la Mata M, Fededa J et al (2004) Multiple links between transcription and splicing. RNA 10:1489–1498.
69. Kozak M (2003) Alternative ways to think about mRNA sequences and proteins that appear to promote internal initiation of translation. Gene 318:1–23.
70. Kretschmer-Kazemi Far R, Sczakiel G (2003) The activity of siRNA in mammalian cells is related to structural target accessibility: a comparison with antisense oligonucleotides. Nucl Acids Res 31:4417–4424.

71. Krogh A (1998) Gene finding: putting the parts together. Guide to human genome computing, M. Bishop, editor 2nd edn. pp. 261–274 Academic Press.
72. Kulp D, Haussler D, Reese M et al (1996) A generalized hidden Markov model for the recognition of human genes in DNA. Proc Int Conf Intell Syst Mol Biol 4:134–142.
73. Lareau L, Green R, Bhatnagar R et al (2004) The evolving roles of alternative splicing. Curr Opin Struct Biol 14:273–282.
74. Lawrence C, Altschul S, Boguski M et al (1993) Detecting subtle sequence signals: a Gibbs sampling strategy for multiple alignment. Science 262:208–214.
75. Lawrence C, Reilly A (1990) An expectation maximization (EM) algorithm for the identification and characterization of common sites in unaligned biopolymer sequences. Proteins 7:41–51.
76. Le SV, Chen JH, Currey KM et al (1988) A program for predicting significant RNA secondary structures. Comput Appl Biosci 4:153–159.
77. LeRoith D, Roberts C (2003) The insulin-like growth factor system and cancer. Cancer Lett 195:127–137.
78. Liang X, Haritan A, Uliel S et al (2003) *trans* and *cis* splicing in trypanosomatids: mechanism, factors, and regulation. Eukaryot Cell 2:830–840.
79. Livny J, Fogel MA, Davis BM et al (2005) `sRNAPredict`: an integrative computational approach to identify sRNAs in bacterial genomes. Nucl Acids Res 33:4096–4105.
80. Mallory A, Vaucheret H (2004) MicroRNAs: something important between the genes. Curr Opin Plant Biol 7:120–125.
81. Margueron R, Trojer P, Reinberg D (2005) The key to development: interpreting the histone code? Curr Opin Genet Dev 15:163–176.
82. Margulies EH, Blanchette M, Haussler D et al (2003) Identification and characterization of multi-species conserved sequences. Genome Res 13:2507–2518.
83. Mathews DH (2004) Using an RNA secondary structure partition function to determine confidence in base pairs predicted by free energy minimization. RNA 10:1178–1190.
84. Mathews DH, Sabina J, Zuker M et al (1999) Expanded sequence dependence of thermodynamic parameters improves prediction of RNA secondary structure. J Mol Biol 288:911–940.
85. Mattick JS (2003) Challenging the dogma: the hidden layer of non-protein-coding RNAs in complex organisms. Bioessays 25:930–939.
86. Mattick JS (2004) RNA regulation: a new genetics? Nature Genetics 5:316–323.
87. McCaskill JS (1990) The equilibrium partition function and base pair binding probabilities for RNA secondary structure. Biopolymers 29:1105–1119.
88. McCutcheon JP, Eddy SR (2003) Computational identification of non-coding RNAs in *Saccharomyces cerevisiae* by comparative genomics. Nucl Acids Res 31:4119–4128.
89. Meisner NC, Hackermüller J, Uhl V et al (2004) mRNA openers and closers: a methodology to modulate AU-rich element controlled mRNA stability by a molecular switch in mRNA conformation. Chembiochem 5:1432–1447.
90. Merino E, Yanofsky C (2002) Regulation by termination-antitermination: a genomic approach in *Bacillus subtilis* and its closest relatives: From Genes to Cells, A. L. Sonenshein, J. A. Hoch, R. Losick, editors pp. 323–336 ASM Press, Washington D.C.

91. Messias AC, Sattler M (2004) Structural basis of single-stranded RNA recognition. Acc Chem Res 37:279–287.
92. Mignone F, Gissi C, Liuni S et al (2002) Untranslated regions of mRNAs. Genome Biology 3 reviews 0004.1-0004.10.
93. Missal K, Rose D, Stadler PF (2005) Non-coding RNAs in *Ciona intestinalis*. ECCB, in press.
94. Missal K, Zhu X, Rose D et al (2006) Prediction of structured non-coding RNAs in the genome of the nematode *Caenorhabitis elegans*. J Exp Zool: Mol Dev Evol, submitted.
95. Mittal V (2004) Improving the efficiency of RNA interference in mammals. Nat Rev Genet 5:355–365.
96. Morgan H, Santos F, Green K et al (2005) Epigenetic reprogramming in mammals. Hum Mol Genet 14 Spec No 1:47–58.
97. Morgenstern B (1999) `DIALIGN 2`: improvement of the segment-to-segment approach to multiple sequence alignment. Bioinformatics 15:211–218.
98. Morillon A, O'Sullivan J, Azad A et al (2003) Regulation of elongating RNA polymerase II by forkhead transcription factors in yeast. Science 300:492–495.
99. Mosig A, Bıyıkoğlu T, Prohaska SJ et al (2005) Discovering cis regulatory modules by optimizing barbeques. Theor Comput Sci, submitted.
100. Mückstein U, Tafer H, Hackermüller J et al (2005) Thermodynamics of RNA-RNA binding. GCB 2005, in press.
101. Nudler E, Mironov AS (2004) The riboswitch control of bacterial metabolism. Trends Biochem Sci 29:11–17.
102. Okazaki Y, Furuno M, Kasukawa T et al (2002) Analysis of the mouse transcriptome based on functional annotation of 60,770 full-length cDNAs. Nature 420:563–573.
103. O'Neill MJ (2005) The influence of non-coding RNAs on allele-specific gene expression in mammals. Hum Mol Genet 14:R113–120.
104. Overhoff M, Alken M, Far RK et al (2005) Local RNA target structure influences siRNA efficacy: A systematic global analysis. J Mol Biol 348:871–881.
105. Pan P, Lieber M, Teale J (1997) The role of recombination signal sequences in the preferential joining by deletion in DH-JH recombination and in the ordered rearrangement of the IgH locus. Int Immunol 9:515–522.
106. Parker JS, Roe SM, Barford D (2005) Structural insights into mRNA recognition from a PIWI domain-siRNA guide complex. Nature 434:663–666.
107. Parsch J, Braverman JM, Stephan W (2000) Comparative sequence analysis and patterns of covariation in RNA secondary structures. Genetics 154:909–921.
108. Pesole G, Mignone F, Gissi C et al (2001) Structural and functional features of eukaryotic mRNA untranslated regions. Gene 276:73–81.
109. Philippakis A, He F, Bulyk M (2005) Modulefinder: a tool for computational discovery of cis regulatory modules. Proc Pac Symp Biocomput 519–30.
110. Prohaska SJ, Fried C, Amemiya CT et al (2003) The shark HoxN cluster is homologous to the human HoxD cluster. J Mol Evol, in press.
111. Prohaska SJ, Fried C, Flamm C et al (2004) Surveying phylogenetic footprints in large gene clusters: applications to Hox cluster duplications. Mol Phyl Evol, in press; doi: 10.1016/j.ympev.2003.08.009.
112. Pudimat R, Schukat-Talamazzini E, Backofen R (2005) A multiple-feature framework for modelling and predicting transcription factor binding sites. Bioinformatics, accepted for publication.

113. Reese M, Hartzell G, Harris N et al (2000) Genome annotation assessment in Drosophila melanogaster. Genome Res 10:483–501.
114. Rehmsmeier M, Steffen P, Hochsmann M et al (2004) Fast and effective prediction of microRNA/target duplexes. RNA 10:1507–17.
115. Rivas E, Eddy SR (2000) Secondary structure alone is generally not statistically significant for the detection of noncoding RNAs. Bioinformatics 16:583–605.
116. Rivas E, Klein, RJ, Jones, TA et al (2001) Computational identification of noncoding RNAs in *E. coli* by comparative genomics. Curr Biol 11:1369–1373.
117. Rodriguez M, Dargemont C, Stutz F (2004) Nuclear export of RNA. Biol Cell 96:639–655.
118. Rogic S, Mackworth A, Ouellette F (2001) Evaluation of gene-finding programs on mammalian sequences. Genome Res 11:817–832.
119. Roth F, Hughes J, Estep P et al (1998) Finding DNA regulatory motifs within unaligned noncoding sequences clustered by whole-genome mRNA quantitation. Nat Biotechnol 16:939–945.
120. Rousseaux S, Caron C, Govin J et al (2005) Establishment of male-specific epigenetic information. Gene 345:139–153.
121. Rueckert RR (1996) Picornaviridae: the viruses and their replication in Virology N. Fields, D. Knipe, P. Howley, editors vol. 1 3rd edn. pp. 609–654 Lippincott-Raven Publishers, Philadelphia, New York.
122. Sætrom P, Sneve R, Kristiansen KI et al (2005) Predicting non-coding RNA genes in *Escherichia coli* with boosted genetic programming. Nucl Acids Res 33:3263–3270.
123. Sandelin A, Pär Engström WA, Wasserman W et al (2004) Jaspar: an open access database for eukaryotic transcription factor binding profiles. Nucl Acids Res 32.
124. Schubert S, Grunweller A, Erdmann V et al (2005) Local RNA target structure influences siRNA efficacy: Systematic analysis of intentionally designed binding regions. J Mol Biol 348:883–93.
125. Schwartz S, Zhang Z, Frazer KA et al (2000) `PipMaker`—a web server for aligning two genomic DNA sequences. Genome Research 4:577–586.
126. Schwarz D, Hutvagner G, Du T et al (2003) Asymmetry in the assembly of the RNAi enzyme complex. Cell 115:99–208.
127. Shabalina S, Ogurtsov A, Kondrashov V et al (2001) Selective constraint in intergenic regions of human and mouse genomes. Trends Genet 17:373–376.
128. Shabalina SA, Kondrashov AS (1999) Pattern of selective constraint in *C. elegans* and *C. briggsae* genomes. Genet Res 74:23–30.
129. Sharan R, Ovcharenko I, Ben-Hur A et al (2003) CREME: a framework for identifying cis-regulatory modules in human-mouse conserved segments. ISMB (Supplement of Bioinformatics) 283–291.
130. Sinha S, van Nimwegen E, Siggia E (2003) A probabilistic method to detect regulatory modules. Bioinformatics 19:i292–i301
131. Stamatoyannopoulos G (2005) Control of globin gene expression during development and erythroid differentiation. Exp Hematol 33:259–271.
132. Stein L (2004) Human genome: end of the beginning. Nature 431:915–916.
133. Sterky F, Lundeberg J (2000) Sequence analysis of genes and genomes. J Biotechnol 76:1–31.
134. Sudarsan N, Barrick JE, Breaker RR (2003) Metabolite-binding RNA domains are present in the genes of eukaryotes. RNA 9:644–647.

135. The *C. elegans* Sequencing Consortium (1998) Genome sequence of the nematode *C. elegans*: a platform for investigating biology. Science 282:2012–2018.
136. Turner D, Sugimoto N, Freier S (1988) RNA structure prediction. Annu Rev Biophys Biophys Chem 17:167–92.
137. van Nimwegen E, Crutchfield JP, Huynen MA (1999) Neutral evolution of mutational robustness. Proc Natl Acad Sci USA 96:9716–9720.
138. Varani G (1997) RNA-protein intermolecular recognition. Acc Chem Res 30:189–195.
139. Vitreschak AG, Rodionov DA, Mironov AA et al (2004) Riboswitches: the oldest mechanism for the regulation of gene expression? Trends Gen 20:44–50.
140. Wagner A, Stadler PF (1999) Viral RNA and evolved mutational robustness. J Exp Zool (Mol Dev Evol) 285:119–127.
141. Washietl S, Hofacker IL (2004) Consensus folding of aligned sequences as a new measure for the detection of functional RNAs by comparative genomics. J Mol Biol 342:19–30.
142. Washietl S, Hofacker IL, Lukasser M et al (2005) Genome-wide mapping of conserved RNA secondary structures predicts thousands of functional noncoding RNAs in human. Nature Biotech, in press.
143. Washietl S, Hofacker IL, Stadler PF (2005) Fast and reliable prediction of noncoding RNAs. Proc Natl Acad Sci USA 102:2454–2459.
144. Weischenfeldt J, Lykke-Andersen J, Porse B (2005) Messenger RNA surveillance: neutralizing natural nonsense. Curr Biol 15:559–562.
145. West A, Fraser P (2005) Remote control of gene transcription. Hum Mol Genet 14 Spec No 1:101–111.
146. West A, Gaszner M, Felsenfeld G (2002) Insulators: many functions, many mechanisms. Genes Dev 16:271–288.
147. Wilusz C, Wilusz J (2004) Bringing the role of mRNA decay in the control of gene expression into focus. Trends Genet 20:491–497.
148. Wray G, Hahn M, Abouheif E et al (2003) The evolution of transcriptional regulation in eukaryotes. Mol Biol Evol 20:1377–1419.
149. Wuchty S, Fontana W, Hofacker IL et al (1999) Complete suboptimal folding of RNA and the stability of secondary structures. Biopolymers 49:145–165.
150. Yeh R, Lim L, Burge C (2001) Computational inference of homologous gene structures in the human genome. Genome Res 11:803–816.
151. Yoshinari K, Miyagishi MKT (2004) Effects on RNAi of the tight structure, sequence and position of the targeted region. Nucl Acids Res 32:691–9.
152. Yuh CH, Bolouri H, Davidson EH (1998) Genomic cis-regulatory logic: Experimental and computational analysis of a sea urchin gene. Science 279:1896–1902.
153. Zuker M (2000) Calculating nucleic acid secondary structure. Curr Opin Struct Biol 10:303–10.
154. Zuker M (2003) Mfold web server for nucleic acid folding and hybridization prediction. Nucl Acids Res 31:3406–15.
155. Zuker M, Stiegler P (1981) Optimal computer folding of large RNA sequences using thermodynamics and auxiliary information. Nucl Acids Res 9:133–148.

10

Dynamic Properties of Cell-Cycle and Life-Cycle Networks in Budding Yeast

Fangting Li, Ying Lu, Tao Long, Qi Ouyang, and Chao Tang

Summary. Dynamic behaviors of protein-protein and protein-DNA interactions in living cells are investigated using the cell-cycle network and life-cycle network in budding yeast as model systems. Our analysis reveals important dynamic properties of the biological networks. In phase space, the resting states of the networks are global attractors, almost all initial protein states evolve to these stationary states; the biological pathways toward the stationary state are globally attracting trajectories. All initial states are through these well-defined sequences. The distributions of attractor size and pathway thickness of biological network are distinct from that of random networks. Perturbation analysis shows that these networks are dynamically stable. These results suggest that protein networks are robustly designed for their functions, and that the global dynamic properties of protein networks are intimately coupled to their functions.

10.1 Introduction

Protein-protein and protein-DNA interactions in living cells constitute molecular dynamic networks that govern various cellular functions. Recently, much attention has been paid to the global or "system" properties of protein networks rather than isolated functions of proteins. Researchers in this field have achieved significant results. For example, Jeong et al [1, 2] showed that biological networks have the same topological power-law scaling properties, and the highly connected proteins with a central role in the networks' architecture are essential in the systems. Albert et al [3] demonstrated that networks with power-law connection distributions are topologically stable; they are highly tolerant against random failures. Milo et al [4] discussed the basic building blocks of most networks. They suggest that different network motifs are used to build different biological networks that perform different functions. However, despite these achievements, we are still in an early stage of understanding the global properties of the networks. Previous studies focused on the topological properties of the biological networks while few studies address their dynamic behavior [5, 6]. Here we report our dynamic analysis of the cell-cycle network and the life cycle network of the budding yeast. Our study reveals important properties of the networks. In the cell-cycle network, almost all (about 96%) initial protein states evolve to the stationary state G1, making it a global attractor; the dynamic paths to G1 for these initial states are through the cell-cycle sequence, making it a globally attracting trajectory. Similar findings are obtained for the life-cycle network, where the cell-cycle network is one of components. We also found that the distributions of the sizes of attractors and trajectories in phase space are very different from that of random networks, and both networks are dynamically stable against random perturbations.

10.2 The Cell-Cycle and Life-Cycle Networks in Budding Yeast

Budding yeast *Saccharomyces cerevisiae* is chosen as a model system in our dynamic study of biological networks because a comprehensive protein-protein interaction network of this single-cell eukaryotic model organism is emerging [7–13]. Budding yeast has a relatively simple life cycle. Yeast cells can exist in either diploid or haploid genetic state. Cells in both forms can proliferate by an ordinary cell-division process under rich nutrition conditions; the process is called the cell cycle. Triggered by starvation, cells in diploid form undergo meiosis to give rise to spores—haploid cells in a dormant state that are resistant to harsh environmental conditions. Under appropriate conditions, pheromone stimulates cells in haploid form to fuse to create a new diploid cell. The whole process is called the life cycle. Figure 10.1 shows the protein interaction network governing the cell cycle and the life cycle processes in budding yeast. The network is based on extensive literature studies, the Chen model [14], and the database at http://mips.gsf.de/ and http://www.proteome.com. Careful document checking and cross-verifying are conducted in order to correct errors in the published draft. The network shown in Fig. 10.1 can be divided into three interacting functional modules: the sporulation and meiosis subnetwork (left), the pheromone and mating cascade (upper right), and the cell-cycle subnetwork (lower right). Depending on the nutrient condition and the genotype, different modules may be activated in response to different signals. We first focus on the dynamic properties of the most comprehensive subnetwork: the module governing the cell cycle process.

The simplified cell-cycle network is shown as the inset of Fig. 10.2. We simplify the network by removing and reconnecting nodes in a way that does not affect the network dynamics; for example, intermediate nodes on

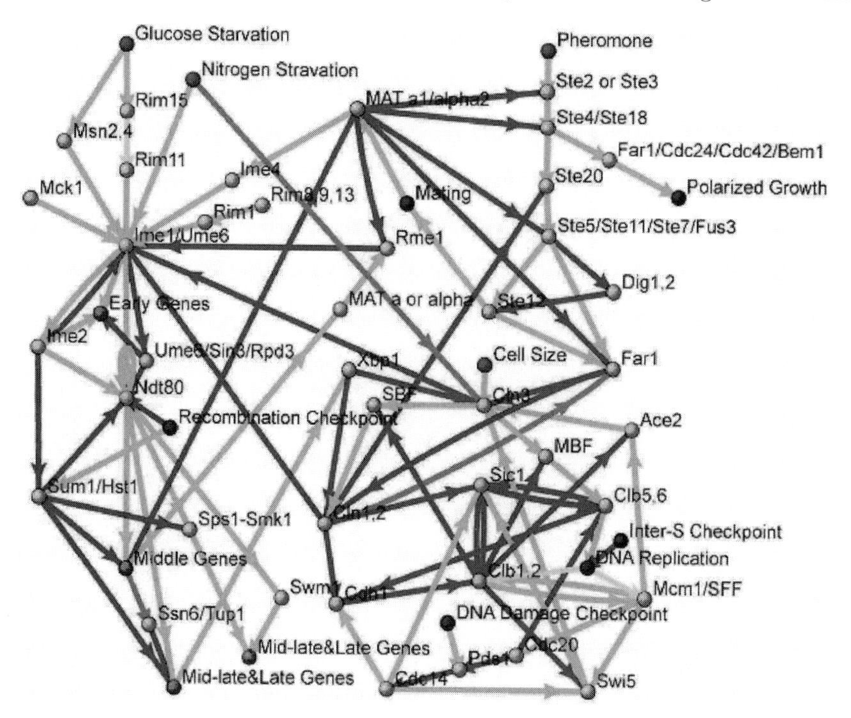

Fig. 10.1. The protein network governing the processes of mating, sporulation, and cell cycle. Yellow nodes represent a single protein, a complex of proteins, or several related proteins. Green nodes represent groups of proteins that are expressed in a special time section. Blue nodes represent a physiological process or checkpoint. Purple nodes represent a signal or corresponding condition. Green and red lines represent positive and negative actions, respectively. Light green and purple lines represent undetermined or indirect actions. All self-degradation actions, yellow lines pointing to node selves, are omitted for clarity. Some of them are displayed in simplified cell-cycle and life-cycle networks inserted in Fig. 10.2 and Fig. 10.3, respectively. The whole net can be divided into three modules: mating (upper right), sporulation (left), and cell cycle (lower right).

a linear path are replaced by an arrow connecting the beginning and the ending nodes. In the simplified network, there are 11 key nodes governing the cell cycle process: Cln3/Cdc28 complex, MBF, SBF, Cln1 or Cln2/Cdc28 complex, Clb5 or Clb6/Cdc28 complex, Sic1, Mcm1/SFF, Clb1 or Clb2/Cdc28 complex, Cdh1, Cdc20&Cdc14, and Swi5. Under rich nutrition conditions, when a yeast cell in either diploid or haploid form grows large enough, the protein Cln3/Cdc28 complex is activated, driving the cell into the excited G1 state. The cell then goes through a sequence of phases of the cell cycle. Activated Cln3/Cdc28 complex activates MBF and SBF, which in turn activate the Cln1 or Cln2 and Clb5 or Clb6, which control the late G1 genes. During the synthesis, or S phase, the cell's DNA is duplicated. After a G2 transition

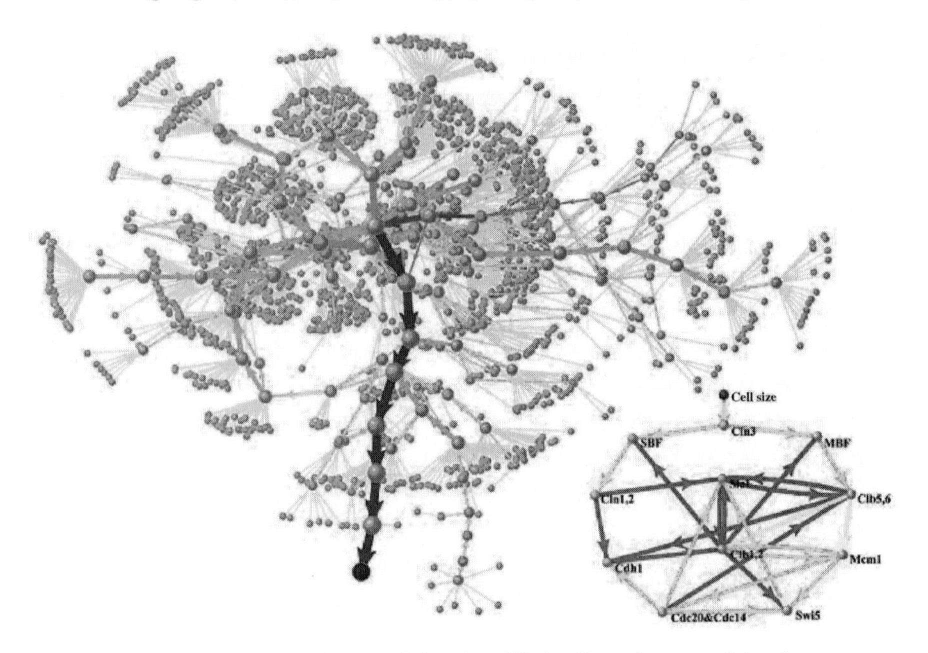

Fig. 10.2. The dynamic pathways of the simplified cell-cycle network in phase space. There are 2048 nodes, representing all $2^{11} = 2048$ possible initial states of the system (signal node cell size is not included in the calculation). An evolution from one state to the next is represented by an arrow. One pathway is defined as an initial state and a series of states that the evolution passes, as well as arrows between them. The real biological pathway and the ground G1 state are colored blue. Thickness of each arrow is in proportion to a logarithm of the number of pathways passing through it, and so is the diameter of each node. The inset of the figure (lower right) is the simplified cell-cycle network.

period, Clb1 or Clb2 is activated; the cell enters the mitotic, or M phase. Mitosis partitions the genome equally to the opposite ends of the cell, and the cell divides into two by budding out a daughter cell. Eventually the cell undergoes the cell cycle arrest period, the ground G1 state, and waits for another dividing signal. Thus, the cell cycle process starts with the excitation by the protein complex Cln3/Cdc28 from the ground G1 state to the excited G1 state, and evolves back to the ground G1 state through a series of well-defined phases.

10.3 Boolean Models and Simulation Results

We now study the dynamic behavior of the cell-cycle network. Because we are only concerned with the overall global dynamic properties of the network consisting of key nodes, we choose to use a simple discrete dynamic model. In the model, each node i has only two states, $S_i = 0$ and $S_i = 1$, representing, respectively, the active and inactive state of the protein. The protein states in the next time step are determined solely by the protein states in the present time step via the following rule:

$$S_i(t+1) = \begin{cases} 1, & \sum_j a_{ij} S_j(t) > 0 \\ 0, & \sum_j a_{ij} S_j(t) < 0 \\ S_i(t), & \sum_j a_{ij} S_j(t) = 0 \end{cases} \tag{10.1}$$

where a_{ij} is the interaction coefficient from protein j to protein i, which depends on the color of the arrow from j to i (see inset of Fig. 10.2). Note that the time steps here are logic steps that represent causality more than actual times. The network dynamics defined by (10.1) is a Boolean network. Boolean networks have been used to model a variety of biological networks and cell states [6, 15, 16]. Referring to general gene expression rules, we take to be 1 and (the minus infinity) for the positive interactions (green arrows) and the negative interactions (red arrows), respectively. The self-degradation (yellow arrows) are modeled as a time-delayed interaction: if a protein with a self yellow arrow is activated at time t ($S_i(t) = 1$) and it receives no further positive or negative inputs from $t+1$ to $t = t + t_d$, it will be degraded at $t = t + t_d$ ($S_i(t + t_d) = 0$). We use $t_d = 4$ in our model.[1] The presence of time-delayed interactions in Boolean networks changes their global dynamic properties significantly. The advantage of using such a discrete and simple dynamic model is that it enables us to analyze a much larger phase space and hence to study the global dynamic properties of the network.

Using the dynamic rules described above on the protein interaction network, we study the evolution of the cell-cycle network as a function of time. First, we set the initial state to the excited G1 state, and observe the system evolve to the ground G1 state. The temporal evolution of protein states is presented in Table 10.1, which follows the cell-cycle sequence—the protein states in each cell-cycle phase are consistent with experimental observations [17–20]. Next, we study the attractors of the network dynamics in phase space. Starting from all possible $2^{11} = 2048$ initial states in the 11-protein network, the system evolves into seven stationary states. No limit cycle is present in the system. We find that 1973 of 2048 initial states (96%) evolve to the ground G1 state—the biological stationary state of the cell—making it a super and the only global attractor. The rest of stationary states attract only less than 4% of the initial state, so that the cell cycle network has no attractors of medium size; only an unusually large global attractor exists. This design of the network ensures the stability of the cell against external and internal noise and perturbations.

[1] For simplicity, we use the same lifetime t_d for all proteins with a self-loop. The results are essentially the same for $t_d > 2$.

Table 10.1. Temporal evolution of protein states from the excited G1 state. The right column indicates the cell cycle phase. Note that the number of time steps in each phase does not reflect its actual duration. Also note that while the on/off of certain nodes sets the start or end of certain cell-cycle phases, for other nodes the precise duration at which they are turned on or off does not have an absolute meaning in this simple model.

Step	Cln3	SBF	MBF	Cln1,2	Cdh1	Swi5	Cdc20	Clb5,6	Sic1	Clb1,2	Mcm1	Phase
1	1	0	0	0	1	0	0	0	1	0	0	Excited G1
2	1	1	1	0	1	0	0	0	1	0	0	G1
3	1	1	1	1	1	0	0	0	1	0	0	G1
4	1	1	1	1	0	0	0	0	0	0	0	G1
5	0	1	1	1	0	0	0	1	0	0	0	S
6	0	1	1	1	0	0	0	1	0	1	1	G2
7	0	0	0	1	0	0	1	1	0	1	1	M
8	0	0	0	1	0	0	1	0	0	1	1	M
9	0	0	0	0	0	0	1	0	0	1	1	M
10	0	0	0	0	1	0	1	0	0	1	1	M
11	0	0	0	0	1	0	1	0	0	0	1	M
12	0	0	0	0	1	1	1	0	1	0	1	M
13	0	0	0	0	1	1	1	0	1	0	0	M
14	0	0	0	0	1	1	0	0	1	0	0	M
15	0	0	0	0	1	0	0	0	1	0	0	Ground G1

Another interesting result of our investigation arose when we asked the following question: How do the initial protein states evolve to their final attractors? Figure 10.2 shows in phase space the temporal evolution from all 2048 initial protein states. The biological path—the cell cycle sequence (Table 10.1)—is labeled in blue. The thickness of each arrow between two nodes is in proportion to a logarithm of the number of pathways passing through it, and so is the diameter of each node. We see that the most probable trajectory from any initial state is the biological path: the system is first attracted to the biological path and then follows the path to the final stationary state. This means not only that the ground G1 state is a globally stable fixed point, but also that the cell cycle sequence is a globally stable dynamic trajectory. This striking dynamic property of the yeast cell cycle network, which is dictated by the network topology that in turn has been shaped by many years of evolution, makes the cell cycle sequence an extremely robust process. We find similar results in cell cycle networks of fission yeast and frog egg, indicating a rather generic behavior.

We next analyse the dynamics of the entire network given by Fig. 10.1, which consists of the cell cycle and the life cycle processes. Again we simplify

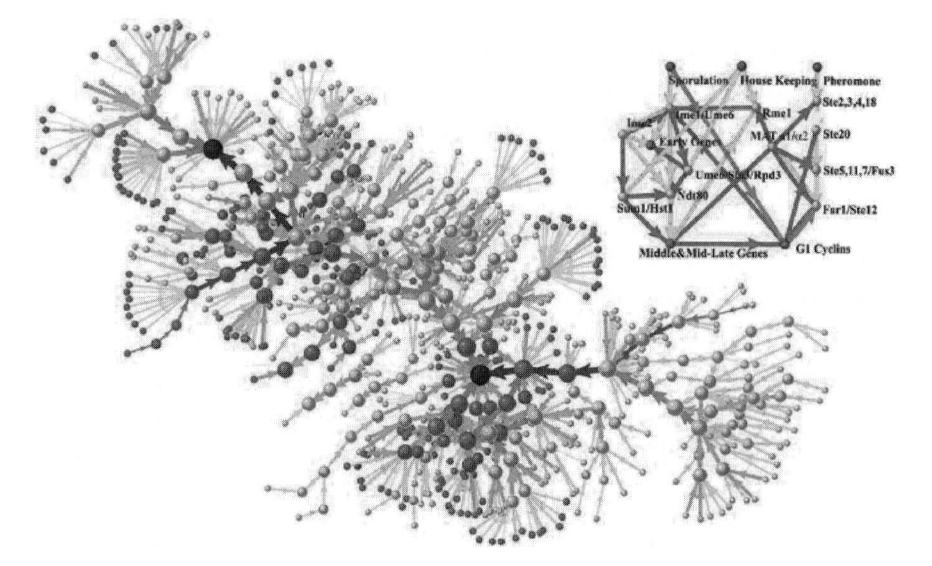

Fig. 10.3. The dynamic pathways of a simplified life-cycle network in phase space. Green and red nodes represent diploid states and haploid states, respectively. Two biological pathways and biological attractors are colored blue. Thick arrows are highlighted by red color. The thickness of arrows and diameter of nodes are applied with the same demonstrating rules as in Fig 10.2. [In principle, there should be $2^{14} = 16384$ nodes (two purple sporulation and pheromone signals are not included in the simulation, while the state of housekeeping is always 1), arrows with value less than 3 are omitted as are corresponding nodes to make the graph clear.] The inset of the figure (upper right) is the simplified life-cycle network.

the network by removing and reconnecting nodes in a way that does not affect the network dynamics. The simplified network is shown as the inset of Fig. 10.3. Note that according to the rules of simplification, a single node (G1 cyclins) can represent the cell-cycle network. To keep certain nodes in the normal cell state, we have introduced a housekeeping node in the simplified network; the state of this housekeeping node is always 1. The dynamic rules are the same as described above, except that the node Ime1 can only be activated with both an active Mat a1/alpha2 (diploid form) and an active sporulation signal (only a diploid cell receiving a sporulation signal can start sporulation). The other green arrows of "AND" logic have already been deleted, so there is no need to interpret it; the simplification rule in the supplement explains this. Starting from all 2^{14} initial states, we trace their evolution trajectories and calculate the number of attractors. There are seven stationary states and no limit cycle. The ground G1 states of haploid and diploid form are the only two global attractors, attracting, respectively, 34.7% and 58.7% of all initial states in this bi-stable biosystem; the other five stationary states only attract about 6.6% of initial states in phase space. The two biological pathways—sporulation and mating—are both attracting trajectories. We calculate all 2^{14} states and delete the arrows whose weight is less than 3 and the corresponding nodes. The evolution trajectories are shown in Fig. 10.3. The two biological paths are very pronounced.

The dynamics of bio-networks (cell cycle and life cycle) own two special features distinct from random networks: they possess super and global attractors and thick biological pathways. Both of the attractors and the pathways have well-defined biological meanings and functions. These features stand out when we compare the dynamic behavior of the bio-networks with that of random networks. Here, we defined random networks to have the same number of nodes and red, green, and yellow lines with the cell-cycle network. As shown in Fig. 10.4A, the attractor size distribution of random network follows a power law, so that they have attractors of all sizes. In comparison, in the cell-cycle network only a single global attractor exists; no medium size attractors are observed. Less than 0.1% of random networks possess this special property. To quantify the thick pathway property, a W value is defined for each pathway to measure the average thickness of the path. Explicitly, for the nth pathway, $W_n = \sum_{i=1}^{L_n-1}, Weight_{i,i+1}/L_n$, with $Weight_{i,j}$ being the number of pathways crossing the arrow from node i to node j, and L_n being the steps of the pathway. A large average W and narrow distribution is the characteristic of bio-networks as shown in Fig. 10.4B, while the W distribution for random networks is quite different, since they lack a main thick pathway corresponding to a narrow W distribution.

10.4 Perturbations of the Cell-Cycle Network

Finally, we study the dynamic stability of the cell-cycle network against perturbations. In the simplified cell-cycle network (see the inset of Fig. 10.2), there are 31 interactions, including 12 negative interactions, 14 positive interactions, and five self-degradations. We give the system a perturbation by deleting one interaction in the network, adding one positive (or negative) interaction, or switching a negative interaction to positive or vice versa, then calculating $\triangle S/S_0$ and $\triangle \overline{W}/\overline{W}_0$. Where $\triangle S/S_0$ and $\triangle \overline{W}/\overline{W}_0$ are, respectively, the change of the biggest attractor size and the change of the average W value after the perturbation, S_0 and $\overline{W}_0$ are, respectively, the size of the biggest attractor and the average W value of the original network. We also study the dynamic stability of 11-node randomly generated networks for comparison. Fig. 10.4C and Fig. 10.4D summarize the results of the perturbation study. We observe that most perturbations have little effect on the size of the biggest attractor and the attractive trajectory, so that the cell-cycle network is dynamically stable. The stability of the cell-cycle network is similar to that of random networks, except that the change of a few connections in the cell-

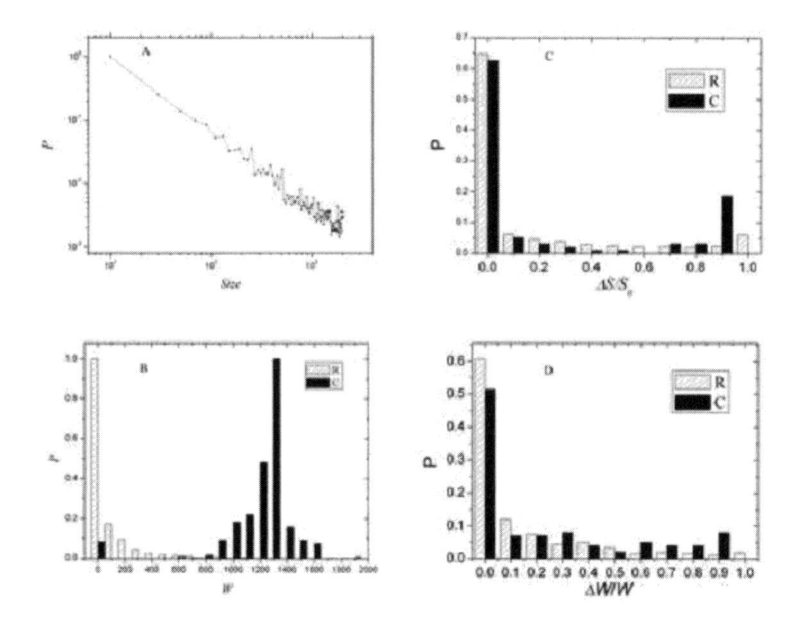

Fig. 10.4. Cell-cycle dynamics in comparison with random networks. (A) Attractor size distribution. The size of an attractor is defined as the number of states that can evolve into it. According to the simulation, the cell cycle is characterized by one ultra-big attractor and several small ones, in contrast with random nets whose attractor size distribution demonstrates the power law. (B) W value distribution. (R: random networks; C: cell-cycle network.) The W value is defined for each pathway to characterize the average thickness of it. The thick-pathway property of the cell cycle corresponds to a relatively narrow W distribution with a big average, while random nets have a much smaller average W value and a quite different distribution. (C) $\triangle S/S_0$ distribution. $\triangle S/S_0$ is the relative change of the attractive basin of the original biggest fixed state. (D) $\triangle\overline{W}/\overline{W}_0$ distribution. $\triangle\overline{W}/\overline{W}_0$ is the relative change of the average W value. The stability of a network is studied with $\triangle S/S_0$ and $\triangle\overline{W}/\overline{W}_0$. The perturbations consist of deleting, adding, or reversing one interaction line in the network. (Ten thousand random nets have been tested.)

cycle network causes a large change in S_0 and $\overline{W}_0$, which hints that these connections play the key roles in the biological functions in the network.

10.5 Conclusion

Our results suggest that besides their topological properties [1–4] biological networks have distinct and robust dynamic properties that are very different from random networks. These properties are intimately related to the underlying biological functions [21]. The study shows that simple and discrete dynamic models can play a complementary role in differential equations in network modeling [22] and can be very useful in analyzing large-scale global properties of the network.

Acknowledgments

This work was partly supported by the National Key Basic Research Project of China (No. 2003CB715900). We thank T. Hua, H. Li, C. M. Zhang, and L.H. Tang for their helpful discussions. The networks and dynamic trajectories are drawn with Pajek (http://vlado.fmf.uni-lj.si/pub/networks/pajek/).

References

1. Jeong H, Tombor B, Albert R, Oltval ZN, Barabasi A-L (2000) The large-scale organization of metabolic networks. Nature 407:651–654.
2. Jeong H, Mason SpP, Barabasi A-L, Oltvai ZN (2001) Centrality and lethality of protein networks. Nature 411:41.
3. Albert R, Jeong H, Barabasi A-L (2000) Attak and tolerance in complex networks. Nature 406:378.
4. Milo R, et al (2002) Network Motifs: Simple Building Blocks of Complex Networks. Science 298:824–827.
5. Tyson JJ, Chen K, Novak B (2001) Network Dynamics and Cell Physiology. Nature Rev Mol Biol 2:908.
6. Huang S, Ingber DE (2000) A Discrete Cell Cycle Checkpoint in Late G_1 That is Cytoskeleton-Dependent and MAP Kinase (Erk)-Independent. Exp Cell Res 261:91.
7. Uetz P, et al (2000) A comprehensive analysis of protein-protein interactions in Saccharomyces cerevisiae. Nature 403:623–627.
8. Schwikowski B, et al (2000) A network of protein-protein interactions in yeast. Nat Biotechnol 18:1257.
9. Ito T, et al (2001) A comprehensive two-hybrid analysis to explore the yeast protein interactome. Proc Natl Acad Sci USA 98:4569.
10. Tong AH, et al (2001) Systematic Genetic Analysis with Ordered Arrays of Yeast Deletion Mutants. Science 294:2364–2368.
11. Gavin AC, et al (2002) Functional organization of the yeast proteome by systematic analysis of protein complexes. Nature 415:141–147.
12. Ho Y, et al (2002) Systematic identification of protein complexes in Saccharomyces cerevisiae by mass spectrometry. Nature 415:180.
13. Mering C, et al (2002) Comparative assessment of large-scale data sets of protein-protein interactions. Nature 417:399–403.

14. Chen K, et al (2000) Kinetic Analysis of a Molecular Model of the Budding Yeast Cell Cycle. Mol Biol Cell 11:369–391.
15. Kauffman SA (1993) The Origins of Order. Oxford University Press, Oxford, p. 191.
16. Shmulevich I, Dougherty ER, Zhang W (2002) Steady-state analysis of genetic regulatory networks modelled by probabilistic Boolean networks. Proc IEEE 90:1778.
17. Lee TI, et al (2002) Transcriptional Regulatory Networks in Saccharomyces cerevisiae. Science 298: 799–804.
18. Spellman PT, et al (1998) Comprehensive Identification of Cell Cycle-regulated Genes of the Yeast Saccharomyces cerevisiae by Microarray Hybridization. Mol Biol Cell 9:3273–3297.
19. Cross FR, et al (2002) Testing a Mathematical Model of the Yeast Cell Cycle. Mol Biol Cell 13:52–70.
20. Simon I, et al (2001) Serial Regulation of Transcriptional Regulators in the Yeast Cell Cycle. Cell 106:697–708.
21. Alon U, Surette MG, Barkai N, Leibler S (1999) Robustness in bacterial chemotaxis. Nature 397:168.
22. Endy D, Brent R (2001) Modelling cellular behaviour. Nature 409:391.

11

Understanding Protein-Protein Interactions: From Domain Level to Motif Level

Huan Yu, Minping Qian, and Minghua Deng

Summary. Understanding protein functional interactions is an important research focus in the post-genomic era. The identification of interacting motif pairs is essential for exploring the mechanism of protein interactions. We describe a word-counting approach for discovering motif pairs from known interactions and pairs of proteins that are putatively not to interact. Our interacting motif pairs are validated by multiple-chain PDB structures and motif pairs extracted from PDB structures. The motif pairs are used to predict interactions between proteins by three different methods. For all the methods used, our predicted protein-protein interactions significantly overlap with the experimental physical interactions. Furthermore, the mean correlation coefficients of the gene expression profiles for our predicted protein pairs are significantly higher than that for random pairs. Supplementary materials are available online at http://ctb.pku.edu.cn/˜yuhuan.

11.1 Introduction

With the finishing of genome sequencing projects, functional annotation of the entire genomes has been a major goal in the post-genomic era. Moreover, the relationships between proteins play a more important role than the function of individual proteins since proteins interact and regulate each other in most biological process. Many experimental and computational methods are developed to identify protein-protein interactions (PPIs).

Traditional Methods

Traditionally, protein-protein interactions have been studied by genetic, biochemical, or biophysical techniques such as protein-protein affinity chromatography, immunoprecipitation, sedimentation, and gel filtration [42]. However, these methods are not suitable for genome-wide interaction detection. High-throughput methods have being developed to identify protein-protein interactions in parallel.

The Yeast Two-Hybrid System

The most widely used large-scale approach for detecting protein interactions is the yeast two-hybrid system introduced by Fields and Song [12]. A eukaryotic transcription factor is split into two distinct functional domains called binding domain (BD) and activating domain (AD). If the bait protein fused to the BD interacts with the prey protein fused to the AD, the transcription machinery will be reconstructed and activate the transcription of the reporter gene. If the two proteins do not interact, the transcription of the reporter gene will not be activated.

Recently, the yeast two-hybrid system was used for high-throughput protein-protein interaction detection from several organisms including *Saccharomyces cerevisiae (yeast)* [26, 49], *Caenorhabditis elegans* [4, 29, 50], *Helicobacter pylori* [43], vaccinia virus [35], mouse [48], and fly [13, 16, 47].

Uetz et al [49] used two different approaches in their experiments. In the first approach, 192 yeast bait proteins fused to the Gal4-DNA-binding domain were screened against nearly 6000 yeast proteins fused to the Gal4-activation domain as prey; 281 interactions were found among 87 proteins. In the second experiment, an interaction sequence tag (IST) approach, 5341 yeast bait proteins were screened against the yeast proteome (nearly 6000 preys) and identified 692 interactions among 817 proteins. Ito et al [26] used IST approach and 4549 interactions among 3278 proteins were reported.

Although the yeast two-hybrid system is suitable for automated large-scale protein-protein interaction screening, the false-positive and false-negative errors are huge. Possible mutations during polymerase chain reaction (PCR) amplification and stochastic activation of reporter gene may lead to the false positives and the inability to detect interactions of membrane proteins or posttranslational modification proteins, and weak interactions may lead to the false negatives [8, 26, 38].

Affinity Purification of Protein Complexes

An alternative way to study protein-protein interactions on a large scale is affinity purification of protein complexes. Tagged bait proteins and associated proteins assemble as protein complexes. The complexes are purified and then separated by gel electrophoresis. The components of complexes are identified by mass spectrometry (MS). MS-based approaches provide protein components in a complex but do not give the direct physical interaction. It is different from the yeast hybrid system, although a relationship does exist since proteins in the same complex are more likely to physically interact with one another. Over 700 different complexes in yeast are identified in two studies [14, 24]. So far the MS-based methods provide the largest protein complex data.

The disadvantage of MS-based approaches is that the interactions may not be detected due to a low-affinity, low-expression level or detection limits of the MS facilities. Most interactions from MS-based methods are by indirect interaction.

Protein Chips

Another direct approach to identify genome-wide protein-protein interactions in parallel is the protein chip method. The purified proteins are printed onto glass slides and then analyzed for binding or enzymatic activities.

MacBeath and Schreiber [32] used protein chips to detect antibody-antigen interactions, protein kinase activities, and protein interactions with small molecules. Zhu et al [55] developed protein chips to identify protein-protein interactions. They identified 39 calmodulin-interactors by adding biotinylated calmodulin to the yeast protein chips. Some of the identified interactions are well-known interactions between calmodulin and calmodulin kinase, but were missed in large-scale yeast two-hybrid and MS-based approaches.

Protein chips can be used to identify not only protein-protein interaction but also protein-lipid, protein-nucleic acid, and protein-drug interactions. The major limitation is the preparation of proteins to analyse, because not all proteins can be purified easily and efficiently.

Protein-Protein Interaction Databases

Since most of the protein interactions were reported in the literature where the information is difficult to manage and compute upon, various protein-protein interaction databases were built to gather the information for further data mining such as MIPS [36] (http://mips.gsf.de), BIND [2] (http://www.bind.ca), DIP [44] (http://dip.doe-mbi.ucla.edu), GRID [5] (http://biodata.mshri.on.ca/grid/servlet/Index), MINT [54] (http://cbm.bio.uniroma2.it/mint), IntAct [23] (http://www.ebi.ac.uk/intact/index.jsp) and STRING [37] (http://string.embl.de). Among these databases, we consider MIPS as the gold standard data since it is a collection of manually curated high-quality PPI data collected from the scientific literature by expert curators.

Furthermore, the PDB database [9] (http://www.rcsb.org/pdb) contains some protein complex structures that are direct evidence for protein-protein interactions. The protein complex structures allows us to determine the protein contact interface by simply calculating the Euclidean distance of atoms. The only problem is that the data size of PDB is too small.

Computational Method for Predicting Protein-Protein Interactions

Although many experimental methods have been developed to identify genome-wide interactions, the overlaps among different data sets are small, suggesting that they contain only a very small fraction of the potential protein-protein interactions [26]. Analysis by computational (*in silico*) methods is required to complete the protein interaction map.

With the completion of genome-sequencing projects, sequence homology search tools such as BLAST [1] (http://www.ncbi.nlm.nih.gov/blast) have been widely used to extend our knowledge of protein functions. However, sequence similarity analysis is applicable only when the function of matched proteins is known and the sequence homology is significant. Several computational methods for predicting genome-wide functional linkages between proteins are developed. These methods can be classified into three categories:

1. Phylogenetic profiles [25, 41]: analysis of the co-presence or co-absence of genes within entire genomes indicates the interactions between those gene products
2. Gene fusion [10, 33]: two separate proteins in one organism are often found as a fusion into a single protein in some other species, which indicates physical interactions
3. Gene neighbourhood [6, 40]: two proteins that are neighbors on the chromosome in several genomes tend to be functionally linked.

To understand protein function, elements of protein that are self-stabilizing and often fold independently of the rest of the protein chain are defined as protein domains. Protein domains have distinct evolutionary origin and function. As a unit of structures, each domain is responsible for a specific interaction with another domain, so it serves as a unit for protein-protein interactions. Several machine-learning approaches have been developed to predict protein interactions on the basis of domain C domain interactions, including support vector machine (SVM) [3], the association method [28, 39, 46], the interacting domain profile [52], the Bayesian method [17, 18, 19], the EM-based MLE method [7, 53], and the linear programming method [22].

Protein-protein interactions are also studied at the motif level. Motif is a short amino acid sequence with a certain functional meaning. Motifs are very important for drug and protein design, often referred to as protein contact interfaces or binding sites. Wang et al [51] identified active motif pairs from interacting protein pairs by an approximate EM algorithm with the motifs from the Prosite database [11]. Li et al [30, 31] discovered the significant motif pairs based on the PDB structures [9] and known protein-protein interactions.

We use a word-counting method to study the protein-protein interactions in the motif level. Starting from protein sequences and protein complexes only, the statistical significant motif pairs are identified as the signature of the protein-protein interactions. By comparing them with PDB structure, we validate our identifications. Furthermore, we predict genome-wide protein-protein interactions from the identified motif pairs, and validate our prediction based on the analysis of overlapping with the MIPS physical interaction and gene expression profiles.

11.2 Data

In our study, yeast sequence data and protein complex data are used to find the signature motif pairs. We use ORF names to identify yeast proteins, resulting in 6294 proteins involved in our experiment; 265 MIPS complexes are downloaded from MIPS to build the protein-protein interaction training data.

To estimate the reliability of our results, protein structures obtained from PDB, MIPS physical interactions, and gene expression data from Spellman et al [45] are used. The protein structures used in this study are a non-redundant subset from PDB where the maximum pairwise sequence identify is 30%, consisting of at least two chains in one entry with resolution 2.0 RA or better. The PDB data contains the three-dimension coordinates of amino acid atoms. The data were obtained on April 8, 2004, containing 544 entries. For MIPS physical interaction data, we exclude those interactions supported only by a high-throughput experiment, resulting in 2579 interactions used in this study. The expression data are obtained from the yeast cell cycle analysis project, containing 6178 genes at 77 time points.

11.3 Methods

11.3.1 Overview of Our Method

We transform the MIPS complexes into pairwise interactions as the positive protein-protein interaction training data. We also generate a negative interacting data set described in the following paragraph. We count all 4-tuple amino acid (AA) pairs occurring in positive and negative data and select statistically significant positive and negative 4-tuple pairs as the word dictionary for further prediction. The identified 4-tuple positive pairs are verified by the tuple pairs on protein complex structures, and are compared with motif pairs discovered by Li et al [30, 31]. Finally, we use an iterative-dependent naive

Bayesian classifier to predict protein-protein interactions based on the identified 4-tuple pairs. The reliability is estimated by comparing the predicted interactions with MIPS physical interactions and gene expression profile correlation coefficients.

11.3.2 Protein-Protein Interaction Training Data

We need both positive and negative protein-protein interaction training data to learn the significant 4-tuple pairs. The positive interactions are extracted from MIPS complexes. We transform the MIPS complexes into pairwise interactions by connecting all possible protein pairs within the complex, resulting in $NP = 10198$ protein pairs.

Several methods have been proposed to construct a negative interaction data set. The two most popular approaches are as follows: In the first approach, proteins in separate subcellular compartments are considered as negatives [34]. The shortcoming of this approach is that there is no information showing whether proteins can bind with each other if they are being put together. Different subcellular locations only indicate that the proteins have no chance to bind. The second approach assumes that two proteins do not interact if there is no positive evidence of interaction [19]. The incomplete and low-quality positive interaction data bring in more false negatives. We use a different approach to generate negative data from positive data. To overcome the shortcoming of the above two approaches, we define a negative PPI pair to be a pair of proteins that have a chance to interact with each other but fail to be observed in the interaction map. To be precise, we first build a protein interaction graph with proteins as nodes and positive interactions as edges. Then all pairs among nodes in a connected graph with the shortest path length greater than a cutoff are treated as putative negative protein-protein interactions. For example, when the cutoff $= 6$, we obtain $NN = 26605$ negative protein-protein interactions.

11.3.3 Significant 4-Tuple Pairs

There are 20^8 possible 4-tuple pairs in the protein pairs. We count all the 4-tuple pairs in the positive and negative protein-protein interaction data. If a 4-tuple pair occurs multiple times in a protein pair, we just count it once. For each 4-tuple pair TP, we denote N_{TP+} and N_{TP-} as the number of protein pairs containing the 4-tuple pair TP in positive and negative protein pairs, respectively.

We select the significant 4-tuple pairs as follows. For a 4-tuple pair TP and the counts N_{TP+} and N_{TP-}, we test whether the 4-tuple pair is over-represented in the positive interacting protein pairs. The negative count N_{TP-}

is used to estimate the background distribution. The approximate background distribution is a Poisson distribution with parameter $\lambda_p = N_{TP-} \times \frac{NP}{NN}$. The p value for the hypothesis can be calculated as

$$P_{raw}(X \geq N_{TP+}) = \sum_{k=N_{TP+}}^{\infty} \frac{\lambda_p^k}{k!} e^{-\lambda_p}.$$

For all $N = 20^8$ 4-tuple pairs, a large number of statistical tests are performed. To avoid the multiple testing problem, the p value p_{raw} is adjusted to $max(1, p_{raw} \times N)$ by the Bonferroni correction. The 4-tuple pairs with adjusted p values lower than .001 are selected as the significant positive 4-tuple pair set S_p resulting in 968,882 4-tuple pairs. Let S_p be the set of these significant positive 4-tuple pairs.

Similarly, the p value for the hypothesis that the 4-tuple pair is underrepresented in the putative negative interacting pairs can be calculated as

$$P_{raw}(X \geq N_{TP-}) = \sum_{k=N_{TP-}}^{\infty} \frac{\lambda_n^k}{k!} e^{-\lambda_n},$$

in which $\lambda_n = N_{p+} \times \frac{NN}{NP}$. Similarly, we select 2,529,775 4-tuple pairs as the significant negative 4-tuple pairs and denote the set of them as S_n.

We call $W = S_p \bigcup S_n$ the significant 4-tuple pair set.

11.3.4 Protein-Protein Interaction Prediction

In the following, we predict protein-protein interactions from the identified significant 4-tuple pair set W. We propose three different prediction models step by step.

Independent Model

We start the prediction from a simple independent classification system. Given a pair of proteins (P_a, P_b), we denote all significant 4-tuple pairs that occur in (P_a, P_b) as $S = \{(LT_i, RT_i, Lpos_i, Rpos_i), i = 1, 2, \cdots, M\}$, where LT_i is the 4-tuple that occurs in P_a and RT_i occurs in P_b, $(Lpos_i, Rpos_i)$ is the occurrence position of (LT_i, RT_i), respectively, and M is the total number of significant 4-tuples that occur in (P_a, P_b). We denote I_{ab} the event that (P_a, P_b) interacts. For 4-tuple pairs TP_A and TP_B in S, we make the following conditional independence assumption:

$$P(TP_A, TP_B|I_{ab}) = P(TP_A|I_{ab}) \times P(TP_B|I_{ab}). \tag{11.1}$$

Following the Bayesian rule, we have

$$\begin{aligned}
P(I_{ab}|S) &= \frac{P(I_{ab}, S)}{P(S)} \\
&= \frac{P(S|I_{ab}) \times P(I_{ab})}{P(S|I_{ab}) \times P(I_{ab}) + P(S|\overline{I_{ab}}) \times P(\overline{I_{ab}})} \\
&= \frac{1}{1 + \frac{P(\overline{I_{ab}})}{P(I_{ab})} \times \frac{P(S|\overline{I_{ab}})}{P(S|I_{ab})}} \\
&= \frac{1}{1 + \frac{P(\overline{I_{ab}})}{P(I_{ab})} \times \prod_{TP \in S} \frac{P(TP|\overline{I_{ab}})}{P(TP|I_{ab})}} \\
&= \frac{1}{1 + \frac{1 - P(I_{ab})}{P(I_{ab})} \times factor(S)},
\end{aligned} \tag{11.2}$$

where $factor(S) = \prod_{TP \in S} \frac{P(TP|\overline{I_{ab}})}{P(TP|I_{ab})}$.

We estimate $factor(S)$ as

$$\widehat{factor}(S) = \prod_{TP \in S} \frac{N_{TP-}/NN}{N_{TP+}/NP}. \tag{11.3}$$

We use an iterative algorithm to estimate the prior probability $P(I_{ab})$. It was estimated in Hazbun and Fields [21] that each protein interacts with about five to 50 proteins. For 6294 yeast proteins and five interactions for each protein, it gives a total of $(6294 \times 5)/2 = 15,735$ real interaction pairs. Given the initial prior probability

$$\widehat{P}_k(I_{ab}) = \frac{2 * 15735}{6293 \times 6294},$$

for step $k = 0$, we estimate $P(I_{ab}|S)$ as

$$\widehat{P}(I_{ab}|S) = \frac{1}{1 + \frac{1 - \widehat{P}_k(I_{ab})}{\widehat{P}_k(I_{ab})} \times \prod_{TP \in S} \frac{N_{TP-}/NN}{N_{TP+}/NP}}.$$

We give a positive interaction prediction of (P_a, P_b) if $P(I_{ab}|S) > 0.999$. We predict all yeast protein pairs and denote the number of positive predictions as K_k. In the next step prior probability is updated by

$$\widehat{P}_{k+1}(I_{ab}) = \frac{2 * K_k}{6293 \times 6294}.$$

The algorithm is iterated until $K_{k+1} = K_k$.

Since we obtain the 4-tuple pairs set S using sliding windows, the assumption of independence (11.1) is not true if TP_A and TP_B overlap. The converged prior probability in the independent algorithm described above is significantly higher than the upper bound of empirical probability 50/6294.

To overcome the shortcoming of the independent model, we present two dependent models to estimate $factor(S)$ in (11.2).

Dominating Model

In contrast to the independent model, the dominating model assumes that only the most significant 4-tuple pairs in S contribute to $factor(S)$. The most significant 4-tuple pairs dominate the interaction. We denote the most significant positive and negative 4-tuple pair in S as $PTP(S)$ and $NTP(S)$. In the dominating model, we modify $factor(S)$ to

$$factor_{dom}(S) = \begin{cases} \frac{P(NTP(S)|\overline{I_{ab}})}{P(NTP(S)|I_{ab})}, & \text{no positive 4-tuple pairs belong to } S \\ \frac{P(PTP(S)|\overline{I_{ab}})}{P(PTP(S)|I_{ab})}, & \text{no negative 4-tuple pairs belong to } S \\ \frac{P(PTP(S)|\overline{I_{ab}})}{P(PTP(S)|I_{ab})} \times \frac{P(NTP(S)|\overline{I_{ab}})}{P(NTP(S)|I_{ab})} \end{cases} \tag{11.4}$$

$factor_{dom}$ is estimated by

$$\widehat{factor}_{dom}(S) = \begin{cases} \frac{N_{NTP(S)-}/NN}{N_{NTP(S)+}/NP}, & \text{no positive 4-tuple pairs belong to } S \\ \frac{N_{PTP(S)-}/NN}{N_{PTP(S)+}/NP}, & \text{no negative 4-tuple pairs belong to } S \\ \frac{N_{NTP(S)-}/NN}{N_{NTP(S)+}/NP} \times \frac{N_{PTP(S)-}/NN}{N_{PTP(S)+}/NP} \end{cases} \tag{11.5}$$

The dominating model runs from one extreme to the other. A majority of information in S is ignored. The prediction is unstable because of the inaccuracy of the selected significant 4-tuple pairs.

Combined Model

The final model is a combination of the independent and the dominating model. The 4-tuple pairs set S is partitioned into disjoint independent sets

$S_1, S_2, ..., S_m$ as following: S is used as nodes to construct a graph, and two 4-tuple pairs $(LT_i, RT_i, Lpos_i, Rpos_i)$ and $(LT_j, RT_j, Lpos_j, Rpos_j)$ are assigned an edge if $|Lpos_i - Lpos_j| \leq 4 \times M$ or $|Rpos_i - Rpos_j| \leq 4 \times M$ ($M = 3$ in this study). Figure 11.1 shows a simple example of the partition. In Figure 11.1, each connected component is clustered as a subset S_i.

The *factor* in the combined model is calculated as

$$factor_{com}(S) = \prod_{i=1}^{m} factor_{dom}(S_i),$$

where $factor_{dom}(S_i)$ is calculated by (11.4) for each individual set S_i. $factor_{com}(S)$ is estimated similarly as (11.5):

$$\widehat{factor}_{com}(S) = \prod_{i=1}^{m} \widehat{factor}_{dom}(S_i).$$

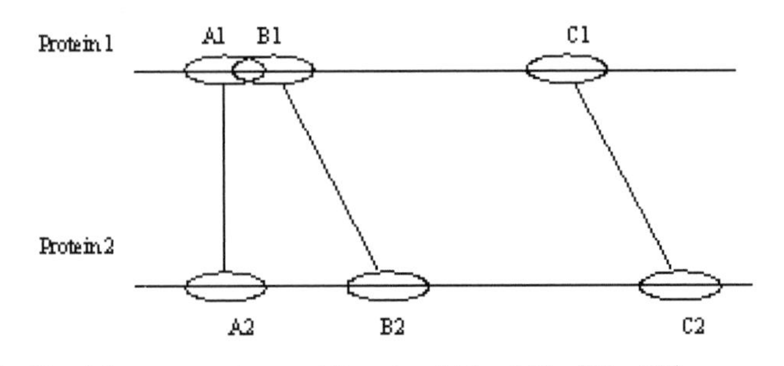

Fig. 11.1. Partition example: motif pairs (A1, A2), (B1, B2) on protein 1 and protein 2 are grouped. Since A1 and B1 are neighbourhood, the isolated motif pair (C1, C2) builds up a group by itself.

11.4 Results

First we verify the overrepresented 4-tuple pair set S_p by comparing it with PDB structure data. The significance of matching is evaluated by random re-sampling. We then compare S_p with the significant motif pairs discovered by Li et al [30]. Finally, we measure the significance of our predicted protein-protein interactions by MIPS physical interactions and gene expression profiles.

Comparing with the PDB Structure

First we give some definitions related to protein structures:

Definition 1. *Atoms contact: Given two protein chains with three-dimensional structural coordinates, $Atom_A$ from one protein chain and $Atom_B$ from the other contact if $dist(Atom_A, Atom_B) < \varepsilon$, where $dist(\cdot,\cdot)$ is the Euclidean distance, and ε is an empirical threshold 5ℝA.*

Definition 2. *Tuple pairs contact: Given two protein chains with three-dimensional structural coordinates and a tuple pair $(Tuple_A, Tuple_B)$, $Tuple_A$ occurs in one protein chain and $Tuple_B$ occurs in the other. $(Tuple_A, Tuple_B)$ contact if one atom from $Tuple_A$ and one atom from $Tuple_B$ contact.*

Definition 3. *Chains verifiable by a set of tuple pairs S: Given two protein chains with three-dimensional structural coordinates, if there exist a tuple pair in S that contact each other.*

Definition 4. *PDB structure verifiable by a set of tuple pairs S: Given a PDB protein complex with three-dimensional structural coordinates, if at least one pair of chains is verifiable by S.*

For the identified overrepresented positive 4-tuple pairs in S_p, 169 of 544 PDB structures are verifiable. Figure 11.2 is an example of verifiable 4-tuple pair. However, when we repeatedly draw the same number (968,882) of 4-tuple pairs randomly as set $R_1, R_2, \cdots, R_{1000}$, and compare each set with PDB structures, the average number of verifiable PDB structures is only 20.9. We use binomial distribution $B(544, p_{binom})$ to approximate the background distribution, in which $p_{binom} = 20.5/544$. The p value of the verifiable performance of set S_p is

$$P(X \geq 169) = \sum_{k=169}^{544} \binom{544}{k} \times p_{binom} \times (1 - p_{binom}) = 1.47e - 144$$

which means a very significant match.

Comparing with Motif Pairs Extracted from PDB

Li et al [30, 31] discovered 930 motif pairs from PDB structures. Our significant tuple pairs is derived from sequence data without any knowledge of protein structures or protein domains. Comparing our identification with their discovery, 496 of 930 motif pairs are confirmed by 4-tuple pairs in S_p and the average match number of the random resampling set is only 22.8. Using the similar background binomial distribution approximation $B(930, p_{binom*})$, where $p_{binom*} = 22.8/930$, the p value of the match performance of set S_p is

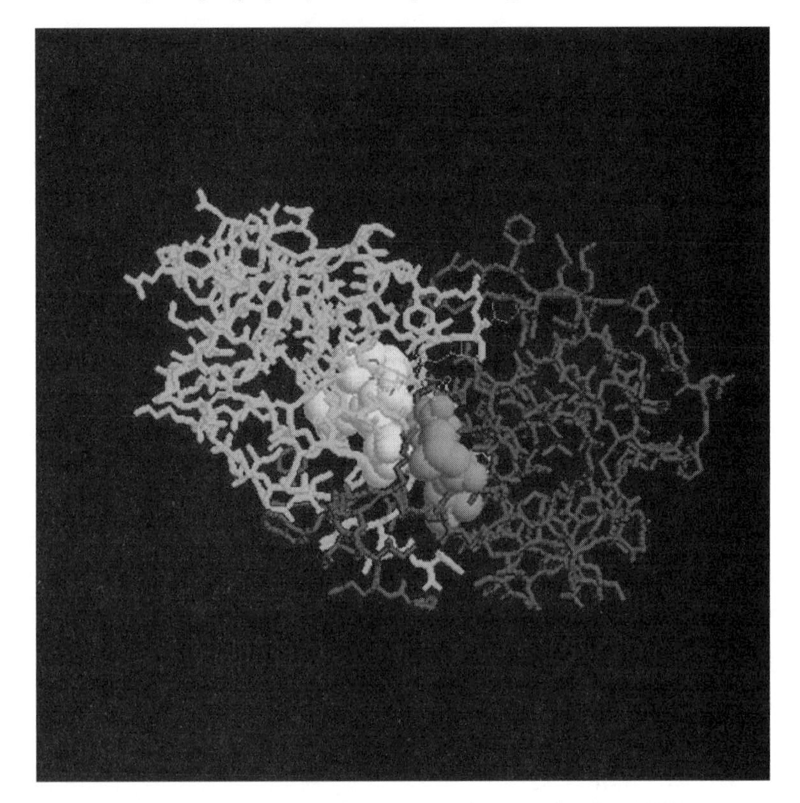

Fig. 11.2. PDB verifying example: PDB1a8g (two molecular: HIV-1 protease) is verified by 4-tuple pair (LLDT LLDT).

$$P(X \geq 496) = \sum_{k=496}^{930} \binom{930}{k} \times p_{binom*} \times (1 - p_{binom*}) < 1.00e - 460$$

which is also a very significant match.

Comparing with MIPS Physical Data

We use MIPS physical interactions to test our protein-protein interaction predictions. There are 2579 MIPS physical interactions. Excluding those interactions in the training set, a test data set of 1964 interactions is obtained. To measure the statistical significance, we use the standard Z score

$$Z = \frac{k_0 - n \times p}{\sqrt{n \times p \times (1 - p)}}$$

where $p = K/L$ (L is the total number of protein pairs), $K = 1964$, n is the number of predictions, and k_0 is the number of matching protein pairs between the 1964 interactions in the test data set and the n predicted interactions. Z has an approximate standard normal distribution under the null hypothesis.

Another measure of the significance is $Fold$ [7], the ratio of the fraction of the predicted protein pairs in the test data set with those in all protein pairs as follows:

$$Fold = \frac{k_0/K}{n/L} .$$

Table 11.1 shows the overlapping numbers between the test set and our predictions and the corresponding statistics. We can see that, although the overlap is not quiet large, it's statistically significant.

Table 11.1. Number of matched protein pairs between the predictions

Model	Prediction	Train	MIPS	MIPS1	$Fold$	Z score	p value
Independent model	4496949	10186	1328	718	1.61	14.72	2.3e-49
Dominating model	38302	10194	618	7	2.51	2.53	5.7e-3
Combined model	68927	10194	630	19	3.26	5.47	2.3e-8

Number of predicted PPIs (Prediction), number of matching PPIs with the training set (Train), number of matching PPIs with the MIPS data (MIPS), and number of matching PPIs with the MIPS excluding the training data (MIPS1), respectively. The corresponding statistics (Fold, Z score, and p value) are also given.

Comparing with Gene Expression Profiles

It has been shown that genes with similar expression profiles are likely to encode interacting proteins [15, 20, 27]. We study the distribution of correlation coefficients for protein pairs with predicted interaction probability. We use the genome-wide cell cycle gene expression data. Figure 11.3 gives the distributions of the pairwise correlation coefficients for all gene pairs, our predicted protein pairs by three models, and the MIPS interaction data.

The statistical significance for the difference between the mean expression correlation coefficient of a putative interaction set and that of random pairs is measured by the T score and the p value for the null hypothesis of no difference between the sample mean and the mean of random gene pairs. The T scores are calculated as the two sample T-test statistic with unequal variances:

$$T_f = \frac{\mu_1 - \mu_2}{\sqrt{\frac{S_1^2}{n_1} + \frac{S_2^2}{n_2}}}$$

with the approximate freedom:

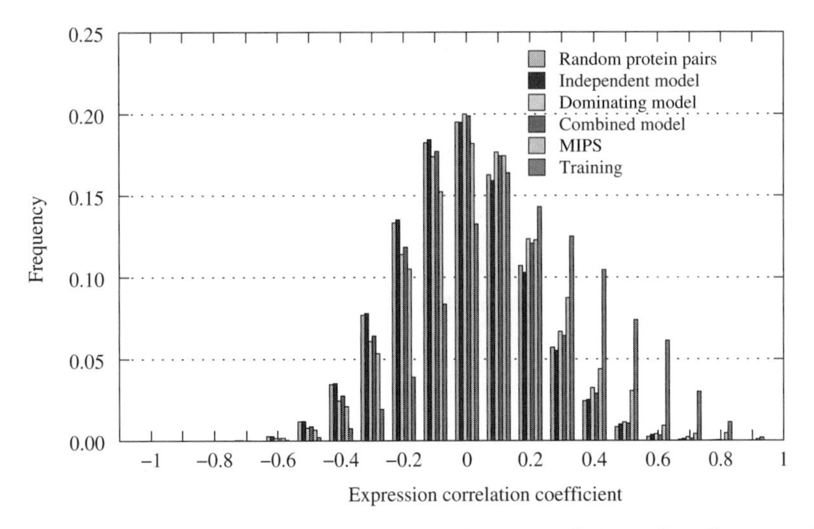

Fig. 11.3. Distributions of the pairwise correlation coefficients for all gene pairs, MIPS interaction data, training data, and our predicted protein pairs excluding the training data.

$$f = \frac{\left(\frac{S_1^2}{n_1} + \frac{S_2^2}{n_2}\right)^2}{\frac{\left(\frac{S_1^2}{n_1}\right)^2}{n_1} + \frac{\left(\frac{S_2^2}{n_2}\right)^2}{n_2}}$$

where μ is the mean of samples, and

$$S^2 = \frac{1}{n-1}\sum_{i=1}^{n}(x_i - \mu)^2$$

is the variance of the samples [7].

Figure 11.3 and Table 11.2 show that the mean correlation coefficient for predicted protein pairs is significantly higher than that for random pairs. Except for the independent model, the mean correlation coefficients for protein pairs are significantly higher than that for random pairs. It should be noted that transforming from the MIPS complexes into pairwise interactions enlarged the mean correlation coefficient of training data. For our predictions, the overlap between training data and our predictions are not used in the calculation of correlation coefficient.

Table 11.2. Distribution of the pairwise correlation coefficients

Pairs	Pairs	Sample mean	Sample variance	T score	p value
All ORFs	18474081	0.031	0.041	0	.5
Independent model	4271174	0.030	0.042	-7.37	1
Dominating model	25067	0.062	0.041	23.93	2.1e-125
Combined model	52079	0.053	0.040	25.33	1.5e-149
MIPS	2415	0.100	0.051	14.92	1.4e-48
Training data	8723	0.253	0.063	82.70	0

Summary statistics of distribution of the correlation coefficient between the gene expression profiles of protein pairs (with gene expression profiles from MIPS), random (all ORFs), training data, and our predictions of three models excluding the training data).

11.5 Discussion

This work describes a general method for predicting protein-protein interactions from sequences. Putative negative protein-protein interactions generated by positive interactions enable us to choose the significant signature pairs by word-counting. The signature pairs are compared with multiple-chain PDB structures and motif pairs extracted from PDB structures. By resampling, it shows that the quality of chosen pairs is significantly better than that of the random 4-tuple pairs, thus validating the theory that protein folding is a consequence of the primary structure, and the functional motifs are highly conserved in the evolution.

At the protein level, protein-protein interactions are separated into attractive and repulsive forces associated with the signatures. We introduce three ways to combine the signatures probabilities into a single protein-protein interaction probability. The significance of our PPI prediction is measured in two ways: (1) comparing the prediction with MIPS physical interactions derived by methods other than the yeast two-hybrid system, and (2) comparing the distribution of gene expression correlation coefficient for the predicted interacting protein pairs with that for random protein pairs. The results show our predictions are statistically significant.

The results also show that using the dependent models to predict protein-protein interactions from the overrepresented and underrepresented 4-tuple pairs indeed improves performance. Generally, the independent assumption is introduced to simplify the computing; it may work if the dependence among features is not too strong. However, because of the existence of long-range correlation between the amino acids after protein folding, the independent model does not fit the current case. In our combined model, the long-range correlation has been taken into account by considering the overlapping between the signatures on the protein pair, so that the predictions by the combined model have a better performance than those by the other two models.

Although our predictions are statistically significant, the overlaps between our predictions and MIPS physical interactions are very small. A possible reason is that the size of the protein interactions network is huge. It is known that every experimental method is biased to certain kinds of proteins and interactions. It is possible that some of our predictions are not suitable to be verified by MIPS physical interaction identifying methods. Another explanation for the small overlaps is that the way to transform the MIPS complexes into pairwise interaction may amplify the data source noise. Moreover, the time and space constraints among protein-protein interactions are ignored in our study. Two proteins may not interact with each other at different expression time or different subcellular location. Also, depending on the environmental conditions such as temperature and others, proteins may have different interaction behaviour with others.

11.6 Acknowledgments

This research is supported by the grants from the National Natural Science Foundation of China (90208022, 10271008, 10329102), the National High Technology Research and Development of China (2002AA234011), and the National Key Basic Research Project of China (2003CB715903).

References

1. Altschul SF, Madden TL, Schaffer AA et al (1997) Gapped BLAST and PSI-BLAST: a new generation of protein database search programs. Nucleic Acids Res 25 (17):3389–3402.
2. Alfarano C, Andrade CE, Anthony K et al (2005) The Biomolecular Interaction Network Database and related tools 2005 update. Nucleic Acids Res 33 (database issue):D418–D424.
3. Bock JR, Gough DA (2001) Predicting protein-protein interactions from primary structure. Bioinformatics 17 (5):455–460.
4. Boulton SJ, Gartner A, Reboul J et al (2002) Combined Functional Genomic Maps of the C. elegans DNA Damage Response. Science 295 (5552):127–131.
5. Breitkreutz BJ, Stark C, Tyers M (2003) The GRID: the General Repository for Interaction Datasets. Genome Biol 4 (3):R23.
6. Dandekar T, Snel B, Huynen M, Bork P (1998) Conservation of gene order: a fingerprint of proteins that physically interact. Trends Biochem Sci 23 (9):324–328.
7. Deng M, Mehta S, Sun F, Chen T (2002) Inferring Domain-Domain Interactions From Protein-Protein Interactions. Genome Res 12 (10):1540–1548.

8. Deng M, Sun F, Chen T (2003) Assessment of the reliability of protein-protein interactions and protein function prediction. Pac Symp Biocomput 2003:140–151.
9. Deshpande N, Addess KJ, Bluhm WF et al (2005) The RCSB Protein Data Bank: a redesigned query system and relational database based on the mmCIF schema. Nucleic Acids Res 33 (database issue):D233–D237.
10. Enright AJ, Iliopoulos I, Kyrpides NC, Ouzounis CA (1999) Protein interaction maps for complete genomes based on gene fusion events. Nature 402 (6757):86–90.
11. Falquet L, Pagni M, Bucher P et al (2002) The PROSITE database, its status in 2002. Nucleic Acids Res 30 (1):235–238.
12. Fields S, Song O (1989) A novel genetic system to detect protein-protein interactions.. Nature 340 (6230):245–246.
13. Formstecher E, Aresta S, Collura V et al (2005) Protein interaction mapping: A Drosophila case study. Genome Res 15 (3):376–384.
14. Gavin AC, Bösche M, Krause R et al (2002) Functional organization of the yeast proteome by systematic analysis of protein complexes. Nature 415 (6868):141–147.
15. Ge H, Liu Z, Church GM, Vidal M (2001) Correlation between transcriptome and interactome mapping data from Saccharomyces cerevisiae. Nat Genet 29 (4):482–486.
16. Giot L, Bader JS, Brouwer C et al (2003) A Protein Interaction Map of Drosophila melanogaster. Science 302 (5651):1727–1736.
17. Gomez SM, Lo SH, Rzhetsky A (2001) Probabilistic Prediction of Unknown Metabolic and Signal-Transduction Networks. Genetics 159 (3):1291–1298.
18. Gomez SM, Rzhetsky A (2002) Towards prediction of complete protein-protein interaction networks. Pac Symp Biocomput 2002:413–424.
19. Gomez SM, Noble WS, Rzhetsky A (2003) Learning to predict protein-protein interactions from protein sequences. Bioinformatics 19 (15):1875–1881.
20. Grigoriev A (2001) A relationship between gene expression and protein interactions on the proteome scale: analysis of the bacteriophage T7 and the yeast Saccharomyces cerevisiae. Nucleic Acids Res 29 (17):3513–3519.
21. Hazbun TR, Fields S (2001) Networking proteins in yeast. Proc Natl Acad Sci USA 98 (8):4277–4278.
22. Hayashida M, Ueda N, Akutsu T (2003) Inferring strengths of protein-protein interactions from experimental data using linear programming. Bioinformatics 19 (suppl 2):ii58–ii65.
23. Hermjakob H, Montecchi-Palazzi L, Lewington C et al (2004) IntAct: an open source molecular interaction database. Nucleic Acids Res 32 (database issue):D452–D455.
24. Ho Y, Gruhler A, Heilbut A et al (2002) Systematic identification of protein complexes in Saccharomyces cerevisiae by mass spectrometry. Nature 415 (6868):180–183.
25. Huynen MA, Bork P (1998) Measuring genome evolution. Proc Natl Acad Sci USA 95 (11):5849–5856.
26. Ito T, Chiba T, Ozawa R, Yoshida M, Hattori M, Sakaki Y (2001) A comprehensive two-hybrid analysis to explore the yeast protein interactome. Proc Natl Acad Sci USA 98 (8):4569–4574.
27. Jansen R, Greenbaum D, Gerstein M (2002) Relating Whole-Genome Expression Data with Protein-Protein Interactions. Genome Res 12 (1):37–46.

28. Kim WK, Park J, Suh JK (2002) Large scale co-evolution analysis of Protein Structural Interlogues using the global Protein Structural Interactome Map (PSIMAP). Genome Inform Ser Workshop Genome Inform 13:42–50.
29. Li S, Armstrong CM, Bertin N et al (2004) A Map of the Interactome Network of the Metazoan C. elegans. Science 303 (5657):540–543.
30. Li H, Li J, Tan SH, Ng SK (2004) Discovery of binding motif pairs from protein complex structural data and protein interaction sequence data. Pac Symp Biocomput 2004:312–323.
31. Li H, Li J (2005) Discovery of stable and significant binding motif pairs from PDB complexes and protein interaction datasets. Bioinformatics 21 (3):314–324.
32. MacBeath G, Schreiber SL (2000) Printing Proteins as Microarrays for High-Throughput Function Determination. Science 289 (5485):1760–1763.
33. Marcotte EM, Pellegrini M, Ng HL, Rice DW, Yeates TO, Eisenberg D (1999) Detecting Protein Function and Protein-Protein Interactions from Genome Sequences. Science 285 (5428):751–753.
34. Martin S, Roe D, Faulon JL (2005) Predicting protein-protein interactions using signature products. Bioinformatics 21 (2):218–226.
35. McCraith S, Holtzman T, Moss B, Fields S (2000) Genome-wide analysis of vaccinia virus protein-protein interactions. Proc Natl Acad Sci USA 97 (9):4879–4884.
36. Mewes HW, Frishman D, Güldener U et al (2002) MIPS: a database for genomes and protein sequences. Nucleic Acids Res 30 (1):31–34.
37. von Mering C, Jensen LJ, Snel B et al (2005) STRING: known and predicted protein-protein associations, integrated and transferred across organisms. Nucleic Acids Res 33 (database issue):D433–D437.
38. Mrowka R, Patzak A, Herzel H (2001) Is There a Bias in Proteome Research? Genome Res 11 (12):1971–1973.
39. Ng SK, Zhang Z, Tan SH (2003) Integrative approach for computationally inferring protein domain interactions. Bioinformatics 19 (8):923–929.
40. Overbeek R, Fonstein M, D'Souza M, Pusch GD, Maltsev N (1999) The use of gene clusters to infer functional coupling. Proc Natl Acad Sci USA 96 (6):2896–2901.
41. Pellegrini M, Marcotte EM, Thompson MJ, Eisenberg D, Yeates TO (1999) Assigning protein functions by comparative genome analysis: Protein phylogenetic profiles. Proc Natl Acad Sci USA 96 (8):4285–4288.
42. Phizicky EM, Fields S (1995) Protein-protein interactions: methods for detection and analysis. Microbiol Rev 59 (1):94–123.
43. Rain JC, Selig L, Reuse HD et al (2001) The protein-protein interaction map of Helicobacter pylori. Nature 409 (6820):211–215.
44. Salwinski L, Miller CS, Smith AJ, Pettit FK, Bowie JU, Eisenberg D (2004) The Database of Interacting Proteins: 2004 update. Nucleic Acids Res 32 (database issue):D449–D451.
45. Spellman PT, Sherlock G, Zhang MQ et al (1998) Comprehensive Identification of Cell Cycle-regulated Genes of the Yeast Saccharomyces cerevisiae by Microarray Hybridization. Mol Bio Cell 9 (12):3273–3297.
46. Sprinzak E, Margalit H (2001) Correlated Sequence-signatures as Markers of Protein-Protein Interaction. J Mol Biol 311 (4):681–692.
47. Stanyon CA, Liu G, Mangiola BA et al (2004) A Drosophila protein-interaction map centered on cell-cycle regulators. Genome Biol 5 (12):R96.

48. Suzuki H, Fukunishi Y, Kagawa I et al (2001) Protein-Protein Interaction Panel Using Mouse Full-Length cDNAs. Genome Res 11 (10):1758–1765.
49. Uetz P, Giot L, Cagney G et al (2000) A comprehensive analysis of protein-protein interactions protein-protein interactions in Saccharomyces cerevisiae in Saccharomyces. Nature 403 (6770):623–627.
50. Walhout AJ, Sordella R, Lu X et al (2000) Protein Interaction Mapping in C. elegans Using Proteins Involved in Vulval Development. Science 287 (5450):116–122.
51. Wang H, Segal E, Ben-Hur A, Koller D, Brutlag DL (2004) Identifying protein-protein interaction sites on a genome-wide scale. In: Saul LK, Weiss Y, Bottou L (eds) Advances in Neural Information Processing Systems 17. MIT Press, Cambridge, MA.
52. Wojcik J, Schachter V (2001) Protein-protein interaction map inference using interacting domain profile pairs. Bioinformatics 17 (Suppl 1):S296–S305.
53. Liu Y, Liu N, Zhao H (2005) Inferring protein-protein interactions through high-throughput interaction data from diverse organisms. Bioinformatics 21 (15):3279–3285.
54. Zanzoni A, Montecchi-Palazzi L, Quondam M, Ausiello G, Helmer-Citterich M, Cesareni G (2002) MINT: a Molecular INTeraction database. FEBS Lett 513 (1):135–140.
55. Zhu H, Bilgin M, Bangham R et al (2001) Global Analysis of Protein Activities Using Proteome Chips. Science 293 (5537):2101–2105.

12

An Efficient Algorithm for Deciphering Regulatory Motifs

Xiucheng Feng, Lin Wan, Minghua Deng, Fengzhu Sun, and Minping Qian

Summary. The identification of transcription factor binding sites (TFBS) by computational methods is very important in understanding the gene regulatory network. Although many methods have been developed to identifying TFBSs, they generally have relatively low accuracy, especially when the positions of the TFBS are dependent. Motivated by this challenge, an efficient algorithm, IBSS, is developed for the identification of TFBSs. Our results indicate that IBSS outperforms other approaches with a relatively high accuracy.

12.1 Introduction

To understand the gene regulatory networks on the genomic scale is one of the great challenges in the post-genomic era. Deciphering the regulatory binding sites (regulatory motifs) of a transcription factor (TF) on DNA sequences is especially fundamental and important for understanding gene regulatory networks. Experimental identification of TF binding sites (TFBS) is slow and laborious, while computational methods for the identification of TFBSs in genomic scale are difficult because binding sites are typically short and degenerate, and the binding mechanisms are not very clear. The availability of high throughput data, such as gene expressions, complete genome sequences of many species, and ChIP-chip (Chromatin Immunoprecipitation) data, has facilitated bioinformatitians to develop new computational methods to identify TFBSs.

There are two basic paradigms of computational identification of TFBSs. One is the weight matrix model (WMM). The basic assumption of the WMM is that it is position specific in the TFBS, and mutations at different positions of the TFBS are independent. However, this is not true in many cases. Short random sequences may be inserted in the TFBS such that the position specificity may be destroyed. There are often significant dependencies between positions within a TFBS [28]. Thus, real TFBSs sometimes do not fit the WMM well, and the results from this model may not be satisfactory. First-order and higher-order Markov models have also been developed to consider the position dependence, but they usually require large data to estimate the exponentially increasing number of parameters in such models [28].

Another paradigm for the identification of TFBSs is based on consensus pattern enumeration (word counting) methods. The main idea of word counting methods is to ascertain the overrepresented words from a set of given samples. The word counting methods do not need any assumptions of position specificity and independence between positions in the TFBS. Because no statistical models for TFBSs are assumed, word counting methods lack the ability to describe the properties of TFs and discrimination on any given sequence to ascertain whether and where they contain TFs. Furthermore, word counting methods sometimes fail to ascertain the TFBS if the TFBS has large variation and the number of training sequences is small. In addition, word counting methods require searching a large candidate consensus pattern space, which can be time consuming.

We have developed an efficient algorithm, IBSS (Intelligent Binding Sites System), for the identification of TFBSs. The procedures and ideas of IBSS can be described as follows (Figure 12.1 gives the schematic view of IBSS):

1. Extracting the Short Conserved Sequences of the TFBS. Suppose that we have a set of sample sequences, which a TF may bind to. Because TFBSs can be degenerate, real TFBSs may not be found to be overrepresented through word-counting methods. On the other hand, short subsequences of the TFBS may be more conserved and be found to be overrepresented through word-counting methods. The main purpose of this step is to extract all these short conserved subsequences. A program, ICOPE (Intelligent Consensus Pattern Enumeration), based on this idea has been developed.

2. Obtaining the TFBS Sample Set. From step 1, a collection of typical short conserved subsequences of the TFBS can be gotten to obtain the TFBS sample sequences for further statistical analysis. We then develop a technique.

3. Modeling the TFBS. In step 1, no statistical models are assumed. Therefore, a proper statistical model is needed for the description and discrimination of TFBSs. Our result indicates that the dependencies between positions in the TFBS sometimes can be very important. Therefore, a model based on the maximal dependence decomposition (MDD) [3] with insertions is developed for the TFBS to model the dependencies between different positions in the TFBS. We derive a statistical inference procedure for the discrimination of TFBSs based on the MDD model.

4. System Evaluation. Cross validation is used to evaluate the performance of our algorithm and the model.

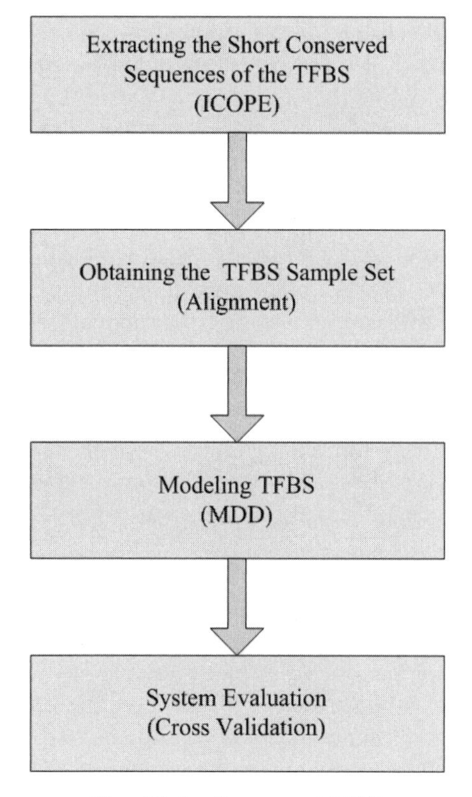

Fig. 12.1. Scheme of IBSS.

In section 12.2.2, we show that ICOPE, which is based on the first step of IBSS, achieves higher accuracy than six well known motif-finding programs based on large-scale ChIP-chip data of transcription factors with known DNA-binding specificities documented in databases.

In section 12.4, we show that the nucleotide bases for the TFBS of TF-YAP7 are highly dependent. We then use the MDD to model the TFBS. The cross-validation results indicate that the MDD model significantly outperforms the weight matrix model (WMM).

12.2 ICOPE

To obtain the set of sample sequences for finding the motifs of the TFBS, usually gene expressions under different conditions or TF mutants are needed. For gene expressions under different conditions, coexpressed genes are clustered. Overrepresented motifs can then be discovered in the promoter regions of these coexpressed genes using motif-finding algorithms. The basic assumption is that the binding sites of the same TFs are similar, and coexpressed genes are likely to be regulated by the same TFs [7]. In the TF mutant microarray, a TF is deleted or overexpressed, and then upregulated or downregulated genes are selected out for motif finding. The basic assumption is that these genes are more likely to be regulated by this TF. However, it is difficult to distinguish the effects of direct or indirect regulation by the TF among these upregulated and downregulated genes. Recently, a functional clustering method was developed to use mutant expression data for motif finding of TFBSs [6].

Motif-finding algorithms can also be used to identify the transcription factors' DNA-binding specificities from ChIP-chip data [18], which provides a way to reveal the genome-wide location of DNA-binding proteins and can be used to construct a genome-wide map of in vivo protein-DNA interactions for any given protein. ChIP-chip has produced a large amount of data [11, 14]. However, the resolution of typical yeast ChIP-chip experiments is only about 1 kb. As a result, computational methods are required to identify the transcription factors' exact DNA binding specificities.

Many computational methods have been developed in the past two decades for discovering motif patterns and TF binding sites in a set of promoter sequences. Most of the motif finding programs can be classified into two groups.

The first group of methods for motif finding uses WMM as the motif model, and usually combines with expectation maximization (EM) or Gibbs sampling to iteratively refine the WMM. Stormo and Hartzell [21] used a heuristic progressive alignment procedure to find motifs. Early WMM-based methods using EM or Gibbs sampling include those by Lawrence and Reilly [12], Lawrence et al. [13], and Bailey and Elkan [1]. By iteratively masking out aligned sites, AlignACE [19] is more effective in finding multiple distinct motifs. BioProspector [16] uses a third-order Markov chain to model the background sequences to improve the motif specificity.

The other group of programs for motif finding uses enumeration methods to find the best consensus pattern; we refer to these methods as word counting methods. Galas et al. [10] developed the earliest motif-finding program using consensus enumeration. Brazma et al. [2] analysed the overrepresentation of regular expression-type patterns. Recent word-counting methods are reported by Wolfertstetter et al. [26], van Helden et al. [25], and Sinha and Tompa [22]. DWE [23] is a word-counting–based tool to discover tissue-specific transcription factor binding sites.

Recently, MDScan [17] combined the two widely adopted motif search strategies, word counting and WMM updating, and incorporated the ChIP-chip ranking information to accelerate searches to enhance its success rates.

The program, ICOPE, was developed based on the first step of IBSS. We searched the ensemble of all short sequences combined with a limited number of degenerate letters and a range of spacers inserted in the set of sample sequences, and selected out all the overrepresented short sequences. These short sequences cannot be too short so that they can contain sufficient information of the TFBS and can be overrepresented in the sample sequences rather than in the background sequences, while they cannot be too long so that their occurrence probabilities will be big enough such that their occurrences can be approximated by their occurrence frequencies. Usually 6 to 7 base pair (bp) informative positions for DNA sequences should serve for practical uses.

It has been shown by several investigators that some TFs, when in low concentration, show a significant preference for multiple occurrences of motifs, whereas in high concentration, they don't have such a preference [4, 11]. ICOPE takes this fact into account. ICOPE uses two different word-counting methods to deal with these two groups of TFs. Obviously, we don't know which group a TF belongs to in advance. So a simple mixture model to combine the two methods is used. Both methods are examined and the method that gives a lower p value will be accepted.

Another character of ICOPE over existing methods lies in the background model. It has been shown that different background models can greatly influence the performance of motif-finding programs [16]. Most other programs use a high-order Markov model as the background model, which is relatively inaccurate. The background model of ICOPE is based directly on resampling from the background sequences.

12.2.1 Systems and Methods

A consensus pattern that occurs significantly more often than chance expectation is said to be overrepresented. Like other consensus enumeration methods, consensus patterns are evaluated based on their overrepresentation in the input sequences. A statistical significance (p value) is given to objectively measure the degree of overrepresentation.

Most conserved patterns of the TFBS are usually in the range of 6 to 10 bp long [22]. ICOPE searches against a very large consensus pattern space. All strings of length 6 over the alphabet {A, C, G, T} and strings of length 7 over the alphabet {A, C, G, T, R, Y, S, W} with a limited number of degenerate letters {R, Y, S, W} are considered. In addition, all the above strings with a spacer inserted are also incorporated. The spacer inserted is a string of 1 to 12 consecutive N's. The spacer can be inserted at any position in the original string. For the search space, our method is found to be similar to the motif method used in YMF [22], except that in YMF the spacer can only be inserted in the center of the motif, while we allow the spacer to occur at any position.

For each consensus pattern S in the search space, we calculate p value p_S^M assuming that the factor has a preference for multiple occurrences of motifs, referred to as factor class I, and calculate p value p_S^S assuming the factor has no such a preference, referred to as factor class II. Then we find the best consensus pattern S_M that minimizes p_S^M and S_S that minimizes p_S^S. If both $p_{S_M}^M$

and $p^S_{S_S}$ are greater than .05, we think the result is not statistically significant and ICOPE outputs nothing. If $p^M_{S_M} < p^S_{S_S}$ and $p^M_{S_M} < .05$, ICOPE assumes the corresponding factor is of factor class I. If $p^S_{S_S} < p^M_{S_M}$ and $p^S_{S_S} < .05$, ICOPE assumes the corresponding factor is of factor class II.

First, we discuss how p^S_S is calculated. This p value is calculated by assuming the factor has a preference of single motif occurrence. For the counting word, only one occurrence of the motif in the promoter region is considered even if a motif occurs several times in a sequence. We count the number of sequences with the motif in the input sequence set. Let N^S_S be the number of sequences in the input sequence set in which pattern S occurs. A good measure of overrepresentation of the consensus pattern must take into account both N^S_S and the background genomic distribution. We merge all the yeast intergenic sequences into one large sequence as the background. Then random sequence sets with the same size as the input sequence set are drawn from the background. The sequences in the resampled random sequence sets have the same sequence length with the input sequence set. For the resampled sequence sets, let n^S_S be the average number of sequences on which the consensus pattern S occurs. Let M be the size of the input sequence set. Then the p value is approximately calculated by using the binominal distribution $B(M, n^S_S/M)$.

The p^M_S can be similarly calculated. This p value is calculated by assuming that the factor has a preference for multiple motif occurrences in the promoter regions. For word counting, every occurrence of motif should be considered. We count N^M_S the number of occurrences of consensus pattern S in the input sequence set. We set an upper limit for the number of occurrences in a single sequence to avoid the interference of repetitive sequence. Currently we set this upper limit to 5. Resampled random sequence sets with the same sequence number and the same sequence length as the input sequence set are generated by a similar method from background sequences and scanned to count the number of occurrences of pattern S. For the resampled random sequence sets, let n^M_S be the average number of occurrences of pattern S. Then the p value is approximately calculated by using the Poisson distribution $P(n^M_S)$.

For every consensus pattern in the search space, we test the hypothesis that the pattern is a binding motif based on p^S_S for factor class I and p^M_S for factor class II. So millions of hypotheses are being tested simultaneously. The p value calculated above by using the binominal distribution or Poisson distribution of each class must be corrected for multiple tests. We use the formula $p_c = 1-(1-p)^N$ to adjust for multiple testing, where p is the p value before correction and p_c is the corrected p value. N is the number of tests conducted, that is, the size of the searched consensus space. Both p^S_S and p^M_S are corrected using this formula. For more advanced and technical discussion of p-value choice in large-scale simultaneous hypothesis testing, see [9].

In practice, the most time-consuming part is the resampling process. This process is done in advance and the result is stored in disk. For every possible length l, let η^S_{Sl} be the probability that a sequence with length l has motif

S if the sequence is drawn randomly from background. Let η_{Sl}^{M} be the average number of occurrences of motif S if the sequence is drawn randomly from background. n_S^S is calculated by summing over η_{Sl}^{S} for every sequence in the input sequence set. n_S^M is calculated by summing over η_{Sl}^{M} for every sequence in the input sequence set. In fact, we do not need and cannot afford to do resampling for every possible length. Because most promoter regions in yeast are less than 1500 bp, we only do this process for length $l \in L$, $L = \{50, 100, 150 \cdots 1450, 1500\}$. If l is not in L, we round it to the nearest length in L.

12.2.2 Results

In Harbison et al. [11], a large set of transcription factors' genome-wide location analysis was conducted under several different conditions. Among those factors, 81 factors have known DNA binding specificities documented in databases (TRANSFAC, YPD, SCPD). Harbison et al. conducted 175 ChIP-chip experiments for these 81 factors. We tested ICOPE on this set of 175 experiments. For 76 of the 175 experiments and for 49 of the 81 factors, the most significant consensus pattern reported by ICOPE match the corresponding database record. Harbison et al. tested six other programs on their data. For these factors with known binding specificities, the best program, MDScan, successfully predicted 37 factors' binding specificity in 59 of 175 experiments. In addition, ICOPE made 135 predictions for those 175 experiments while MDScan made 156 predictions. It is obvious that ICOPE has a higher accuracy. The five other programs used by Harbison et al. can successfully predicted only 20 to 33 of the factors' binding pattern. Detailed information is given in Table 12.1. A comprehensive result of ICOPE's prediction on these 175 experiments and comparison with database record is given in Table 12.2.

We did a similar evaluation on a subset of the 81 transcription factors, which contains 52 factors for which at least one program's prediction matches the known specificity. We dropped those that have only a database record but have no program's prediction matching the record for three reasons. First, the database might have errors, although the chance is small. Second, the ChIP-Chip experiment is rather noisy. Third, and the most important, transcription factors may take effect only in certain specific condition. The condition under which ChIP-Chip experiment is conducted might not be that exact condition. So if none of the programs can predict the binding specificity, the condition under which ChIP-Chip experiment is conducted might be incorrect. The result for this set of factors is given in Table 12.1 and summarized in Table 12.2.

As mentioned in section 12.2.1, ICOPE uses a very simple strategy to combine the two p values. We find that such a strategy to combine those two p values is much better than using one of the two p values alone. Using one p

Table 12.1. Comparison of ICOPE with six other programs on a set of 52 factors with database documented binding specificity.

FA	KN	I	A	C	D	K	M	N
Abf1	rTCAyTnnnnACGw	+	+	+	+	+	+	+
Ace2	GCTGGT					+		
Aft2	AAAGTGCACCCATT	+	+	+	+		+	+
Azf1	TTTTTCTT							+
Bas1	TGACTC	+	+			+	+	+
Cad1	TTACTAA	+	+	+	+		+	+
Cbf1	rTCACrTGA	+	+	+	+	+	+	+
Cin5	TTACrTAA	+	+	+	+			
Fkh1	GGTAAACAA	+	+	+	+	+	+	+
Fkh2	GGTAAACAA	+	+	+	+	+	+	+
Gal4	CGGnnnnnnnnnnnCCG	+	+			+		
Gat1	GATAA	+				+		
Gcn4	ArTGACTCw	+	+	+	+	+	+	+
Gcr1	GGCTTCCwC	+			+			
Gln3	GATAAGATAAG	+		+	+	+		
Hap1	CGGnnnTAnCGG	+		+			+	
Hap3	CCAAT	+						
Hap4	YCNNCCAATNANM	+	+	+	+		+	+
Hsf1	TTCTAGAAnnTTCT	+	+	+	+	+	+	+
Ino2	ATTTCACATC	+		+	+		+	+
Ino4	CATGTGAAAT	+	+	+	+	+	+	+
Leu3	yGCCGGTACCGGyk	+	+		+	+	+	+
Mac1	GAGCAAA	+			+			
Mbp1	ACGCGT	+	+	+	+	+	+	+
Mcm1	wTTCCyAAwnnGGTAA	+	+	+	+		+	+
Mot3	yAGGyA	+						
Msn2	mAGGGG	+					+	
Nrg1	GGaCCCT	+	+	+	+		+	+
Pdr1	CCGCGG						+	
Pho4	cacgtkng	+			+	+		+
Put3	CGGnnnnnnnnnnCCG	+						
Rap1	wrmACCCATACAyy	+	+	+	+		+	+
Rcs1	AAmTGGGTGCAkT	+		+	+		+	+
Reb1	TTACCCGG	+	+	+	+	+	+	+
Rph1	CCCCTTAAGG	+						
Rpn4	GGTGGCAAA	+	+	+	+	+	+	+
Sip4	yCGGAyrrAwGG	+			+			
Skn7	ATTTGGCyGGsCC	+	+	+	+		+	+
Sko1	ACGTCA	+						
Stb5	CGGnstTAta	+			+			+
Ste12	ATGAAAC	+	+	+	+	+	+	+
Sum1	AGyGwCACAAAAk	+	+	+	+		+	+

Sut1	CGCG	+	+		+		+	
Swi4	CnCGAAA	+	+	+	+	+	+	+
Swi6	CnCGAAA / ACGCGT	+	+	+	+		+	+
Tec1	CATTCy	+		+				
Tye7	CAnnTG	+	+	+	+		+	
Ume6	wGCCGCCGw	+	+	+	+	+	+	+
Xbp1	CTTCGAG	+						
Yap1	TTAsTmA	+	+	+	+		+	
Yap7	TTACTAA	+	+	+	+		+	+
Zap1	ACCCTAAAGGT	+			+			+

FA = factor name, KN = known specificity, I = ICOPE, A = AlignACE, C = CONVERGE, D = MDscan, K = Kellis et al., M = MEME, N = MEME_c.

Table 12.2. Summary of Table 12.1

	I	A	C	D	K	M	N
MA	49	30	31	37	20	33	31
Percent_1	60.5%	37.0%	38.3%	45.7%	24.7%	40.7%	38.3%
Percent_2	94.2%	57.7%	59.6%	71.2%	38.4%	63.5%	59.6%

MA = the number of factors for which each program's prediction match the database record. Percent_1 = MA/81, 81 is total number of factors with database documented DNA binding specificity. Percent_2 = MA/52, 52 is number of factors with database documented DNA binding specificity and with at least one program's prediction matching the database record.

value alone will miss some of the factors belonging to the other group. Figure 12.2 compares the results of considering different p values. As illustrated in Figure 12.2, ICOPE_M is the program that uses p value p_S^M, and ICOPE_S is the program that uses p value p_S^S. Both programs can correctly predict 36 factors' DNA binding specificity. Seven factors (CAD1, GCR1, PUT3, RPH1, STB5, XBP1, ZAP1) can be correctly predicted by ICOPE_M and cannot be correctly predicted by ICOPE_S. Seven other factors (GAT1, HAP2, HAP3, MSN2, SUM1, SUT1, TEC1) can be correctly predicted by ICOPE_S and cannot be correctly predicted by ICOPE_M. ICOPE can correctly predict DNA binding specificity for 49 factors whose DNA binding specificity can be correctly predicted by either ICOPE_M or ICOPE_S. There is only one exception, HAP2. For this factor, both ICOPE_M and ICOPE_S make predictions. ICOPE_M gives a lower p value, but its prediction is incorrect, whereas the prediction of ICOPE_S is correct. So ICOPE takes the result of ICOPE_M and makes a mistake.

We sorted the ICOPE's predictions on the 175 experiments based on the p values given by ICOPE. The sorted results are given in Table 12.3. We

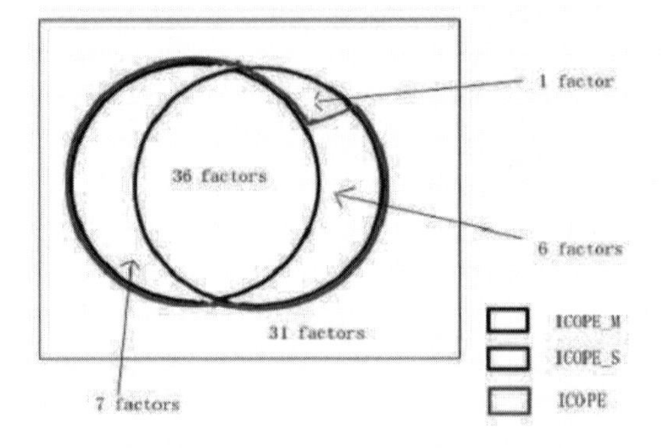

Fig. 12.2. ICOPE has a significantly higher accuracy than both ICOPE_M and ICOPE_S. In the rectangle are the 81 factors with known DNA binding specificity. The black circle stands for the factors whose DNA binding specificity can be correctly predicted by ICOPE_M. The blue circle stands for the factors whose DNA binding specificity can be correctly predicted by ICOPE_S. The red curve stands for the factors whose DNA binding specificity can be correctly predicted by ICOPE.

can see that the p value is a good indicator of whether the prediction is reliable. ICOPE makes 39 predictions with $\log 10\ p$ values less than -14, among which 37 (95%) predictions are correct. ICOPE makes 23 predictions with $\log 10\ p$ values between -14 and -7, among which 18 (78%) predictions are correct. ICOPE makes 73 predictions with $\log 10\ p$ values between -7 and -1.3, among which 21 (29%) predictions are correct. If the p value is very low, as low as 10^{-14}, the prediction is highly reliable. As the p value goes higher, the prediction becomes less reliable.

Table 12.3. The p value given by ICOPE indicates the reliability of its prediction

p-value range	N	N_C	$P(\%)$
$\log_{10} p \leq -14$	39	37	95%
$-14 \leq \log_{10} p \leq -7$	23	18	78%
$-7 \leq \log_{10} p \leq -1.3$	73	21	29%

N = number of predictions, N_C = number of correct predictions, $P(\%)$ = percentage of correct predictions

We run ICOPE on the 135 ChIP-chip experiment for 101 factors without known DNA binding specificity. The predictions is given in Table 12.4. The result sorted by p value is given in Table 12.5. The program made 87 predictions for these 135 experiment. Six of the predictions have very low p value (less then 10^{-14}). Based on the prediction accuracy on factors with known DNA binding specificity, we estimate that these predictions may be highly reliable (95% accurate). Sixteen predictions have very low p value (between 10^{-14} and

10^{-7}). Based on the prediction accuracy on factors with known DNA binding specificity, we estimate that these predictions might have an accuracy of 78%. The predictions with p value less than 10^{-7} are given in Table 12.4.

Table 12.4. The predictions made by ICOPE on factors with unknown DNA binding specificity (only predictions with a p value lower than 10^{-7} are shown)

Experiment	Prediction	p value
FHL1_YPD	TGTNNGGRT	−54.0149
FHL1_RAPA	TGTNNGGRT	−53.7184
FHL1_SM	TGTNNGGRT	−49.6275
RDS1_H2O2Hi	CGGCCG	−15.7688
SFP1_SM	AYCNNTACA	−15.6778
GAT3_YPD	GCTNAGCS	−14.4517
DIG1_Alpha	YGTTTCA	−13.1135
YFL044C_YPD	TANNNATWTA	−12.0532
PHD1_YPD	AGGCAC	−11.735
SNT2_YPD	GGSGCTA	−11.4775
STB1_YPD	CGCSAAA	−11.2591
HOG1_YPD	ACNNNNNCRCAC	−10.9645
DAT1_YPD	GCTNAGCS	−9.77751
RDS1_YPD	CGGCCG	−9.60608
MET4_YPD	CACGTG	−9.46732
SPT23_YPD	TTAWTAA	−9.42808
DIG1_BUT90	YGTTTCA	−8.70506
MET4_SM	CACGTG	−8.61795
YDR026c_YPD	TASCCGG	−8.54541
MGA1_H2O2Hi	CGGCCG	−8.50797
RGM1_YPD	GCTNAGCS	−7.38143
YDR520C_YPD	CCGSCGG	−7.17487

Because we did the resampling process in advance for almost every possible length of input sequence and stored the result in disk, ICOPE runs pretty fast. For an input sequence set with 10 sequences and average length of 500, about 1.5 minutes is required. For an input sequence set with 100 sequences and average length of 500, about 9 minutes is required.

ICOPE is implemented using C++ programming language and a Web interface is implemented using PHP. The URL for the Web interface is at http://ctb.pku.edu.cn/˜xcfeng/icope/.

Supplementary information, including data and omitted results are available at http://ctb.pku.edu.cn/˜xcfeng/icope/supp/.

12.3 Obtaining the TFBS Sample Set

Although ICOPE's prediction on transcription factors' DNA binding specificity has a high accuracy, the sequences generated by ICOPE are typical sequences of the TFBS. Real TFBS sample sequences should be obtained for further statistical analysis by MDD. The technique to obtain the TFBS sample set is described as follows.

Let C be the collection of overrepresented short sequences generated from ICOPE. Align each short sequence in C, to all the sample sequences mentioned in step A of IBSS, with a limited number of mismatches, and highlight the aligned part. Then get all the connected segments that are highlighted, and denote the set of all segments by D_O. After multiple alignment on all segments in D_O, we choose the aligned consensus sequences as the TFBS sample set, denote by D. Gaps are usually contained in sequences of D since insertions are kept from multiple alignment.

12.4 Modeling the TFBS

Because no statistical models for TFBSs are assumed in ICOPE, it is not possible to test whether there are some binding sites and where they are for a given sequence.

The weight matrix model (WMM) introduced by Staden [20] is a more accurate representation, but the weight matrix assumes independence between different positions in the motif. For some transcription factors, it has been shown that there are strong dependencies between different positions of their binding motif [28]. So weight matrix is not a good choice to represent such transcription factors' DNA binding specificity, although they are good enough for other transcription factors. The weight array model (WAM) [27], which was applied by Zhang and Mar, could detect dependencies between adjacent positions by using first-order (nonstationary) Markov models. Higher-order WAMs can be used to model higher-order dependencies in the TFBS sequences, but there are often insufficient data available to estimate the exponentially increasing number of parameters in such models [28].

Recently, some new methods have been developed to model such long distant dependence. A maximal dependence decomposition (MDD) [3] based model, with the insertion, denoted by O, added in as a symbol, is developed for the TFBS. The MDD model has the advantage of capturing and modeling the dependencies between different positions in the TFBS. The main idea of MDD is that it splits up the training data to fit different WMMs to suitably define subsets of the data based on most dependent positions, which are measured by the χ^2 test.

12.4.1 Using χ^2 Test to Measure Dependence Between Positions

For a given TFBS sample set D obtained from section 12.3, we use the χ^2 test to measure dependencies between positions of the sequences in D. Suppose D contains N aligned sequences of length k. For any given position i and j ($1 \le i, j \le k$, $i \ne j$), and for a given nucleotide b ($b \in${A,C,T,G,O}),[1] we use the $\chi^2(C_{ib}, X_j)$ test, to measure the dependency between the consensus indicator variable, C_{ib} (1 if the nucleotide b at position i matches the consensus at i, 0 otherwise) and the nucleotide indicator X_j identifying the nucleotide at position j.

For a given position i, nucleotide b, and position j, we have

	$C_{ib} = 0$	$C_{ib} = 1$	
$X_j = A$	n_{A0}	n_{A1}	$n_{A\cdot}$
$X_j = C$	n_{C0}	n_{C1}	$n_{C\cdot}$
$X_j = G$	n_{G0}	n_{G1}	$n_{G\cdot}$
$X_j = T$	n_{T0}	n_{T1}	$n_{T\cdot}$
$X_j = O$	n_{O0}	n_{O1}	$n_{O\cdot}$
	$n_{\cdot 0}$	$n_{\cdot 1}$	

where

$$n_{c\cdot} = \sum_{t=0,1} n_{ct}, \qquad c \in \{A, C, G, T, O\},$$

$$n_{\cdot t} = \sum_{c \in \{A,C,G,T,O\}} n_{ct}, \qquad t = 0, 1,$$

$$\sum_{\substack{c \in \{A,C,G,T,O\} \\ t=0,1}} n_{ct} = N.$$

Define

$$\hat{p_c} = \frac{n_{c\cdot}}{N}, \qquad c \in \{A, C, G, T, O\},$$

$$\hat{q_t} = \frac{n_{\cdot t}}{N}, \qquad t = 0, 1.$$

We use the statistic

$$\chi^2(C_{ib}, X_j) = \sum_{\substack{c \in \{A,C,G,T,O\} \\ t=0,1}} \frac{(n_{ct} - N\hat{p_c}\hat{q_t})^2}{N\hat{p_c}\hat{q_t}}.$$

$\chi^2(C_{ib}, X_j)$ obeys the $(2-1) \times (5-1) = 4$ d.f. χ^2-distribution [5], and a significant χ^2 value should be great than 13.3 at the relatively level of $p < .01$, indicating a significant dependent position in the TFBS.

Based on the χ^2 test, we find that some cases of the transcription factors' binding motif show strong dependencies between different positions.

[1] O stands for one gap.

12.4.2 Maximal Dependence Decomposition (MDD)

For a given TFBS sample sequence set D, the MDD model is to build a conditional probability model, based on the significant dependence between positions (both nonadjacent and adjacent dependencies are considered), instead of the unconditional WMM probabilities model [3].

The algorithm of the MDD model is described as follows: For a given sequence set D, assume that it contains N aligned sequences of length k.

1. Check whether there exist i, j, b that satisfy $\chi^2(C_{ib}, X_j) \geq 13.3$, that is, whether there exist significant dependent positions. If yes, continue to the next step; otherwise, stop.

2. For each position i and nucleotide b, calculate $S_{ib} = \sum_{j \neq i} \chi^2(C_{ib}, X_j)$, which measures the amount of dependencies between the variable C_{ib} and the nucleotides at the remaining positions. Choose i_1 and b_1 that maximize S_{ib},
$$S_{i_1 b_1} = \max_{i,b} S_{ib}.$$

3. Divide D into two subsets based on whether or not position i_1 has nucleotide b_1. Check whether each of the two subsets contains enough sequences to estimate WMM frequencies for further subdivision. If each of the two subsets contains enough sequences to estimate WMM frequencies, we divide set D into $D_{i_1 b_1}$ and $\overline{D}_{i_1 b_1}$. $D_{i_1 b_1}$ contains all the sequences that has nucleotide b_1 at position i_1, and $\overline{D}_{i_1 b_1}$ contains no nucleotide b_1 at position i_1.

4. Repeat steps 1, 2, and 3 on the subsets $D_{i_1 b_1}$ and $\overline{D}_{i_1 b_1}$, and from then on, position i_1 will be not considered.

The MDD model splits the set D into subsets and constructs a binary tree structure based on the above algorithm. We use the WMM model to describe each leaf of subsets. Figure 12.3 shows the procedures for splitting data set D of TF-YAP7.

To make the algorithm clearer, we give an example of YAP7. We use ChIP-chip data from [11] of YAP7 under the condition H2O2Lo. ICOPE is used to predict the binding sites of YAP7. ICOPE predicts 100 words with $p \leq 10^{-5}$. We highlight the 123 ChIP-chip selected sequences that can be aligned, with no mismatch, by the 100 words; 117 (95.12%) of the 123 ChIP-chip selected sequences could be highlighted. We then obtain all the 598 sequence segments that have been highlighted. The 598 sequence segments are aligned by ClustalW [24] with its default parameters, and the most consensus aligned sequences of 10 bp length from 598 sequence segments are picked out as the TFBS sample set D.

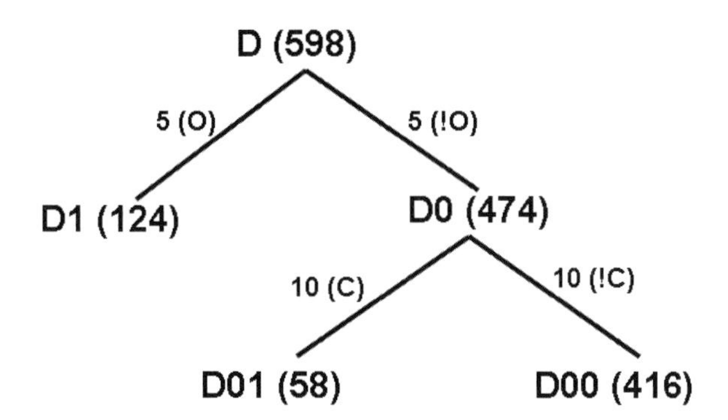

Fig. 12.3. Tree result of YAP7. D, $D1$, $D0$, $D01$, $D00$ stand for the node names of tree. The number in the parentheses behind the node name stands for the number of sequences. The annotation on the edge represents how to split the set of the parent node into two subset. For example, D splits into $D1$, and $D0$, based on position 5, has or has no O. O stands for gap, ! stands for NOT.

Using χ^2 test, we find significant dependencies between the positions of the sequences in D. The TFBS sample set D is used to train the MDD model, and the results are summarized in Figures 12.3 and 12.4. Figure 12.3 shows the binary tree constructed by the MDD model. From it we learn that other positions have significant dependencies on whether position 5 has a gap. Conditional on position 5 having no gap, the other positions have significant dependencies on whether position 10 has C. Figure 12.4 shows the sequence logo [7] plots of the data set in each node of the binary tree in Figure 12.3. From the sequence logo can be seen strong dependencies between the positions of YAP7.

12.4.3 Statistical Model and Discrimination

The MDD model splits the training data set into subsets and constructs a binary tree based on the most significant dependent positions. Any sequences could find a corresponding unique leaf on the binary tree. For each leaf l, the MDD model assigns a probability $P_{mdd}(L = l)$ for reaching this leaf from the root. Separate WMM models are estimated for each subset on the leaf of the binary tree. For a given leaf l, it has weight matrix $\mathbf{W}^l$, and W^l_{ib} stands for the frequency of nucleotide b in position i on leaf l.

Suppose that we have a sequence $S = s_1 s_2 \dots s_k$, and it is on leaf l from the MDD model. Condition on leaf l, sequence S has the probability

$$P(S = s_1 s_2 \dots s_k | L = l) = \prod_{i=1,2,\dots,k} W^l_{is_i}.$$

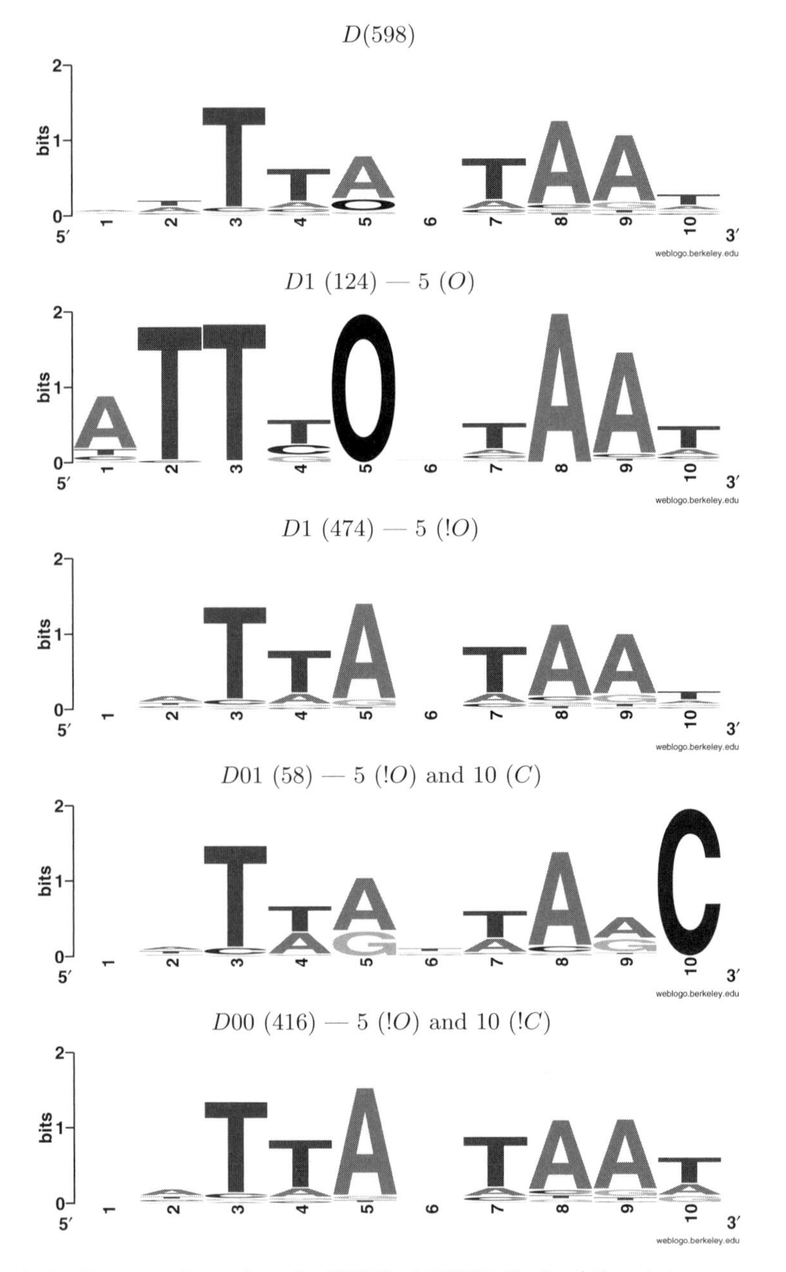

Fig. 12.4. Sequence logo plots for TFBS of YAP7. Each of the plots corresponds to the node of the binary tree in Figure 12.3. The positions dependencies can been seen from sequence logo plots.

From the MDD model, the probability of obtaining such a sequence S is

$$P_{mdd}(S = s_1 s_2 \dots s_k) = P_{mdd}(L = l) \cdot P(S = s_1 s_2 \dots s_k | L = l).$$

Next we give a discriminant method to identify whether or not a sequence, $S = s_1 s_2 \dots s_k$, is a transcription factor's binding sides. Let $\Omega = 1$ indicate that a given sequence is a TFBS, and $\Omega = 0$ otherwise. Suppose that for any sequence S with length k,

$$P_{mdd}(S|\Omega = 1) \sim N(\mu_1, \sigma_1),$$

$$P_{mdd}(S|\Omega = 0) \sim N(\mu_2, \sigma_2).$$

From [8], the threshold μ^* is derived as

$$\mu^* = \frac{\mu_1 \sigma_2 + \mu_2 \sigma_1}{\sigma_1 + \sigma_2}.$$

The discriminant rule for S is

$$\begin{cases} \text{S is TFBS}, & \text{when } P_{mdd}(S) > \mu^*, \\ \text{S is not TFBS}, & \text{when } P_{mdd}(S) \leq \mu^*. \end{cases}$$

We use the sequence set D generated from the procedures described in section 12.3 as the positive sample, and randomly select the same number of sequences with the same length from background sequences as our negative sample, denoted as B. We use the positive sequences to train the MDD model, and use both the positive and negative sequences to estimate the threshold μ^* for further discrimination.

To test the efficiency of the MDD model, we use the 598 TFBS sample sequences of YAP7 described in section 12.4.2 as our positive sample, and obtain the negative sample B from the background sequences. We use 10-fold cross-validation to evaluate the MDD model. To compare with the WMM model, we also use the same positive and negative sample to do the 10-fold cross-validation on the WMM model. Figure 12.5 gives the receiver operating characteristic (ROC) curve of the results, which shows their performances in terms of TPr and FPr, where TPr is the sensitivity of a classifier measuring the fraction of positive cases that are classified as positive, and FPr is the false alarm measuring the fraction of incorrectly classified negative cases: $TPr = \frac{TP}{TP+FN} = \frac{TB}{Pos}$, $FPr = \frac{FP}{TN+FP} = \frac{FP}{Neg}$.

From Figure 12.5, we learn that the MDD model outperforms the WMM model because the MDD model can capture and model the dependencies between positions while the WMM model cannot.

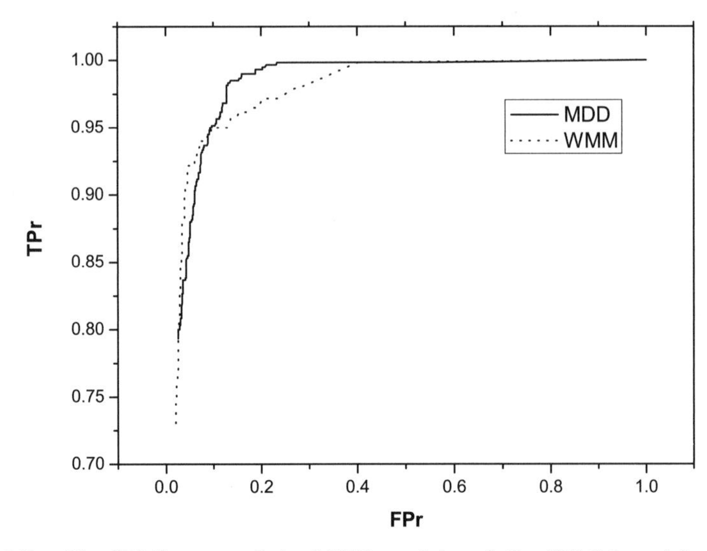

Fig. 12.5. The ROC curve of the MDD model and the WMM model of YAP7. The X-axis is FPr (false positive rate), and the Y-axis is TPr (true positive rate).

12.5 Discussion

The identification of transcription factor binding sites by computational methods is an essential step in understanding the gene regulatory network. Although many methods have been developed to identifying TFBSs, they generally have relatively low accuracy especially when the positions of the TFBS are dependent. Motivated by this challenge, an efficient algorithm, IBSS, is developed for the identification of TFBSs. Our results indicate that IBSS outperforms other approaches with a relatively high accuracy.

In this study, the background model of ICOPE is only based on resampling. The binominal and Poisson distributions are only an approximation. Further study on the background model may help to improve the accuracy of ICOPE.

Combinatorial regulation is an essential feature of transcriptional regulation. Identification of TFBSs provides the materials for further understanding the combinational regulation. TFs usually form a protein complex to regulate the gene, so the distance between the TFBSs of the cooperative TFs should be limited; sometimes the TFBSs of the cooperate TFs may even overlap. This biological knowledge gives us a clue to find out the cooperative TFs by the distances of TFBS pairs.

An inverse problem of TFBS finding is to detect the protein motif sequences that bind to the DNA sequences. A natural question is to ask whether or not the protein motif sequences that bind to the common binding sites on DNA sequences are conserved. With knowing the TFBSs of many TFs, the protein sequences of the TFs that share common binding sites on the DNA sequences may be used for further statistical analysis.

On the statistical model of TFBSs, the MDD-based model is used instead of the traditional WMM-based model. The advantage of the MDD model is that it considers the position dependencies of the TFBS. In the future, a more accurate model of TFBSs should be developed.

Acknowledgment

This work benefits greatly from the research environment and the supercomputer at the Center for Theoretical Biology of Peking University. This work is supported in part by grants from the National Natural Sciences Foundation of China (90208002, 10271008), State 863 High Technology R&D Project of China (2002AA234011) and the Special Funds for Major State Basic Research of China (2003CB715903).

References

1. Bailey TL, Elkan C (1994) Fitting a mixture model by expectation maximization to discover motifs in biopolymers. In: Altman R, Brutlag, D, Karp P, Lathrop R, Searls D (eds) Proceedings of the Second International Conference on Intelligent Systems for Molecular Biology. AAAI Press, Menlo Park, CA.
2. Brazma A, Jonassen I, Vilo J, Ukkonen E (1998) Predicting gene regulatory elements in silico on a genomic scale. Genome Res 8:1202–1215.
3. Burge C, Karlin S (1997) Prediction of complete gene structures in human genomic DNA. J Mol Biol 268:78–94.
4. Bussemaker HJ, Li H, Siggia ED (2001) Regulatory element detection using correlation with expression. Nat Genet 27:167–171.
5. Casella G, Berger RL (2001) Statistical Inference, 2nd ed. Duxbury Press.

6. Chen GX, Hata N, Zhang MQ (2004) Transcription factor binding element detection using functional clustering of mutant expression dat. Nucleic Acids Res 32:2362–2371.
7. Crooks GE, Hon G, Chandonia JM, Brenner SE (2004) WebLogo: a sequence logo generator. Genome Res 14:1188–1190.
8. Duda RO, Hart PE, Stork DG (2000) Pattern Classification, 2nd ed. Wiley-Interscience.
9. Efron B (2004) Large-scale simultaneous hypothesis testing: the choice of a null hypothesis. J Am Statistical Assoc 99:97–104.
10. Galas DJ, Eggert M, Waterman MS (1985) Rigorous pattern-recognition methods for DNA sequence: analysis of promoter sequences from Escherichia coli. J Mol Biol 186:117–128.
11. Harbison CT et al (2004) Transcriptional regulatory code of a eukaryotic genome. Nature 431:99–104.
12. Lawrence CE, Reilly AA (1990) An expectation maximization (EM) algorithm for the identification and characterization of common sites in unaligned biopolymer sequences. Proteins 7:41–51.
13. Lawrence CE, Altschul SF, Boguski MS, Liu JS, Neuwald AN, Wootton J (1993) Detecting subtle sequence signals: a Gibbs sampling strategy for multiple alignment. Science 262:208–214.
14. Lee TI et al (2002) Transcriptional regulatory networks in Saccharomyces cerevisiae. Science 298:799–804.
15. Li H, Wang W (2003) Dissecting the transcription networks of a cell using computational genomics. Curr Opin Genet Dev 13:611–616.
16. Liu XS, Brutlag DL, Liu JS (2001) BioProspector: discovering conserved DNA motifs in upstream regulatory regions of co-expressed genes. Pac Symp Biocomput 6:127–138.
17. Liu XS, Brutlag DL, Liu JS (2002) An algorithm for finding protein-DNA binding sites with applications to chromatin-immunoprecipitation microarray experiments. Nature Biotech 20:835–839.
18. Ren B, Robert F, Wyrick J et al (2000) Genome-wide location and function of DNA binding proteins. Science 290:2306–2309.
19. Roth FP, Hughes JD, Estep PW, Chruch GM (1998) Finding DNA regulatory motifs within unaligned noncoding sequences clustered by whole-genome mRNA quantitation. Nature Biotech 16:939–945.
20. Staden R (1984) Computer methods to locate signals in nucleic acid sequences. Nucleic Acids Res 13:505–519.
21. Stormo GD, Hartzell GW (1989) Identifying protein-binding sites from unaligned DNA fragments. Proc Natl Acad Sci USA 86:1183–1187.
22. Sinha S, Tompa M (2002) Discovery of novel transcription factor binding sites by statistical overrepresentation. Nucleic Acids Res 30:5549–5560.
23. Sumazin P, Chen GX, Hata N, Smith AD, Zhang T, Zhang MQ (2004) DWE: Discriminating Word Enumerator. Bioinformatics 21:31–38.
24. Thompson JD, Higgins DG, Gibson TJ (1994) ClustalW: improving the sensitivity of progressive multiple sequence alignment through sequence weighting, position-specific gap penalties and weight matrix choice. Nucleic Acids Res 22:4673–4680.
25. van Helden J, Andre B, Collado-Vides J (2000) Discovering regulatory elements in non-coding sequences by analysis of spaced dyads. Nucleic Acids Res 28:1808–1818.

26. Wolfertstetter F, Frech K, Herrmann G, Werner T (1996) Identification of functional elements in unaligned nucleic acid sequences by a novel tuple search algorithm. Bioinformatics 12:71–81.
27. Zhang MQ, Marr TG (1993) A weight array method for splicing signal analysis. Computer Application in the Biosciences (CABIOS) 9 (5):499–509.
28. Zhao XY, Huang HY, Speed T (2004) Finding short DNA motifs using permuted Markov models. Proceeding of RECOMB 4:68–75.

13

The Stochastic Model and Metastability of the Gene Network

Yuping Zhang and Minping Qian

Summary. There are large numbers of molecules, including proteins, DNA, RNA and so on, with complicated motions in the living cells. The interactions between them constitute molecular networks that carry out various cellular functions. These molecular networks should be dynamically stable against various fluctuations that are inevitable in the living world. For such large complex systems, what is the mechanism to regulate them functioning reliably and stably? Many works have contributed to this problem, from the dynamic point of view or the probabilistic point of view. In this chapter, we address this issue from the metastability point of view of a dynamic system perturbed by noises.

We model the network regulating the cell cycle of the budding yeast by a system of differential equations with noises, and explain how such a system can lead to a discrete model, a finite stationary Markov chain, which well matches the observation by biologists.

A continuous dynamic model of a system of ODEs can reasonably model the basic average dynamic behaviour of the interacting system of particles. But a discrete model on n particles of "on" and "off" type meets the biological observation well. We explain how the dynamic system can lead to a discrete state Markov chain on $\{0, 1\}^n$, by metastability theory, with the behaviour well matching the biological observation.

13.1 Introduction

Numerous molecules, including proteins, DNA, RNA, and small molecules, have complex interactions in the cell. The emergence and development of many high-throughput data-collection techniques, such as microarrays [62], protein chips or yeast two-hybrid screens [84], automated reverse-transcriptase polymerase chain reaction (RT-PCR) and two-dimensional (2D) gel electrophoresis help us to simultaneously obtain the expression profiles of a cell's components at any given time and find how and when these molecules interact with each other. It presents an opportunity to construct the real gene networks from experimental observation. Gene network inference is a hot research field of contemporary molecular biology.

Complex interactions between the cell's numerous constituents determine not only the structure of biological networks but also most other biological characteristics, such as complexity, versatility, dynamics, and robustness [6–10, 12, 41, 42]. Another major challenge of contemporary molecular biology is to systematically investigate the complex molecular processes underlying biological systems—how these molecules and their interactions determine the functions of all kinds of complex living systems, as well as other system-scale biological characteristics of cellular networks. Much research in network biology indicates that cellular networks, such as an organism's metabolic network and genetic interaction network, are driven by self-organizing processes and governed by universal laws [14–21].

13.1.1 From Experiment Data to Gene Network

Increment and accumulation of biological data, such as expression profiles data, CHIP-chip data, and DNA sequences data, make it possible to infer gene network and understand the functioning of organisms on the molecular level. One should select candidate genes that show significant expression changes, because large amounts of data can be simultaneously routinely generated by large-scale gene screening technologies, such as mRNA hybridization micro-arrays and RT-PCR. To identify these genes of interest, the simplest methods are straightforward scoring methods, according to whether a significant change exists at one or all conditions or whether the fluctuation pattern shows high diversity according to Shannon entropy [63].

Besides selecting genes that show significant expression changes, we sometimes further need to find coexpression genes, assuming they are caused by co-regulation. We can classify gene expression patterns to find coexpression genes for exploring shared functions and regulations. Many clustering methods can be used to accomplish classification. Before clustering the gene expression data, we should choose the distance measure first, which is used to quantify the difference in expression profiles between two genes, and may be as important as the choice of clustering algorithm. Different distance measures emphasize different regularities presented within the data. Distance measures can be divided into at least three classes, according to different types of regularities in the data [58]:

1. Similarity according to positive correlations, which may identify similar or identical regulation

2. Similarity according to positive and negative correlations, which may also help identify control processes that antagonistically regulate downstream pathways

3. Similarity according to mutual information, which may detect even more complex relationships.

When the distance measure has been chosen, one can choose the preferable clustering method. Many clustering algorithms have been proposed. As a whole, clustering algorithms can be divided into hierarchical and nonhierarchical methods. Hierarchical methods return a hierarchy of nested clusters, where each cluster typically consists of the union of two or more smaller clusters. Nonhierarchical methods typically cluster N objects into K groups in an iterative process until certain goodness criteria are optimized [58], such as the following:

1. The K-means algorithm [2] can be used to partition N genes into K clusters, where K is predetermined by the user (see Tavazoie et al [64] for an application to yeast gene expression).

2. The self-organized map (SOM) method is closely related to K-means and has been applied to mRNA expression data of yeast cell cycles as well as hematopoietic differentiation of four well-studied model cell lines [65].

3. The expectation-maximization (EM) algorithm [66] for fitting a mixture of Gaussians (also known as fuzzy K-means; [3]) is very similar to K-means, and has been used by Mjolsness et al [67] to cluster yeast data.

4. Autoclass is also related to EM, in that a mixed probability distribution has been found. In addition, Bayesian methods have been used to derive the maximum posterior probability classification, and the optimum number of clusters [4].

Different clustering methods can work well in different applications. It's hard to say which clustering method is the best.

Clustering of gene expression data can only help elucidate the regulation (or co-regulation) of individual genes, not what is regulating what. The eventual goal of gene network inference is to understand the integrated behaviour of networks of regulatory interactions—construct a coarse-scale model of the network of regulatory interactions between the genes. To deduce the unknown underlying regulatory network from a large amount of data, one requires inference of the causal relationships among genes, that is, reverse engineering the network architecture from its activity profiles [58].

Reverse engineering is the process of elucidating the structure of a system by reasoning backward from observations of its behaviour [59]. In reverse engineering biological networks, a complex genetic network underlies a massive set of expression data, and the task is to infer the connectivity of the genetic circuit. Many methods can be used, such as directed graph, Bayesian network, and Boolean network and their generalizations; ordinary and partial differential equations; qualitative differential equations; stochastic equations; and so on.

Complexity of the model used to infer a gene network determines the number of needed data points, such as the data requirement of a fully connected Boolean network model, $O(2^N)$, while a continuous fully connected model with additive restrictions requires at least $N + 1$ data points [85, 86].

Different models can help us to analyse the biological systems at different levels—from the very coarse and abstract to the very concrete. One can choose models like Boolean networks to handle large-scale data in a global fashion with examination of very large systems (thousands of genes), or choose fine-grained quantitative stochastic models, such as full biochemical interaction models with stochastic kinetics in Arkin et al [60], to analyse biological systems in the very concrete scale.

13.1.2 Topological Properties

Interactions between numerous molecules in cells produce various types of molecular interaction networks including protein-protein interaction, metabolic, signaling, and transcription-regulatory networks. Most of these networks are proved to have system-scale behaviour.

In 1960, Paul Erdös and Alfrëd Rënyi [24] initiated the study of the mathematical properties of random networks. The Erdös-Rënyi (ER) model of a random network starts with N nodes and connects independently each pair of nodes with probability p. The node degrees follow a Poisson distribution, which indicates that most nodes have roughly the same number of links, approximately equal to the network's average degree. The clustering coefficient $[C_I = 2n_I/k(k-1)$, where n_I is the number of links connecting the k_I neighbours of node I to each other$]$ is independent of a node's degree. The mean path length ($< l >$, which represents the average over the shortest paths between all pairs of nodes and offers a measure of a network's overall navigability) is proportional to the logarithm of the network size $< l >\sim \log N$, which indicates that it is characterized by the small-world property. Recently, a series finding indicates that many real networks share common architectural features that deviated from the random network. The most striking property is that, in contrast to the Poisson degree distribution, for many real networks, from social networks to cellular networks [15–19, 21, 25–30], the number of nodes with a given degree follows a power law. That is, the probability that a chosen node has exactly k links follows $P(k) \sim k^{\gamma}$, where γ is the degree exponent, with its value for most networks being between 2 and 3 [23]. Networks with a power degree distribution are called scale-free [23, 32, 33], and are highly nonuniform; most of the nodes have only a few links, and a few nodes have a very large number of links. Scale-free networks with degree exponents $2 < \gamma < 3$, a range that is observed in most biological and nonbiological networks, are ultrasmall [31, 33], with the average path length following $< l >\sim \log\log N$, which is significantly shorter than $\log N$, which characterizes random small-world networks. Gene duplication is likely to be a key mechanism for generating the scale-free topology [14, 34–40].

In contrast to the scale-free features of complex networks emphasizing the organizing principles that determine the network's large-scale structure, the concept of modularity is introduced to characterize specific networks. It starts from the bottom and looks for highly representative patterns of interactions. Small regulatory interaction patterns, called subgraphs and motifs, occupy distinct positions in and between organs, offering insights into their dynamic role in information processing [48]. Milo et al [49] identified small subgraphs (motifs) that appear more frequently in a real network than in its randomized version. In general, modularity refers to a group of physically or functionally linked molecules (nodes) working together to achieve a different function [6, 41–43]. Biological functions are carried by discrete functional modules. For example, temporally co-regulated groups of molecules are known to govern various stages of the cell cycle [44–46], or to convey extracellular signals in bacteria or the yeast pathways. Natural selection aims to maintain function, which, however, is rarely carried by single components, but rather by a network of interacting subunits. Therefore, we should see a tendency toward the evolutionary conservation of subnetworks that are capable of carrying biological function. Motifs aggregate into motif clusters, which is likely a general property of most real networks [47, 49]. However, modularity and scale-free property seem to be contradictory. The definition of modules seems to indicate the existence of some groups of nodes relatively isolated from the rest of the system, whereas in a scale-free network hubs are in contact with a high fraction of nodes, which makes the existence of relatively isolated modules unlikely [14]. Because of the coexistence of clustering and hubs, topological modules are not independent, but rather hierarchical [43, 51]. In fact, many real biological systems, including all examined metabolic [51] and protein interaction networks, demonstrate that a hierarchical relationship among modules universally exists.

13.1.3 Robustness and Dynamics

Networks with power-law distributed degrees are robust to random perturbations. Upon removal of randomly chosen nodes, the mean distance (network diameter) between network nodes that can still be reached from each other increases only very little, while in graphs with other degree distribution, network diameter can increase substantially [52]. Cellular networks can be subject to random errors as a result of mutations or protein misfolding, as well as harsh external conditions eliminating essential metabolites [53]. Many real complex systems have a key feature of robustness, which refers to the system's ability to maintain relatively normal behaviour responding to changes in the external conditions or internal organization [14]. Jeong et al found that metabolic network graphs with power-law distributed degrees are robust against perturbations [15]. A considerable amount of attention has been paid to the quantitative modeling and understanding of the budding yeast cell cycle regulation [21, 25, 68–77], and similar robust results have been obtained for the protein network of yeast as well [21, 25].

On a molecular level, functions in time, that is, causality and dynamics, are manifested in the behaviour of complex networks. The dynamics of these networks resemble trajectories of state transitions, which correspond to temporal gene expression. The concept of attractors is what really lends meaning to these trajectories; that is, the attractors are the high-dimensional dynamic molecular representations of stable phenotypic structures such as differentiated cells and tissues, either healthy or diseased [1, 55]. In higher metazoa, each gene or protein is estimated on average to interact with four to eight other genes [56], and to be involved in ten biological functions [57]. The global gene expression pattern is therefore the result of the collective behaviour of individual regulatory pathways. In such highly interconnected cellular signaling networks, gene function depends on its cellular context; thus understanding the network as a whole is essential [58].

One available method to research the principles of network behaviour is to radically simplify the individual molecular interactions, and focus on the collective outcome. Kauffman [54] represented Boolean networks, in which each gene is considered as a binary variable (either ON or OFF) and regulated by other genes through logical or Boolean functions [61]. Even the construction of Boolean network is very simple; the network behaviour is already extremely rich [1]. Many useful concepts naturally emerge from such a simple mathematical model. Take, for example, the budding yeast cell cycle regulation system. Li et al [70] introduced a deterministic Boolean network [54] model and investigated its dynamic and structural properties. Their main result is that the network is both dynamically and structurally stable. The biological stationary state is a big attractor of the dynamics; the biological pathway is a globally attracting dynamic trajectory. These properties are largely preserved with respect to small structural perturbations to the network, for example, adding or deleting links.

Boolean networks provide a useful conceptual tool for investigating the principles of network organization and dynamics. We can study the role of various constraints on global behaviour in terms of network complexity, stability, and evolvability. Investigations into abstract models will help us understand the cybernetic significance of network features, and provide meaningful questions for targeted experimental exploration.

13.2 Discrete Stochastic Model

As referred to in section 13.1.3, Li et al [70] introduced a deterministic Boolean network model and investigated its dynamic and structural properties. However, one crucial point left unaddressed in their study is the effect of stochasticity or noise, which inevitably exists in a cell and may play important roles [78]. We investigated the stochastic effect on the deterministic network model of Li et al [79]. We found that both the biological stationary state and the biological pathway are well preserved under a wide range of noise level. When the noise is larger than a value of the order of the interaction strength, the network dynamics quickly become noise dominating and lose their biological meaning. In Li et al, six attractors without biological meanings are produced. Our result indicates that six other attractors without biological meanings are unstable under a real, noisy environment. All states converge to the most stable state (stationary G1 state—the attractor with biological meaning in [70]).

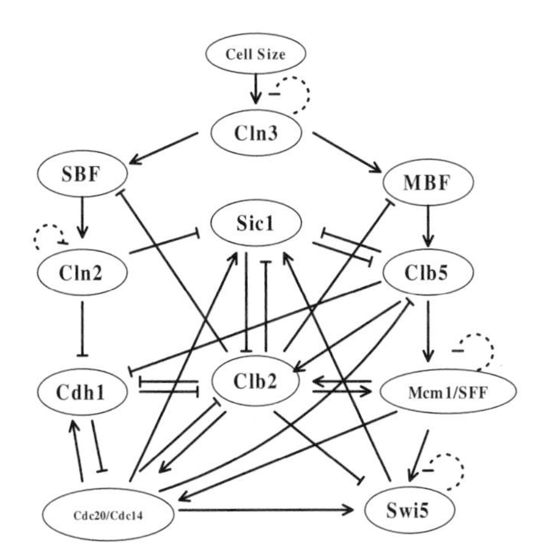

Fig. 13.1. The cell-cycle network of the budding yeast. Each node represents a protein or a protein complex. Arrows are positive regulation, "T"-lines are negative regulation, dotted "T"-loops are degradation.

In our model, the 11 nodes in the network shown in Fig. 13.1, namely, Cln3, MBF, SBF, Cln2, Cdh1, Swi5, Cdc20, Clb5, Sic1, Clb2, and Mcm1, are represented by variables $(s_1, s_2, ..., s_{11})$, respectively. Each node i has only two values: $s_i = 1$ and $s_i = 0$, representing the active state and the inactive state of the protein i, respectively. Mathematically we consider the network evolving on the configuration space $S = \{0, 1\}^{11}$; the $2^{11} = 2048$ "cell states" are labeled by $\{n = 0, 1, ..., 2047\}$. The statistical behaviour of the cell state at the next time step is determined by the cell state at the present time step. That is, the evolution of the network has the Markov property [5]. The time steps here are logic steps that represent causality rather than actual times. The stochastic process is assumed to be time homogeneous. Under these assumptions and considerations, we define the transition probability of the Markov chain as follows:

$$P_r(s_1(t+1), ..., s_{11}(t+1)|s_1(t), ..., s_{11}(t)) = \prod_{i=1}^{11} P_r(s_i(t+1)|s_1(t), ..., s_{11}(t)), \tag{13.1}$$

where

$$P_r(s_i(t+1) = \sigma_i|s_1(t), ..., s_{11}(t)) = \frac{\exp(\beta(2\sigma_i - 1)\sum_{j=1}^{11} a_{ij}s_j(t))}{\exp(\beta\sum_{j=1}^{11} a_{ij}s_j(t)) + \exp(-\beta\sum_{j=1}^{11} a_{ij}s_j(t))},$$

if $\sum_{j=1}^{11} a_{ij}s_j(t) \neq 0, \sigma_i \in \{0, 1\}$; and

$$P_r(s_i(t+1) = s_i(t)|s_1(t), ..., s_{11}(t)) = \frac{1}{1+e^{-\alpha}}, \tag{13.2}$$

if $\sum_{j=1}^{11} a_{ij}s_j(t) = 0$. We define $a_{ij} = 1$ for a positive regulation of j to i and $a_{ij} = -1$ for a negative regulation of j to i. If the protein i has a self-degradation loop, $a_{ii} = -0.1$. The positive number β is a temperature-like parameter characterizing the noise in the system [80]. To characterize the stochasticity when the input to a node is zero, we have to introduce another parameter α. This parameter controls the likeliness for a protein to maintain its state when there is no input to it. Notice that when β, $\alpha \to \infty$, this model recovers the deterministic model of Li et al [70]. In this case, they showed that the G1 state ($n = 68$) is a big attractor, and the path $1092 \to 836 \to 964 \to 896 \to 904 \to 907 \to 155 \to 51 \to 55 \to 53 \to 116 \to 100 \to 68$ is a globally attracting trajectory. Our study focuses on the stochastic properties of the system.

We first study the property of the biological stationary state G1 and define an "order parameter" as the probability for the system to be in the G1 state, π_{G1}. Plotted in Fig. 13.2A is the value of the order parameter as a function of the control parameter β. At large β (low "temperature" or small noise level), the G1 state is the most probable state of the system and π_{G1} has a significant value. (Note that for a finite α there are "leaks" from the G1 state and the maximum π_{G1} is less than 1.) When β is lowered, one observes a sharp transition at around $\beta_c \approx 1.1$ where π_{G1} drops to a very small value, indicating a "high temperatures" phase in which the network dynamics cannot converge to the biological steady state G1. The system, however, is rather resistant to noise. The transition "temperature" is quite high—the value of

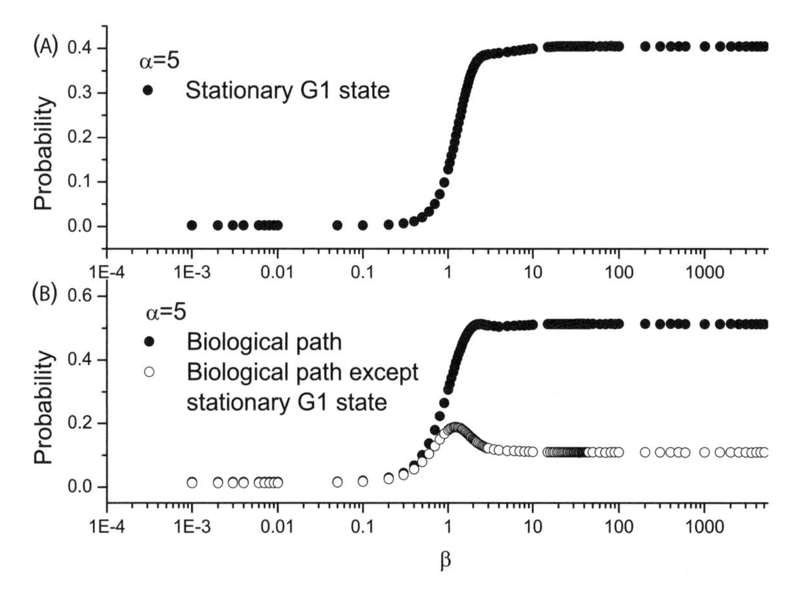

Fig. 13.2. The probability of the stationary G1 state and the biological pathway. (A) The order parameter π_{G1} as a function of β. (B) The sum of the probability for the states on the bio-pathway with and without G1 being included.

$\beta_c \approx 1.1$ implies that the system will not be significantly affected by noise until a fraction of $e^{-1.1}/(e^{1.1} + e^{-1.1}) \approx 0.1$ of the updating rules is wrong.

We next study the statistical properties of the biological pathway of the cell-cycle network. In Fig. 13.2B, we plot the probability for the system to be in any of the biological states along the biological pathway, as a function of β. We also plot the same probability with the G1 state excluded. One observes a similar transition as before. The jump of the probability of the states along the biological pathway in the low temperature phase is due to the fact that in this phase the probability flux among different states in the systems is predominantly the flow along the biological pathway. To visualize this, we show in Fig. 13.3 an example of the probability flux among all 2048 states. Each node in Fig. 13.3 represents one of the 2048 states. The size of a node reflects the stationary distribution probability of the state. If the stationary probability of a state is larger than a given threshold value, the size of the node is in proportion to the logarithm of the probability. Otherwise, the node is plotted with the same smallest size. The arrows reflect the pure probability flux (only the largest flux from any node is shown). The probability flux is divided into seven grades, which are expressed by seven colors: light-green, canary, goldenrod, dandelion, apricot, peach and orange. The warmer the color is, the wider the arrow is, and the larger the probability flux. The width of an arrow is in proportion to the logarithm of the probability flux it carries. The arrow representing the probability flux from the stationary G1 state to the excited G1 state (the start of the cell-cycle) is shown in dashed lines. One

Fig. 13.3. The probability flux. For a node, only the largest flux from it is shown. The nodes on the biological pathway are denoted with different colors: purple, the stationary G1 state; blue, the other G1 states; olive-green, S state; dandelion, G2 state; and red, M states. All other states are noted in normal green. The simulations were done with $\alpha = 5$ and $\beta = 6$.

observes that once the system is "excited" to the start of the cell cycle process [here by noise (α) and in reality mainly by signals like "cell size",] the system will essentially go through the biological pathway and come back to the G1 state. Another feature of Fig. 13.3 is that the probability flux from any states other than those on the biological pathway is convergent with the biological pathway. Notice that this diagram is similar to the Fig. 2 of [70]. For $\beta < \beta_c$, this feature of a convergent high flux bio-pathway disappears.

We also define a "potential" function and study the change of the "potential landscape" as a function of β. Specifically, we define

$$S_n = -\log \pi_n = \beta E_n, \tag{13.3}$$

where E_n is the pseudo-energy defined by

$$E_n = -\frac{\log \pi_n}{\beta}. \tag{13.4}$$

Fig. 13.4 shows four examples of $\Delta S_n = S_n - S0$ distribution, where the reference potential $S0$ in each plot is set as the highest potential point in the system.

One observes that far from the transition point ($\beta = 0.01$, Fig. 13.4A), the potential values are high (around -4), and the landscape is flat. Near to but below the transition point ($\beta = 0.6$, Fig. 13.4B), some local minima (blue points) become more pronounced, but the landscape still remains rather flat. We notice that these minimum points do not correspond to the biological pathway. Right after the transition point ($\beta = 1.5$, Fig. 13.4C), the system quickly condenses into a landscape with deep valleys. The state with the lowest

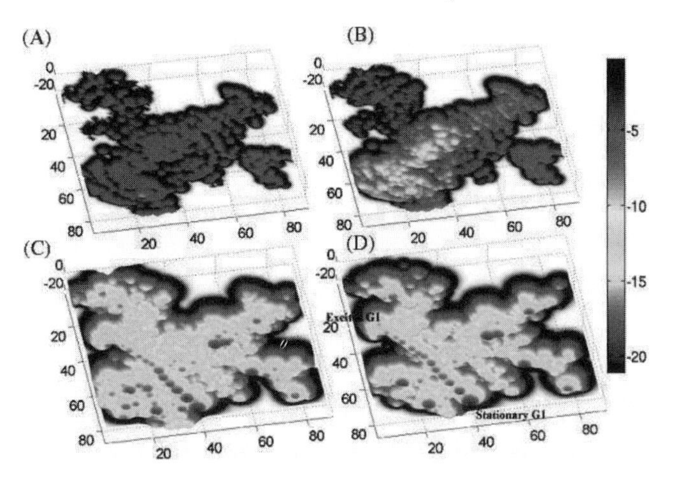

Fig. 13.4. The "potential" landscape of the system before and after the transition. (A) $\beta = 0.01$, (B) $\beta = 0.6$, (C) $\beta = 1.5$, (D) $\beta = 6.0$, all for $\alpha = 5$. The color code gives the relative value of the potential function.

potential value corresponds to the stationary G1 state. A linear line of blue dots from up-left to down-middle corresponds to the biological pathway, which forms a deep valley. Some deep blue dots out of the biological pathway are local attractors in [70]. Notice that although their potential values are low, they attract only a few nearby initial states—all these points are more or less isolated. After the transition point, the potential landscape does not change qualitatively ($\beta = 6$, Fig. 13.4D). As $\beta, \alpha \to \infty$, the landscape becomes seven deep holes, each corresponding to an attractor of the system [70].

13.3 Continuous Stochastic Model

We have introduced stochastic noises into the discrete model and have given a microscopic statistical understanding of the system. The model is still an "on-off" model, which do not match very well the real interaction between genes and proteins. The Boolean approximation assumes highly cooperative binding (very "sharp" activation response curves) or positive feedback loops to make the variables saturate in ON or OFF positions. However, examining real gene expression data, it seems clear that genes spend a lot of their time at intermediate values; gene expression levels tend to be continuous rather than binary. We need to introduce variables, characterizing the quantities of each kind of gene and protein, which are crucial for the chance of binding.

13.3.1 Interacting Particle System Model

Assume r kinds of proteins or genes in all. N_i^M is the total number of the ith kind of particle. $N_i(t)$ $(0 \leq N_i(t) \leq N_i^M)$ is the number of the ith kind of particle that is active at time t. Given that there are currently $N(t)$ particles that are active in the system, as $h \to 0$, let

$$P(N_i(t+h) - N_i(t) = k|\mathbf{N}(t)) = \begin{cases} \lambda_i(\mathbf{N}(t))h + o(h) & , \text{ if } k = 1 \\ \mu_i(\mathbf{N}(t))h + o(h) & , \text{ if } k = -1 \\ o(h) & , \text{ if } |k| > 1 \end{cases}$$

which assumes only one increasing or decreasing event occurs in a very small interval of time. While the probability is not exactly zero for more than one event, it is negligible. The above conditions imply that

$$P(N_i(t+h) - N_i(t) = 0|\mathbf{N}(t)) = 1 - \lambda_i(\mathbf{N}(t))h - \mu_i(\mathbf{N}(t))h + o(h). \quad (13.5)$$

We get a Markov chain to characterize the system. The transition probability of the Markov process is

$$P(\mathbf{N}(t+h)|\mathbf{N}(t)) = \prod_{i=1}^{r} P(N_i(t+h)|\mathbf{N}(t)). \quad (13.6)$$

The biological observation can only indicate which kind of particles is mostly active or inactive.

13.3.2 Approximate SDE for Particle Systems

Set $X_t^i = \frac{N_i(t)}{N_i^M}$, $t \geq 0$. This system nests on $[0,1]^r$. $\mathbf{X_t}$ is the proportion of all kinds of particles that are active or in a working situation at time t. Each dimension X_t^i $(i \in [1,r])$ stands for the proportion of a kind of particles that are active, which behaves like birth and death processes.

This can be approximated by a stochastic differential equation (SDE) taking the value $[0,1]^r$, where r is the number of kinds of particles.

$$dX_t^i = \varphi_i(\mathbf{X_t})dt + \varepsilon \Sigma_{i=1}^n X_t^i dw_t \quad (13.7)$$

In our system, the attractors in the determinate model all are located at the vertices of points (or sets).

The motion of the system moving from one to the other metastable state can be approximated by the Markov chain on $\{0,1\}^r$, which is like the Markov chain of the "on-off" model.

The biological path reported is the transition path between high-level attractors, which can be reached in the real observation time scale.

13.4 Metastability

Metastability was a phenomenon first reported in the physics literature. Robert Schomann was the first to bring the subject to the attention of probabilists, to discuss the metastable behaviour of the contact process and two-dimensional Ising model. Chen, Feng, and Qian introduced the large deviation method to solve this problem mathematically [80, 81].

Roughly speaking, metastability is a property of the system that persists in its existing equilibrium when undisturbed (or only slightly disturbed) but is able to pass to a more stable equilibrium when sufficiently disturbed.

13.4.1 Metastability of Markov Chain

Let the state space $\mathbf{S} = \{\xi_1, ..., \xi_N\}$ be finite. Consider a family of Markov chains on $\mathbf{S}$ with transition matrices $\{\mathbf{P}^\beta(\xi, \eta); \beta \in [0, \infty]\}$, which satisfy

$$\lim p^\beta(\xi, \eta) = p^\infty(\xi, \eta), \forall \xi, \eta \in \mathbf{S} \tag{13.8}$$

as $\beta \to \infty$ (the temperature $\to 0$). Assume that the following limit exists:

$$\lim_{\beta \to \infty} -\frac{1}{\beta} \log p^\beta(\xi, \eta) = C_{\xi\eta}, \tag{13.9}$$

$$\text{if } p^\infty(\xi\eta) = 0 \text{ (with convention that } \log 0 = -\infty), \text{then } C_{\xi\eta} > 0. \tag{13.10}$$

The family of Markov chains with transition probability matrices $\{\mathbf{P}^\beta(\xi, \eta); \beta \in [0, \infty]\}$ is an exponential perturbation of the Markov chain with transition probability matrix $\mathbf{P}^\infty$. If $\{\mathbf{P}^\infty(\xi, \eta)\}$ is degenerate and irreducible, the stochastic model becomes the deterministic dynamical system. In Qian et al [81], the metastability of exponentially perturbed Markov chains is exploited by the following theorems. Before describing the theorems, we first give some definitions. For a subset $K \subset S$, the exit time of Markov chain $\{X_n\}$ is denoted by

$$\tau(K) = inf\{n; X_n \notin K\}, \tag{13.11}$$

and the hitting time by

$$\sigma(K) = inf\{n; X_n \in K\}. \tag{13.12}$$

Let $\{A_1, ..., A_s\}$ be the set of all recurrent classes of $\mathbf{P}^\infty$. We assume that $\mathbf{P}^\infty_{A_i}$ is aperiodic with every $i \in \{1, ..., s\}$, while $\mathbf{P}^\infty_{A_i}$ is the transition probability matrix restricted to recurrent class A_i. For each A_i, which corresponds to an attractor of dynamical systems, define the attractive basin B_i to be

$$B_i = \{\xi | \mathbf{P}^\infty(\sigma(A_i) < \infty | X_0 = \xi) > 0\}). \tag{13.13}$$

Obviously $\cup_i B_i = S$. However, for the stochastic version we are considering now, $B_i \cap B_j \neq \emptyset$ is well possible. The behaviour of the Markov chain (such as exit time, exit distribution, ergodicity, etc.) with very large β is determined by $C_{\xi\eta}$. Define

$$T(B_i) = min\{\Sigma_{k=1}^{l} C_{\xi_{k-1}\xi_k} | l \geq 1, \xi_0 \in A_i, \xi_1, \xi_2, ..., \xi_{l-1} \in B_i, \xi_l \notin B_i\} \tag{13.14}$$

and for $\xi \in B_i, \eta \notin B_i$, $T_{\xi\eta}(B_i)$ can be similarly defined.

Theorem 1. *Suppose that $\{X_n\}$ is a Markov chain starting from $\xi \in B_i$.*
i) Exit time

$$\lim_{\beta\to\infty} \frac{1}{\beta} \log E_\xi^\beta \tau(B_i) = T(B_i). \tag{13.15}$$

ii) Exit distribution: For $\eta \notin B_i$,

$$\lim_{\beta\to\infty} \frac{-1}{\beta} \log P^\beta(X_{\tau(B_i)} = \eta | X_0 = \xi) = -T(B_i) + T_{\xi\eta}(B_i), \tag{13.16}$$

$$\lim_{\beta\to\infty} P_\xi^\beta(X_{\tau(B_i)} \in \{\eta \notin B_i : T_{\xi\eta}(B_i) = T(B_i)\}) = 1. \tag{13.17}$$

Theorem 2. *If the initial state $\xi \in B_i \setminus \cup_{j\neq i} B_j$, then $\frac{\tau(B_i)}{E_\xi^\beta \tau(B_i)}$ converges in distribution to the exponential random variable with mean 1, as $\beta \to \infty$. In particular, for $\delta > 0$ small enough, we have*

$$\lim_{\beta\to\infty} P_\xi^\beta(e^{(T(Bi)-\delta)\beta} < \tau(B_i) < e^{(T(Bi)+\delta)\beta}) = 1, \tag{13.18}$$

$$\lim_{\beta\to\infty} \frac{-1}{\beta} \log P_\xi^\beta(\sigma(\zeta) > e^{\delta\beta}) \geq \delta, \forall \zeta \in A_i. \tag{13.19}$$

Theorem 3. *Suppose that $\{\upsilon_i(\zeta)\}$ is the invariant measure of $\mathbf{P}_{A_i}^\infty$. For $\zeta \in A_i$, $\xi \in B_i \setminus \cup_{j\neq i} B_j$, $0 < \delta < T(B_i)$, then*

$$\lim_{\beta\to\infty} E_\xi^\beta [\frac{1}{N_\beta} \sum_{k=1}^{N_\beta} I_{\{\zeta\}}(X_k) - \upsilon_i(\zeta)]^2 = 0, \tag{13.20}$$

where I is the indicator function and N_β is the integral part of $e^{(T(B_i)-\delta)\beta}$.

Demonstration of the above three theorems and more details can be found in Qian et al [81]. The phenomena characterized by the theorems are collectively called metastability. This kind of metastability can also be observed at higher levels. In Qian et al the higher-level metastability is also described.

The transition from one attractive basin to another attractive basin also exhibits metastability. Some attractive basins constitute an attractive basin of the second level. The Markov chain stays in a certain attractive basin of the second level for an asymptotically exponential random time, and then enters another attractive basin of the second level with a definite hitting probability. Repeating this process, we can get attractive basins of the third level, and so on.

Let us explain the metastability of the Markov chain with exponential perturbation more distinctly. For β is infinite, the state space of the system breaks down into several subspaces. Every state of each subspace never gets to the other subspaces. In other words, every state can only evolute in its own subspace. If β is large but finite, the system will be ergodic; the stable states (or sets) become "metastable" states (or sets). Noise makes the attractors stronger or weaker. How hard it is to leave one attractor and reach another attractor is also affected by the noise level. The evolutional distance between different states will become closer or further apart compared to the situation when β is infinite. There is a hierarchical structure, like a pyramid, between these metastability states. Considering a metastable state b, the time that the system stays is in the order of $e^{\beta T(b)}$ (as $\beta \to \infty$), where $T(b)$ is a constant. In the time less than the scale of $e^{\beta T(b)}$, the state b looks stable; when the time scale is larger than $e^{\beta T(b)}$, b becomes unstable, and goes to a metastable state of a higher level. That's why we call b a metastable state. As $\beta \to \infty$, the system spends most of its time at the "most stable" (the highest level of) metastable state(s), and spends less time in the next level of metastable state(s). In a fixed short time scale, the system almost never reaches the metastable states of low-enough levels [81]. Thus in biological observation these states never can be seen. When the system is going from a metastable state to another metastable state, usually there is a most possible path, with probability of almost 1 [81]. Therefore, in our observation we can only see the system going on the most possible path.

In our example [79] (see Fig. 13.4), the states with cooler colors are metastable states of higher levels. When $\beta \to \infty$, the system almost stays at the bluest state. Transition from one metastable state to another metastable state goes along the transition path of Fig. 13.3 (arrows), with probability of almost 1.

13.4.2 Metastability of SDE

We introduced the metastability theory of discrete Markov chains above. In fact, a similar metastability theory of stochastic differential equations exists. Suppose that we have a stochastic differential equation as follows:

$$dx_t = b(x_t)dt + \varepsilon dw_t \tag{13.21}$$

where $\varepsilon = \gamma/\beta$ and γ is independent of β. Then $\beta \to \infty$ corresponds to $\varepsilon \to 0$. When $\varepsilon \to 0$, the system become a deterministic dynamic system with several attractors. These attractors can also move from one to the others under random perturbation. Attractors (points) are not stable sets (points) forever, but become metastable sets (states) at different levels. A hierarchical structure like a pyramid between these metastability states also exists.

13.5 Conclusion

We introduced some stochastic models for the yeast cell cycle network. In a discrete stochastic model, we have found that there exists a transition point as the noise level is varied. With a lot of noise, the network behaves randomly; it cannot carry out the ordered biological function. When the noise level drops below a critical value, which is of the same order as the interaction strength ($\beta_c \approx 1.1$), the system becomes ordered: the biological pathway of the cell cycle process becomes the most probable pathway of the system and the probability of deviating from this pathway is very small. So in addition to the dynamic and the structural stability [70], this network is also stable against stochastic fluctuations. Metastability theory makes continuous stochastic models connect with discrete stochastic models. By dint of metastability theory, we interpret why the simulation results of discrete models match with real biological situations well. We used a pseudo-potential function to describe the dynamic landscape of the system. In this language, the biological pathway can be viewed as a valley in the landscape [82, 83]. This analogy to equilibrium systems may not be generalizable, but it would be interesting to see if one can find more examples in other biological networks, which are very special dynamic systems.

References

1. Kauffman SA (1993) The Origins of Order, Self-Organization and Selection in Evolution. Oxford University Press, Oxford.
2. MacQueen J (1967) Some methods for classification and analysis of multivariate observation. In: Le Cam L M, Nyeman J (eds) Proceedings of the Fifth Berkeley Symposium on Mathematical Statistics and Probability, vol I. University of California Press.
3. Bezdek JC (1981) Pattern Recognition with Fuzzy Objective Function Algorithms. Plenum Press, New York.
4. Cheeseman P, Stutz J (1996) Bayesian classification (autoclass): theory and results. In: Fayyad UM, Piatetsky-Shapiro G, Smyth P, Uthurusamy R

(eds) Advances in Knowledge Discovery and Data Mining. AAAI Press/MIT Press, Cambridge, MA. http://icwww.arc.nasa.gov/ic/projects/bayes-group/images/kdd-95.ps.
5. Chung KL (1967) Markov Chains with Stationary Transition Probability. Springer, New York.
6. Hartwell LH, Hopfield JJ, Leibler S, Murray AW (1999) From molecular to modular cell biology. Nature 402:C47–C52.
7. Hasty J, McMillen D, Collins JJ (2002) Engineered gene circuits. Nature 420:224–230.
8. Kitano H (2002) Computational systems biology. Nature 420:206–210.
9. Koonin EV, Wolf YI, Karev GP (2002) The structure of the protein universe and genome evolution. Nature 420:218–223.
10. Oltvai ZN, Barabasi AL (2002) Life's complexity pyramid. Science 298:763–764.
11. Wall ME, Hlavacek WS, Savageau MA (2004) Design of gene circuits: lessons from bacteria. Nature Rev Genet 5:34–42.
12. Bray D (2003) Molecular networks: the top-down view. Science 301:1864–1865.
13. Alon U (2003) Biological networks: the tinkerer as an engineer. Science 301:1866–1867.
14. Barabsi AL, Oltvai ZN (2004) Network biology: understanding the cell's functional organization. Nature Rev Genet 5:101–113.
15. Jeong H, Tombor B, Albert R, Oltvai ZN, Barabasi AL (2000) The large-scale organization of metabolic networks. Nature 407:651.
16. Eisenberg E, Levanon EY (2003) A. Phys Rev Lett
17. Jeong, Neda, B AL (2003) A. (author names incomplete) Europhys Lett
18. Li et al (2004) Science
19. Giot et al (2003) Science
20. Zhou H (2003) Network landscape from a Brownian particle's perspective. Phys Rev E 67:041908.
21. Vogelstein B, Lane D, Levine A J (2000) Surfing the p53 network. Nature 408:307–310.
22. Watts D J, Strogatz S H (1998) Collective dynamics of 'small-world' networks. Nature 393:440.
23. Barabasi AL, Albert R (1999) Emergence of scaling in random networks. Science 286:509–512.
24. Erdös P, Rënyi A (1960) On the evolution of random graphs. Publ Math Inst Hung Acad Sci 5:17–61.
25. Jeong H, Mason SP, Barabasi AL, Oltvai ZN (2001) Lethality and centrality in protein networks. Nature 411:41–42.
26. Wagner A (2001) The yeast protein interaction network evolves rapidly and contains few redundant duplicate genes. Mol Biol Evol 18:1283–1292.
27. Yook SH, Oltvai ZN, Barabasi AL (2004) Functional and topological characterization of protein interaction networks. Proteomics 4:928–42.
28. Uetz P et al (2000) A comprehensive analysis of protein—protein interactions in Saccharomyces cerevisiae. Nature 403:623–627.
29. Ito T et al (2001) A Comprehensive two-hybrid analysis to explore the yeast protein interactome. Proc Natl Acad Sci USA 98:4569–4574.
30. Featherstone D E, Broadie K (2002) Wrestling with pleiotropy: genomic and topological analysis of the yeast gene expression network. Bioessays 24:267–274.
31. Chung F, Lu L (2002) The average distances in random graphs with given expected degrees. Proc Natl Acad Sci USA 99:15879–15882.

32. Barabasi AL, Bonabeau E (2003) Scale-free networks. Sci Am 288:60–69.
33. Cohen R, Havlin S (2003) Scale-free networks are ultrasmall. Phys Rev Lett 90:058701.
34. Rzhetsky A, Gomez S M (2001) Birth of scale-free molecular networks and the number of distinct DNA and protein domains per genome. Bioinformatics 17:988–996.
35. Qian J, Luscombe N N, Gerstein M (2001) Protein family and fold occurrence in genomes: power-law behaviour and evolutionary model. J Mol Biol 313:673–681.
36. Bhan A, Galas D J, Dewey T G (2002) A duplication growth model of gene expression networks. Bioinformatics 18:1486–1493.
37. Pastor-Satorras R, Smith E, Sole RV (2003) Evolving protein interaction networks through gene duplication. J Theor Biol 222:199–210.
38. Vazquez A, Flammini A, Maritan A, Vespignani A (2003) Modeling of protein interaction networks. Complexus 1:38–44.
39. Kim J, Krapivsky PL, Kahng B, Redner S (2002) Infinite-order percolation and giant fluctuations in a protein interaction network. Phys Rev E Stat Nonlin Soft Matter Phys 66:055101.
40. Wagner A (2003) How large protein interaction networks evolve. Proc R Soc Lond B 270:457–466.
41. Wall ME, Hlavacek WS, Savageau MA (2004) Design of gene circuits: lessons from bacteria. Nature Rev Genet 5:34–42.
42. Alon U (2003) Biological networks: the tinkerer as an engineer. Science 301:1866–1867.
43. Ravasz E, Barabasi AL (2003) Hierarchical organization in complex networks. Phys Rev E Stat Nonlin Soft Matter Phys 67:026112.
44. Simon I et al (2001) Serial regulation of transcriptional regulators in the yeast cell cycle. Cell 106:697–708.
45. Tyson JJ, Csikasz-Nagy A, Novak B (2002) The dynamics of cell cycle regulation. Bioessays 24:1095–1109.
46. McAdams HH, Shapiro L (2003) A bacterial cell-cycle regulatory network operating in time and space. Science 301:1874–1877.
47. Shen-Orr SS, Milo R, Mangan S, Alon U (2002) Network motifs in the transcriptional regulation network of escherichia coli. Nat Genet 31: 64–68.
48. Balázsi G, Barabasi AL, Oltvai ZN (2005) Topological units of environmental signal processing in the transcriptional regulatory network of Escherichia coli. Proc Natl Acad Sci 102:7841–7846.
49. Milo R et al (2002) Network motifs: simple building blocks of complex networks. Science 298:824–827.
50. Bhalla US, Ram PT, Iyengar R (2002) MAP kinase phosphatase as a locus of flexibility in a mitogen-activated protein kinase signaling network. Science 297:1018–1023.
51. Ravasz E, Somera AL, Mongru DA, Oltvai ZN, Barabasi AL (2002) Hierarchical organization of modularity in metabolic networks. Science 297:1551–1555.
52. Albert R, Jeong H, Barabasi AL (2000) Error and attack tolerance of complex networks. Nature 406:378.
53. Albert R, Barabasi AL (2002) Statistical mechanics of complex networks. Rev Mod Phys 74:47–97.
54. Kauffman SA (1969) Metabolic stability and epigenesis in randomly constructed genetic nets. J Theoret Biol 22:437–467.

55. Somogyi R, Sniegoski CA (1996) Modeling the complexity of genetic networks: understanding multigenic and pleitropic regulation. Complexity 1(6):45–63.
56. Arnone MI, Davidson EH (1997) The hardwiring of development: organization and function of genomic regulatory systems. Development 124:1851–1864.
57. Miklos GL, Rubin GM (1996) The role of the genome project in determining gene function: insights from model organisms. Cell 86:4 521–9.
58. D'haeseleer P, Liang S, Somogyi R (2000) Genetic network inference: from co-expression clustering to reverse engineering. Bioinformatics 16(8):707–26.
59. Hartemink AJ (2005) Reverse engineering gene regulatory networks. Nat Biotech 23:554–555.
60. Arkin A, Ross J, McAdams HH (1998) Stochastic kinetic analysis of developmental pathway bifurcation in phage λ-infected Escherichia coli cells. Genetics 149:1633-1648.
61. Yuh CH, Bolouri H, Davidson EH (1998) Genomic cis-regulatory logic: experimental and computational analysis of a sea urchin gene. Science 279:1896–1902.
62. Zweiger G (1999) Knowledge discovery in gene-expression microarray data: mining the information output of the genome. Trends Biotech 17:429–436.
63. Fuhrman S, Cunningham MJ, Wen X, Zweiger G, Seilhamer JJ, Somogyi R (2000) The application of shannon entropy in the identification of putative drug targets. Biosystems 55 (1-3):5–14.
64. Tavazoie S, Hughes JD, Campbell MJ, Cho RJ, Church GM (1999) Systematic determination of genetic network architecture. Nat Genet 22:281–285.
65. Tamayo, P, Slonim D, Mesirov J, Zhu Q, Kitareewan S, Dmitrovsky E, Lander ES, Golub TR (1999) Interpreting patterns of gene expression with self-organizing maps: methods and application to hematopoietic differentiation. Proc Natl Acad Sci USA 96(6):2907–2912.
66. Dempster AP, Laird NM, Rubin DB (1977) Maximum likelihood estimation from incomplete data. J R Stat Soc B 39:1–38.
67. Mjolsness E, Mann T, Castaño R, Wold B (1999) From Co-expression to Co-regulation: An Approach to Inferring Transcriptional Regulation among Gene Classes from Large-Scale Expression Data. Technical Report JPL-ICTR-99-4, Jet Propulsion Laboratory Section 365. http://www-aig.jpl.nasa.gov/public/mls/papers/emj/GRN99prprnt.pdf.
68. Chen KC, Csikasz-Nagy A, Gyorffy B, Val J, Novak B, Tyson JJ (2000) Kinetic analysis of a molecular model of the budding yeast cell cycle. Mol Biol Cell 11:369–391.
69. Cross FR, Archambault V, Miller M, Klovstad M (2002) Testing a mathematical model of the yeast cell cycle. Mol Biol Cell 13:52–70.
70. Li FT, Long T, Lu Y, Ouyang Q, Tang C (2004) The yeast cell-cycle network is robustly designed. Proc Natl Acad Sci 101:4781–4786.
71. Chen HC, Lee HC, Lin TY, Li WH, Chen BS (2004) Quantitative characterization of the transcriptional regulatory network in the yeast cell cycle. Bioinformatics 20:1914–1927.
72. Chen KC, Calzone L, Csikasz-Nagy A, Cross FR, Novak B, Tyson JJ (2004) Integrative analysis of cell cycle control in budding yeast. Mol Biol Cell 15:3841–3862.
73. Cross FR, Schroeder L, Kruse M, Chen KC (2005) Quantitative characterization of a mitotic cyclin threshold regulating exit from mitosis. Mol Biol Cell 16(5): 2129–38.

74. Futcher B (2002) Transcriptional regulatory networks and the yeast cell cycle. Curr Opin Cell Biol 14(6):676–83.
75. Murray AW (2004) Recycling the cell cycle: cyclins revisited. Cell 116:221–234.
76. Ingolia NT, Murray AW (2004) The ups and downs of modeling the cell cycle. Curr Biol 14:R771–R777.
77. Tyers M (2004) Cell cycle goes global. Curr Opin Cell Biol 16(6):602–13.
78. Rao CV, Wolf DM, Arkin AP (2002) Control, exploitation and tolerance of intracellular noise. Nature 420:231.
79. Zhang YP, Qian MP, Ouyang Q, Tang C et al (To appear).
80. Albeverio S, Feng J, Qian MP (1995) Role of noises in neural networks. Phys Rev E 52:6593–6606.
81. Qian MP, Chen DY, Feng JF (1996) The metastability of exponentially perturbed Markov chains. Sci China Ser A 39:7–28.
82. Ao P (2004) Potential in stochastic differential equations: novel construction. J Phys A 37:L25–L30.
83. Zhu XM, Lan L, Hood L, Ao P (2004) Calculating biological behaviors of epigenetic states in the phage lambda life cycle. Funct Integr Genomics 4:188–195.
84. Ito T, Chiba T, Ozawa R, Yoshida M, Hattori M, Sakaki Y (2001) A comprehensive two-hybrid analysis to explore the yeast protein interactome. Proc Natl Acad Sci 98(8):4569–4574.
85. D'haeseleer P, Wen X, Fuhrman S, Somogyi R (1999) Linear modeling of mRNA expression levels during CNS development and injury. Pacific Symposium on Biocomputing 4:41–52. http://www.smi.stanford.edu/projects/helix/psb99/Dhaeseleer.pdf.
86. Weaver DC, Workman CT, Stormo GD (1999) Modeling regulatory networks with weight matrices. Pacific Symposium on Biocomputing 4:112–123. http://www.smi.stanford.edu/projects/helix/psb99/Weaver.pdf.

14

Messenger RNA Information: Its Implication in Protein Structure Determination and Others

Liaofu Luo and Mengwen Jia

Summary. Three problems on mRNA information in protein-coding regions are discussed: first, how the mRNA sequence information (tRNA gene copy number) is related to protein secondary structure; second, how the mRNA structure information (stem/loop content) is related to protein secondary structure; third, how the specific selection for mRNA folding energy is made among genomes. From statistical analyses of protein sequences for humans and *E. coli* we have found that the m-codon segments (for $m = 2$ to 6) with averagely high tRNA copy number (TCN) (larger than 10.5 for humans or 1.95 for *E. coli*) preferably code for the alpha helix and that with low TCN (smaller than 7.5 for humans or 1.7 for *E. coli*) preferably code for the coil. Between them there is an intermediate region without structure preference. In the meantime, we have demonstrated that the helices and strands on proteins tend to be preferably "coded" by the mRNA stem region, while the coil on proteins tends to be preferably "coded" by the mRNA loop region. The occurrence frequencies of stems in helix and strand fragments have attained 6 standard deviations more than the expected. The relation between mRNA stem/loop content and protein structure can be seen from the point of mRNA folding energy. Both for *E. coli* and humans, the mRNA folding energy in protein regular structure is statistically lower than that in randomized sequence, but for irregular structure (coil) the Z scores are near their control values. We also have studied the folding energy of native mRNA sequence in 28 genomes from a broad view. By use of the analysis of covariance, taking the covariable G+C content or base correlation into account, we demonstrate that the intraspecific difference of the mRNA folding free energy is much smaller than the interspecific difference. The distinction between intraspecific homogeneity and interspecific inhomogeneity is extremely significant ($p < .0001$). This means the selection for local mRNA structure is specific among genomes. The high intraspecific homogeneity of mRNA folding energy as compared with its large interspecific inhomogeneity can be explained by concerted evolution. The above result also holds for the folding energy of native mRNA relative to randomized sequences. This means the robustness of the distinction between intraspecific homogeneity and interspecific inhomogeneity of mRNA folding under the perturbation of sequential and structural variation.

DNA, RNA, and protein are three basic elements of a life system. A Chinese ancient philosopher named Laozhi said: "One generates two, two generates three, and three generates all things in the Universe." So, three means infinity. The interaction network of these three elements makes the genetic language complex enough to represent life. Although the importance of DNA and protein interaction is well known the role of RNA in the network has not been analysed thoroughly. We shall discuss some aspects of mRNA information in the determination of protein structure and in the specific selection among genomes.

14.1 mRNA Sequence Information (tRNA Gene Copy Number) Related to Protein Secondary Structure

There have been three generations in the empirical prediction of protein secondary structure. The first generation of the empirical prediction is based on single-residue statistics. The second generation is based on segment statistics (typically 11 to 21 adjacent residues were taken from a protein). In the third generation of prediction the evolutionary information is used through the method of multiple sequence alignment. A typical example is the particular neural network-based method PHD. Up to now the highest prediction accuracy currently attained is about 76% by use of the support vector machine method. So, "the dinosaurs of secondary structure prediction are still alive" [1]. We feel that the relatively low accuracy of secondary structure prediction has its deep origin in the formation of secondary structure. The problem is twofold. The first is related to the importance of long-range information and environmental information in determining the secondary structure. The second is the possible role of mRNA sequence on the formation of protein structure. The latter is more fundamental since it is a challenge to Anfinsen's [2] sequence-structure principle. The possible structural signals in mRNA sequence were analyzed by several authors [3–6]. Based on di-peptide frequencies we have studied the influences of codon usage on protein secondary structure [7]. It is demonstrated that for humans, the structural preferences of codons in 45 (or 79) di-peptides are different from those of amino acids, and they could not be explained by stochastic fluctuations at the 95% (or 90%) confidence level, and for *E. coli* the number is 36 (or 60). So the codon usage may influence protein secondary structure at the level of 10% or lower for different species, from higher mammals to prokaryotes. Apart from the possible structural signal in mRNA sequence, the influence of messenger RNA on protein secondary structure may occur through two other approaches, namely, the codon tRNA abundance and the stem-loop structure of mRNA [8–10].

Since the regular secondary structure (alpha helix and beta strand) occurs in the very early epoch of protein folding, and the tRNA molecule is the adaptor of the mRNA sequence to the amino acid sequence, we ask if the tRNA molecule can exert some influence on the formation of protein secondary structure? Consider m-codon segment, $m = 2, 3, 4, 5$, or 6 (hereafter called m-mer). The average of tRNA gene copy number (TCN) values over codons in an m-mer is denoted as v [9]. Consider a sliding window of width of m codons shifted along the mRNA sequence by 1 codon each step and count the number of m-codon segments corresponding to a definite protein secondary structure α, β, or c (the m-mer that corresponds to two structures will not be taken into account). The m-mer frequency in the kth interval of v that codes for protein secondary structure α, β, or c is denoted as n_k^j ($j = \alpha, \beta$, or c). The total number of m-mers in the kth interval is denoted by n_k, $n_k = \sum_j n_k^j(obs)$. Set

$$q_j = \frac{\sum_k n_k^j(obs)}{\sum_{kj} n_k^j(obs)} \qquad (j = \alpha, \beta, \text{or } c)\,. \tag{14.1}$$

Theoretically, the distribution n_k^j ($j = \alpha, \beta$, or c) in three structures is a stochastic variable, obeying multinomial distribution

$$n_k^j(exp) = n_k \cdot q_j \tag{14.2}$$

and the corresponding deviation

$$\sigma_k^j = \sqrt{n_k \cdot q_j \cdot (1 - q_j)}\,. \tag{14.3}$$

Calculating the parameter of codon preference for protein secondary structure

$$F_k^j = \frac{n_k^j(obs) - n_k^j(exp)}{\sigma_k^j} \tag{14.4}$$

in each interval of v we obtain $F_k^j - v$ relations for three structures α, β, and c.

We count the number of m-mers in a window of width v_0, $v_0 = (v_{max} - v_{min})/20$, and shift the window by steps of $v_0/10$. The resulting $F_k^j(m) - v$ relations for humans and *E. coli* are shown in scattered point diagrams, Figure 14.1 and Figure 14.2. The data have been filtered by $|F_k^j| \geq 3$.

From Figures 14.1 and 14.2 we find the following:

1. The mRNA sequences consisting of m-codons ($m = 2$ to 6) with an averagely high copy number of tRNA, namely v larger than around 10.5 for humans or v larger than around 1.95 for *E. coli*, preferably code for α helix but less commonly code for coil.

2. The structural preference/avoidance turns out to be contrary to the codons with low tRNA copy number. As the average TCN v smaller than around 7.5 for humans or v smaller than around 1.7 for *E. coli*, the m-mers preferably code for coil but less commonly code for α helix.

3. There exists a clear transition region between high v and low v regions of different preferences. Detailed studies show that the location of the intermediate region slightly shifts for different m.

4. For beta strand the preference/avoidance tendency is not obvious. In most cases $-2 < F_k^{\beta} < 2$.

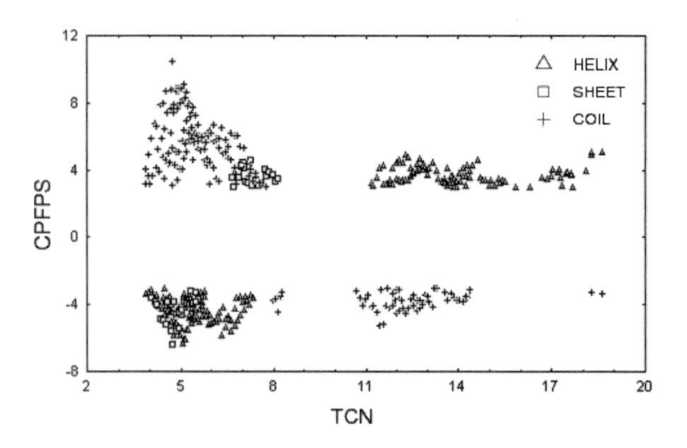

Fig. 14.1. $F_k^j(m)$ - v relations for humans.
$m = 2, \ldots, 6$. The interval number between v_{max} and v_{min} is supposed to be 20.

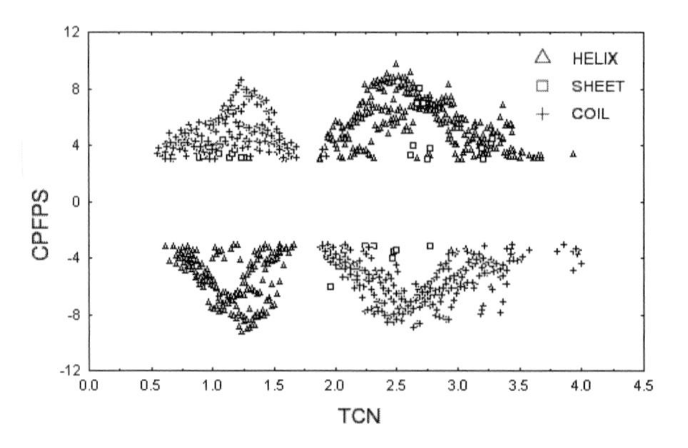

Fig. 14.2. $F_k^j(m)$ - v relations for *E. coli*.
$m = 2, \ldots, 6$. The interval number between v_{max} and v_{min} is supposed to be 20.

All strong preference/avoidance modes with $F_k^j(m) \geq 3$ and ≤ -3 in non-overlapping TCN intervals are listed in Tables 14.1 and 14.2 for humans and *E. coli*, respectively. These modes are arranged in a line following the order of TCN. The mode notation, for example, kc,Nhe means in the kth interval of average TCN the m-mer preferably codes for the random coil (denoted as c) but less commonly code for α helix (denoted as h) and β strand (denoted as e), etc.

The preference of TCN for protein secondary structure is essentially a problem of the existence of mutual interaction between two factors. That is, the stochastic variable n_k^j, the number of m-mer in the kth TCN interval coding for protein structure j ($j = 1, \ldots, a$; $k = 1, \ldots, b$), depends on two factors: factor A, protein secondary structure, and factor B, the interval of codon TCN v. Define Q_A as A-induced deviation, and Q_B as B-induced deviation. The total deviation Q_{tot} can be calculated by

$$Q_{tot} = Q_A + Q_B + Q_{A \times B} \tag{14.5}$$

where $Q_{A \times B}$ describes the mutual effect of two factors. Decompose $Q_{A \times B}$ into two parts, Q_N (1 degree of freedom) and Q_{error} ($(a-1)(b-1)-1$ degrees of freedom). Define

$$Q_R = \frac{Q_N}{Q_{error}}((a-1)(b-1)-1), \tag{14.6}$$

Q_Rs for humans and *E. coli* are calculated and listed in Table 14.3. Tukey [11] proved that there exists mutual interaction between two factors—factor A and B at significance level α if $Q_R > F_{1-\alpha}(1, df)$ where $F_{1-\alpha}(1, df)$ is $100(1-\alpha)$th percentiles of the F distribution with numerator degree of freedom 1 and denominator degree of freedom $df = (a-1)(b-1)-1$. From Table 14.3 we find that Q_R is always much larger than $F_{0.999}$ $(1, df)$ for all m-mers in humans and *E. coli*. Therefore, the mutual interaction between protein secondary structural type and codon TCN is very marked.

Table 14.1. Strong structure-preference/avoidance modes of codon TCN (human) (9 TCN intervals) ($F(m) \geq 3$ or ≤ -3)

m-mers	Low TCN region			High TCN region			
2mers	1c	2c,Nhe	3e,Nc			7h	8h
3mers	1c,Nh	2c,Nhe	3e		6h	7h	
4mers		2c,Nhe	3Nh	5h	6h,Nc		
5mers		2c,Nh	3c,Nh	5h,Nc	6h		
6mers		2c,Nh	3c,Nh	5h,Nc	6h,Nc		

Finally, to study the origin of structure preference we compare

$$\left|\frac{n_k^\alpha(obs)}{n_k} - q_\alpha\right| + \left|\frac{n_k^\beta(obs)}{n_k} - q_\beta)\right| + \left|\frac{n_k^c(obs)}{n_k} - q_c\right| = R_k \tag{14.7}$$

Table 14.2. Strong structure-preference/avoidance modes of codon TCN (*E. coli*) (19 TCN intervals) ($F(m) \geq 3$ or ≤ -3)

m-mers	Low TCN region					High TCN region							
2mers	1c	2Nh	3c,Nh	4c,Nh	5c	7Ne	8h	9h,Nc	10h,Nc	11h,Nc	12h,Nc		15Nc
3mers		2c,Nh	3ec,Nh	4c,Nh	5c,Nh	7h	8h,Nc	9h,Nc	10h,Nc	11h,Nc	12h,Nc	13h,Nc	15Nc
4mers	1c	2c,Nh	3c,Nh	4c,Nh	5c,Nh	7h,Nc	8h,Nc	9h,Nc	10h,Nc	11h,Nc	12h,Nc	13h	
5mers		2c,Nh	3c,Nh	4c,Nh	5Nh	7h,Nc	8h,Nc	9h,Nc	10h,Nc	11h,Nc			
6mers	1c	2c,Nh	3c,Nh	4c,Nh	5Nh	7h,Nc	8h,Nc	9h,Nc	10h,Nec	11h			

with

$$\left|\frac{n_k^\alpha(stoch)}{n_k(stoch)} - q_\alpha\right| + \left|\frac{n_k^\beta(stoch)}{n_k(stoch)} - q_\beta)\right| + \left|\frac{n_k^c(stoch)}{n_k(stoch)} - q_c\right| = R_k^{stoch} \tag{14.8}$$

where *stoch* means the quantity calculated in a codon-randomized sequence. If the structure preference results from an amino acid sequence, then one should have $R_k = R_k^{stoch}$. However, by direct statistics we find

$$\left(1 - \frac{R_k^{stoch}}{R_k}\right) > 0.5 \quad \text{or} \quad < -0.5 \tag{14.9}$$

for more than half TCN intervals. So the contribution to the structural preference does not come from the amino acid sequence alone. Accompanying the amino acid sequence the nonuniform codon usage gives an important contribution to the structural preference.

Table 14.3. Assessment of mutual interaction between protein secondary structure and codon TCN

		3-mers	4-mers	5-mers	6-mers
	df	29	25	21	19
Humans	Q_R	90.11	82.27	63.76	62.83
	$F_{0.999}$	13.39	13.88	14.59	15.08
	df	23	19	17	15
E. coli	Q_R	84.00	53.50	50.30	50.18
	$F_{0.999}$	14.19	15.08	15.72	16.59

The average TCN (v) of m-mer is divided into 21 intervals for humans and 19 intervals for *E. coli*. $df = (a-1)(b-1) - 1 = 2b - 3$ ($a = 3, b =$ effective interval number). $F_{0.999}(1,df)$ is the percentile of F distribution, which gives the threshold of Q_R at significance level .001. Q_R is a measure of mutual interaction between two factors—protein secondary structure and codon TCN, which is calculated from their statistical deviations. It shows $Q_R > F_{0.999}(1,df)$ for all m-mers in humans and *E. coli*. So, the mutual interaction between two factors does exist.

14.2 mRNA Structure Information (Stem-Loop Content) Related to Protein Secondary Structure

The mRNA secondary structure is deduced from nucleotide sequence by use of RNA structure 3.6 [12]. We fold the mRNA sequence through base pairing by use of RNA structure 3.6 in a window of 100 nucleotides, and shift the window along the sequence. The unpairing part in the tail of the first 100 nucleotides is put into the shifted window and participates in the next folding. Based on the above model, we postulate the secondary structure of mRNA as a number of hairpins or more complex units, constructed by loops and stems (pairing bases). The nucleotide in the loop is denoted by 0 and that in the stem by 1. So, the secondary structure of an mRNA is depicted by a sequence written by two symbols, 0 and 1.

To study the relation between mRNA structure and protein secondary structure, we align the amino acid sequence for a given secondary structure (helix, denoted as H; extended strand, denoted as E; turn, denoted as T; or coil, denoted as C) [13] and the corresponding mRNA sequence written by 0 and 1. Set the observed base number of mRNA structure j occurring in the kth protein secondary structure denoted by $n_k^j(obs)$. Define F_k^j as in (14.4) but with its meaning changed. It gives a measure of preference-avoidance of protein secondary structure k for the mRNA structure j.

Based on the IADE database, which contains 2269 protein sequences [8], we calculate the mRNA folding and study the relation between RNA stem-loop frequencies and protein secondary structure. We find that the regular secondary structures—helices and strands—on proteins are strongly related to the stems of the corresponding mRNA structure. These regular structures tend to be preferably "coded" by the mRNA stem region, while the coil on proteins tend to be preferably "coded" by the mRNA loop region (Table 14.4).

Table 14.4. Preference of protein secondary structural types for the mRNA stems

	$k = H$(helix)	$k = E$(strand)	$k = T$(turn)	$k = C$(coil)
F_k^1	3.40	4.21	3.43	−9.25
σ_k^1	186	136	104	169

$(F_k^0 = -F_k^1)$

To obtain better statistics, we define a four-amino-acid-fragment that shows pronounced secondary structural propensity as "structural word" (SW). We study the following types of SW: H-type, the secondary structures corresponding to four-amino-acid-fragment are HHHH; E-type, the secondary structures corresponding to four-amino-acid-fragment are EEEE; T-type, the secondary structures corresponding to four-amino-acid-fragment are TTTT; and boundary type, the secondary structures corresponding to four-amino-acid-fragment are T and other structures (H,E,C), for example, HHHT, TTEE, TCCC, HHTC, etc. An i-type structural word means that the occurrence of the word in structure i is not at random with 95% confidence level, and this word is a characteristic word of the structure i.

Based on SWs, we are able to manifest the relation between protein secondary structure and mRNA stem-loop structure more clearly. We calculate the occurrence frequency of stem or loop in each kind of SW. In calculations, the double count should be avoided. For successive SWs of the same type, the overlapping part should be counted only once. We deduce the preference of protein secondary structures for the mRNA stems/loops and find the tendencies are more obvious if we observe the structural words. The preferences of the H-type SWs (H) and E-type SWs (E) for the mRNA stem structure are very marked. As seen from Table 14.5, the occurrence frequencies of stems in H and E words have attained 6 standard deviations more than expected. The result is understandable since H and E words are amino acid fragments characteristic of helix and strand. Their preferences for stems more clearly reflect the essential connections of helix and strand to the stem structure of mRNA.

Table 14.5. Preference of protein structural words for the mRNA stems

	H-type word	E-type word	T-type word	T-related boundary word
F_k^1	5.73	6.83	−1.52	9.21
σ_k^1	151	109	77	132

($F_k^0 = -F_k^1$)

The n-nucleotide fragment analyses (up to $n = 6$) further prove the above conclusion. All n-mers solely composed of loops very scarcely occur in helices and strands (H-type and E-type words) with a high confidence level, but they tend to code for nonregular secondary structures. However, the strands (E-type SWs) preferably tend to be coded by n-mers solely composed of stems. The H-type words also preferably tend to be coded by n-mers mainly composed of stems but the tendency is not so obvious as in E-type words (Table 14.6).

Note: The secondary structures of protein are usually classified into three categories, namely, helix, strand, and coil. The class of coil includes all structures not assigned as helix or strand [13]. But we have found in coil that the turns and turn-related boundaries are of special interest. They are preferably coded by stems, different from other coils. So, we have separated them as an independent class, different from the other coils, in the stem-loop preference analysis.

The relation between mRNA stem/loop content and protein structure can be seen from the point of mRNA folding energy. The folding energy is calculated by use of the RNAfold program from the Vienna RNA package [12, 14, 15]. The energy Z-score is defined by

$$Z = \frac{E_{native} - \langle E_{random} \rangle}{STD} \tag{14.10}$$

where $\langle E_{random} \rangle$ means the energy of a codon-randomized sequence averaged over a large number of samples (larger than 50) generated from the native sequence, and STD means its standard deviation. Simultaneously, we define

Z^{reg} — when the regular secondary structure segments are randomized

Z^{coil} — when the coil segments are randomized

Z^{reg}_{ctrl} — when part codons are randomized with the same percentage as regular structure in native sequence, which is served as a control of Z^{reg}

Z^{coil}_{ctrl} — when part codons are randomized with the same percentage as coil segments, which is served as a control of Z^{coil}

The above Z scores are free energy Z scores. If energy E is changed to the percentage of nucleotides in stem, then the score calculated is called stem-loop score Z_{SL}. The energy Z score is oppositely correlated with stem content Z score. Figure 14.3 gives an example.

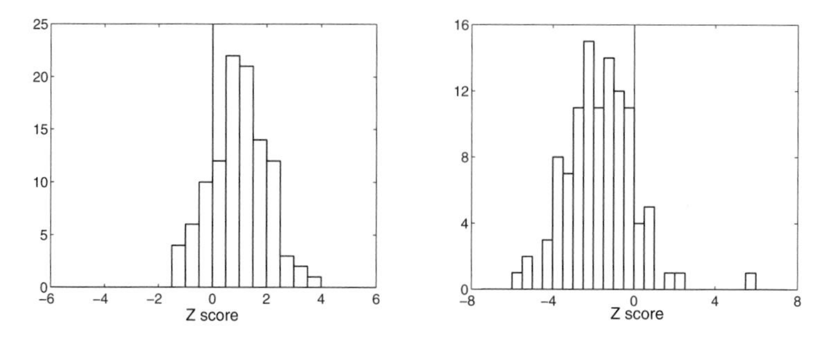

Fig. 14.3. The histograms of stem-loop and energy Z score distribution (*E. coli*). The left figure gives the stem-loop Z score (Z_{SL}) distribution for 107 *E. coli* genes, and the right figure the energy Z score distribution for 107 *E. coli* genes.

Table 14.6. Preference of protein structural words (A) and structural types (B) for the stem-loop structure in mRNA oligonucleotide fragments (n-mers with $n \leq 6$) (all modes with $F_k^j \geq 3$ or ≤ -3 are listed)

A

Strc	0	1	00	11	000	101	111	0000	1011	1111	00000
H.W.	−5.7	5.7	−6.9	3.8	−7.7	3.2	—	−8.6	3.1	—	−9.5
E.W	−6.8	6.8	−6.9	6.5	−6.6	—	7.7	−6.0	—	8.4	−5.0
B.W.	−9.2	9.2	−10.3	7.6	−12.1	—	6.8	−13.2	—	5.7	−13.6
T.W.	—	—	—	—	—	—	—	—	—	—	—

Strc	00111	10000	10111	11101	11111	000000	100000	100001	110000	111101	111111
H.W.	—	—	3.1	—	—	−10.1	—	—	—	—	—
E.W.	—	−3.1	—	3.1	8.9	−3.6	−3.4	—	−3.7	3.9	8.5
B.W.	3.3	—	—	—	4.9	−13.9	—	3.3	—	—	4.0
T.W.	—	—	—	—	—	—	—	—	—	—	—

B

Strc	0	1	00	11	000	001	111	0000
helix	−3.4	3.4	−4.2	—	−4.9	—	—	−5.7
strand	−4.2	4.2	−4.0	4.6	−3.6	—	6.0	−3.2
turn	−3.4	3.4	—	—	−3.3	4.2	—	—
other	9.2	−9.2	8.8	−6.1	8.5	—	−6.1	7.9

Strc	0011	1111	00000	00111	11111	000000	000110	111111
helix	—	—	−6.0	—	—	−6.5	—	—
strand	—	6.3	—	—	6.4	—	—	6.5
turn	3.6	—	—	4.2	—	—	3.1	—
other	—	−4.5	7.0	—	−3.0	6.8	—	—

The Z scores for mRNA sequences in different protein secondary structures are calculated and shown in Table 14.7. It gives the average energy Z scores of 107 *E. coli* genes and 125 human genes. From Table 14.7 we find for *E. coli* the mean Z score of regular structure is −1.38 with control −1.01 (difference 0.37), and for humans, the two values are −1.71 and −1.24, respectively (difference 0.47). So, both for *E. coli* and humans, the Z scores of regular structure are explicitly lower than their control values. However, the case is different for coil region. For *E. coli*, the average Z score of coil region is −0.93, very near the control value −0.86 (difference 0.07), and for humans, the Z score of coil is −1.03, even slight larger than its control −1.20 (difference −0.17). So, we conclude the mRNA folding energy in protein regular structure (helix and strand) is statistically lower than that in randomized sequence, but for irregular structure (coil) no such conclusion can be deduced. The Z scores for irregular structure are near their control values.

The detailed difference between Z scores in regular and irregular structures can be plotted in a diagram. We study the distributions of $Z^{reg} - Z^{reg}_{ctrl}$ and $Z^{coil} - Z^{coil}_{ctrl}$ for 107 *E. coli* and 125 human genes. The results are shown in Figure 14.4. Evidently, both for *E. coli* and humans the maximum distributions of Z score difference in regular structure are located at some values smaller than zero, that is, the distributions shifted toward the left as compared with those of irregular structure. Note that in spite of the maximum of Z score difference for coil taking a value slightly larger than zero, its average value (the average Z^{coil} minus its control) is near zero due to the bias-to-left of the shape of the coil distribution curve. These results indicate obviously that the mRNA sequence coding for protein regular structure has more negative free energy (relative to randomized samples) than the sequence segment coding for coil.

A more detailed comparison of Z scores for regular structure and irregular structure can be established through a statistical test. Consider 107 (125) Z^{reg}'s and Z^{reg}_{ctrl}'s for *E. coli* (humans) as two sets of variables separately. By use of the Aspin-Welch t test, we find the Z scores of regular structure, Z^{reg}, are significantly different from the control set both for *E. coli* and humans. So the folding free energy of native mRNA sequence segment coding for protein regular structure (relative to the energy of randomized sequence) is significantly different from the control set (Table 14.8). But for irregular structure, the difference between two Z scores, Z^{coil} and Z^{coil}_{ctrl}, is not significant.

Table 14.7. The dependence of energy Z-score on protein structure

	Z^{reg}	Z^{reg}_{ctrl}	Z^{coil}	Z^{coil}_{ctrl}
E. coli	−1.38	−1.01	−0.93	−0.86
Humans	−1.71	−1.24	−1.03	−1.20

We have indicated in previous paragraphs that stems tend to code for protein regular structure while loops tend to code for irregular structure. So, the mRNA sequence coding for protein regular structure has more negative folding energy due to the hydrogen bonds that existed between stems and that decrease the folding energy. The conclusion is consistent with the present energy Z score analysis. We studied the Z score in protein regular structure and that in irregular structure separately. We have deduced that the Z score in protein regular structure (Z^{reg}) is more negative than that in coil (Z^{coil}). This shows again the mRNA sequence coding for protein regular structure has more negative folding energy.

Table 14.8. Aspin-Welch t test and F test for the difference of Z scores

	t test				F test			
	Regular struc		Irregular struc		Regular struc		Irregular struc	
	t-value	df	t-value	df	F-value	df_1, df_2	F-value	df_1, df_2
Human	-2.0^*	240	1.02	247	1.43^+	124	1.10	124
E. coli	-2.04^*	200	-0.44	193	1.66^+	106	1.91^+	106

$*$ The calculated t-value (absolute value) $> t_{0.975}$ and the difference of average Z scores is very significant between Z^{reg} and Z^{reg}_{ctrl}. $+$ means the calculated F-value $> F_{0.975}$ and the difference of deviations of Z scores is very significant between Z^{reg} and Z^{reg}_{ctrl} (or Z_{coil} and Z^{coil}_{ctrl}). df = degree of freedom. When F-value is large enough, the Aspin-Welch t-test for average Z scores should be used (as shown in the left panel).

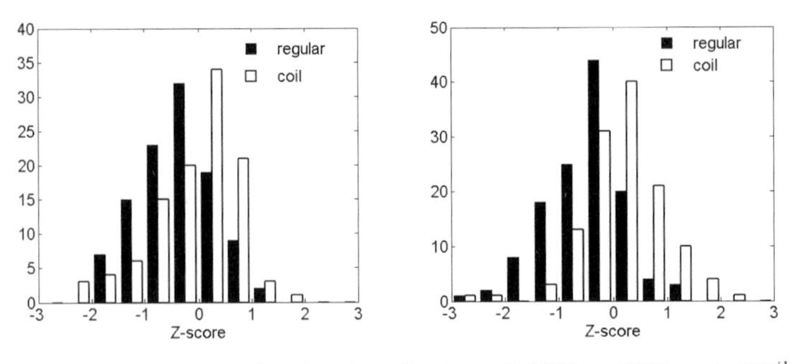

Fig. 14.4. The histograms for the distribution of $Z^{reg} - Z^{reg}_{ctrl}$ and $Z^{coil} - Z^{coil}_{ctrl}$ for *E. coli* and humans. The left figure gives the distribution of $Z^{reg} - Z^{reg}_{ctrl}$ and $Z^{coil} - Z^{coil}_{ctrl}$ for 107 *E. coli* genes and the right figure gives the distribution of $Z^{reg} - Z^{reg}_{ctrl}$ and $Z^{coil} - Z^{coil}_{ctrl}$ for 125 human genes.

Two Factors—tRNA Gene Copy Number and Stem/Loop Content—Are Coupled

We have proposed a phenomenological model on the relation between structure-preference and translational accuracy [9]. The model assumes the protein structure preference of codons is dependent on translational accuracy. Then the translational accuracy is assumed to be proportional to tRNA abundance (tRNA copy number, TCN). Apart from TCN we introduce R to describe other translational accuracy-related factors that influence structure preference. R factor is a matrix that, on the one hand, depends on three protein structural types, and on the other hand, depends on three TCN regions of codons. We have established equations that should be satisfied by R matrix elements, and demonstrated that the stem/loop content of mRNA is a possible solution for R factor. The result suggests that there may exist an adjusting factor on translation accuracy that is expressed in three TCN regions and correlates with stem/loop distribution. This factor plays a role of positive regulation for helix but negative regulation for random coil. So, the interaction between mRNA structure and tRNA abundance may play an important role in an associated way in protein secondary structure formation.

14.3 mRNA Folding Energy—Specific Selection Among Genomes

This subsection addresses the mRNA folding energy of genes in a wider range of species (28 genomes including eight archaea, 14 eubacteria, and six eukaryota) from a broad view that will give us some new insights into the role of mRNA local folding in the genome evolution. For each species about 120 genes (coding regions) are chosen stochastically. Using the RNAfold program from the Vienna RNA package [14, 15], we fold each mRNA sequence in a local window pattern, namely, the sequence is folded in short regions of 50 bases and shifted by 10 bases [16]. The averaged local folding energy E_{native} of the jth native mRNA sequence in the ith genome is denoted by y_{ij}. We study the statistical property of folding energies by use of the analysis of covariance. Set $y_{ij} = (E_{native})_{ij}$ (i=1, ..., t treatments or classes and for each treatment i, j=1, ..., n_i observations). In the general linear model one assumes y_{ij} depending on another set of covariables x_{ij}, with these two sets of variables obeying a regression equation in the form of

$$y_{ij} \approx a + bx_{ij} \tag{14.11}$$

for each class. Set

$$\begin{aligned} x_{ij} &= (G+C)\% \qquad \text{(model } GC\text{)} \\ &= D_1 + D_2 \qquad \text{(model } D_1 + D_2\text{)} \end{aligned} \tag{14.12}$$

D_1 and D_2 are informational redundancies and D_1+D_2 describes the base correlation in a DNA sequence [17]. Define the error sum of squares

$$\begin{aligned} SSe &= \sum_{ij}(y_{ij} - a - bx_{ij})^2 \\ &= \sum_{ij}(y_{ij} - \overline{y}_i)^2 - b^2\sum_{ij}(x_{ij} - \overline{x}_i)^2, \end{aligned} \tag{14.13}$$

$$b = \frac{\sum_{ij}(y_{ij} - \overline{y}_i)(x_{ij} - \overline{x}_i)}{\sum_{ij}(x_{ij} - \overline{x}_i)^2},$$

where $\overline{y}_i(\overline{x}_i)$ is the average of $y_{ij}(x_{ij})$ in a given class i, and the mean error sum of squares

$$MESS = \frac{SSe}{df(SSe)} = \frac{SSe}{\sum_i n_i - t - 1}. \quad (14.14)$$

Define the class sum of squares

$$CLS = \sum_{ij}((y_{ij} - \overline{y})^2 - b'^2(x_{ij} - \overline{x})^2) - \sum_{ij}((y_{ij} - \overline{y}_i)^2 - b^2(x_{ij} - \overline{x}_i)^2),$$

$$b' = \frac{\sum_{ij}(y_{ij} - \overline{y})(x_{ij} - \overline{x})}{\sum_{ij}(x_{ij} - \overline{x})^2} \quad (14.15)$$

where $\overline{y}(\overline{x})$ is the overall average of $y_{ij}(x_{ij})$, and the mean class sum of squares

$$MCSS = \frac{CLS}{df(CLS)} = \frac{CLS}{t-1}. \quad (14.16)$$

The covariant F-test is defined by

$$F = \frac{CLS/(t-1)}{SSe/(\sum_i n_i - t - 1)} \quad (14.17)$$

which gives a criterion for examining the difference between classes. Evidently, it includes the usual F-test as a special case when x variable vanishes. We put 28 species into three classes. For class 1 (archaea) $t = 8$, for class 2 (eubacteria) $t = 14$, and for class 3 (eukaryota) $t = 6$. Through the analysis of covariance we obtain mean error sum of squares ($MESS$) and mean class sum of squares ($MCSS$) of folding free energy of native mRNA. The former describes the square deviation of mRNA folding energy in a genome and the latter describes the square deviation of this energy among genomes. By comparing the calculated F value with F-distribution the significance level (p value) is deduced. The results are summarized in Table 14.9.

The results in Table 14.9 are given in two regression models (model GC and model $D_1 + D_2$). All results show $MCSS$ much larger than $MESS$ at the significance level $< .0001$. So, the square deviation of mRNA folding energy among genomes is definitely higher than that within a genome. Simultaneously, the results show that the linear relation between energy variable y_{ij} and covariable x_{ij} (GC content or $D_1 + D_2$) existed.

To understand its meaning we study the energy difference between native and randomized sequence. That is, we set

$$y_{ij} = (E_{native} - \langle E_{random} \rangle)_{ij} \quad (14.18)$$

Three types of random sequences are produced based on different randomization methods. The first procedure, called Codonrandom (CODRAN), preserves the same encoded amino acid sequence of mRNA sequence under the random codon choice. The second procedure, called Codonshuffle (CODSHU), shuffles synonymous codon in given mRNA sequence and preserves the same encoded amino acid sequence and codon usage (base composition) of mRNA sequence. The third procedure called Dicodonshuffle (DICODSH) was developed by Katz and Burge [16]. This procedure preserves the dinucleotide composition, codon usage, and encoded amino acid sequence of the mRNA sequence.

Table 14.9. Square sum of folding energy of native mRNA sequence

	$MCSS$	(df_1)	$MESS$	(df_2)	F	p value	$\|R\|$
Model GC							
CLS 1	14.61	7	1.120	949	13.04	<.0001	0.959
CLS 2	16.66	13	0.602	1652	27.66	<.0001	0.940
CLS 3	19.75	5	0.883	705	22.37	<.0001	0.941
Model D_1+D_2							
CLS 1	1295.1	7	3.336	949	388.2	<.0001	0.873
CLS 2	178.5	13	1.541	1652	115.8	<.0001	0.861
CLS 3	440.4	5	2.614	705	168.5	<.0001	0.814

$MCSS$ = mean class sum of squares, $MESS$ = mean error sum of squares, df = degree of freedom, $df_1 = t-1$, $df_2 = \sum n_i - t - 1$ (t, number of species; n_i, sequence number taken in this study for species i), $F = MCSS/MESS$, R = correlation coefficient in linear regression, CLS1 including eight archaea—*A. fulgidus* (sequence number 120), *A. pernix* (119), *H. sp.* (119), *M. thermoautotropicum* (120), *P. abyssi* (120), *S. solfataricus* (120), *S. tokodaii* (120), and *T. volcanium* (120), CLS2 including 14 eubacteria—*A. aeolicus* (120), *B. burgdorferi* (120), *B. subtills* (120), *C. acetobutylicum* (120), *E. coli* (107), *H. influenzae* (120), *H. pylori* (120), *M. pneumoniae* (120), *R. prowazekii* (120), *S. pcc6803* (120), *S. typhi* (120), *T. maritima* (120), *T. pallidum* (120), and *V. cholerae* (120), CLS3 including six eukaryota—*A. thaliana* (120), *C. elegans* (114), *D. melanogaster* (120), *Homo sapiens* (125), *S. cerevisiae* (120), and *P. falciparum* (113). The p value column gives the significance level.

By use of the same method we find the extremely significant difference between $MCSS$ and $MESS$ also existed for folding energy of native sequence relative to randomized sequence (Table 14.10). This is a universal law: whether the folding energy of native sequence is lower or higher than the randomized sequence [16, 18, 19], whether the folding is energy-favorable or not, all species exhibit the same trend—the energy deviation in different genomes is

Table 14.10. Square sum of folding energy of native mRNA relative to randomized sequence

	$MCSS$	(df_1)	$MESS$	(df_2)	F	p value	$\|R\|$
CODRAN	Model GC						
CLS 1	14.27	7	0.981	949	14.55	<.0001	0.896
CLS 2	14.06	13	0.540	1652	26.05	<.0001	0.812
CLS 3	50.17	5	0.943	705	53.22	<.0001	0.803
CODRAN	Model D_1+D_2						
CLS 1	420.9	7	1.395	949	301.8	<.0001	0.849
CLS 2	64.20	13	0.860	1652	94.36	<.0001	0.755
CLS 3	192.4	5	1.187	705	162.1	<.0001	0.744
DICODSH	Model GC						
CLS 1	1.16	7	0.206	949	5.62	<.0001	0.268
CLS 2	1.76	13	0.160	1652	10.95	<.0001	0.291
CLS 3	0.79	5	0.130	705	6.09	<.0001	0.204
DICODSH	Model D_1+D_2						
CLS 1	1.866	7	0.208	949	8.96	<.0001	0.252
CLS 2	1.827	13	0.160	1652	11.40	<.0001	0.292
CLS 3	0.704	5	0.131	705	5.37	<.0001	0.192

See table legend given below Table 14.9.

always higher than that in a genome. Further, the above law is irrespective of the kingdoms of species and irrespective of the randomization method that has been adopted [16, 18, 19]. Another important result summarized in the two tables is the approximate linear relation between energy variable y_{ij} and covariable x_{ij} (GC content or D_1+D_2). The correlation coefficients $|R|$ are always large in the CODRAN randomization procedure both in model GC and model D_1+D_2. But in the DICODSH randomization the linear relation between energy variable and covariable (G+C)% or D_1+D_2 does not exist.

The above results show that the selection for mRNA folding is specific among genomes. From the point of evolution the large interspecific difference occurs due to the rapid accumulation of mutations. However, if we assume that, by certain mechanisms of concerted evolution (coincidental evolution) the mutation, and in turn, the change of mRNA folding energy can spread horizontally to all gene members in the same genome, then the high intraspecific homogeneity of mRNA folding energy as compared with its large interspecific inhomogeneity can be explained naturally. Numerous examples of concerted evolution of multigene families have been proposed. A large body of data obtained by restriction enzyme analysis and DNA sequencing techniques has attested to the generality of concerted evolution [20]. We suggest that the concept of concerted evolution of multigene families may be generalized to mRNA folding of genes in the whole genome. Under this assumption the high specific selection for mRNA folding among genomes compared with high intraspecific homogeneity is a result of concerted evolution. The deeper meaning of gene interaction in a genome at RNA level and the evolutionary conservation of mRNA folding should be clarified in a future investigation.

The intraspecific homogeneity and interspecific inhomogeneity of mRNA folding can be visualized as the clustering of folding free energy of genes in energy space for each species. To study the relation between folding energy of the native sequence (E_{native}) and the corresponding relative energy, we have made linear regression analysis and find there exists a good linear relation between these two energies for each genome (in 27 genomes for CODRAN case and 24 genomes for DICODSH case). Further, setting the folding free energy averaged over genes of species i as

$$\langle (E_{native})_{ij} \rangle_j = E_i \tag{14.19}$$

and the energy difference averaged over genes as

$$\langle (E_{native} - \langle E_{random} \rangle)_{ij} \rangle_j = D_i \tag{14.20}$$

we find E_i and D_i linearly related to each other for 28 species with correlation coefficient $R = 0.963$ in CODRAN and 0.386 in DICODSH randomization. These R values explicitly exceed the critical correlation coefficient at significance .05 ($R_{0.05} = 0.374$). So the clustering of genes in energy space leads to the clustering of genes in relative energy space. Therefore, the assumed concerted evolution may lead to not only the homogeneity of mRNA folding energy among genes in a genome, but also the homogeneity of folding free energy of native mRNA relative to randomized sequence. Since the randomization can be regarded as a perturbation applied to the native mRNA sequence, the above result shows that the homogeneity of folding energy among genes in a genome should remain valid even under certain perturbation of mRNA sequence and structure. This means the distinction between intraspecific homogeneity and interspecific inhomogeneity of mRNA folding free energy is robust.

References

1. Rost B, Sander C (2000) Third generation prediction of secondary structure. In: Webster DM (ed) Methods in Molecular Biology vol 143. Humana Press, New Jersey.
2. Anfinsen CB (1973) Principles that govern the folding of protein chains. Science 181:223–230.

3. Brunak S, Engelbrecht J (1996) Protein structure and the sequential structure of mRNA: alpha-helix and beta-sheet signals at the nucleotide level. Proteins 25:237–252.
4. Oresic M, Shalloway D (1998) Specific correlations between relative synonymous codon usage and protein secondary structure. J Mol Biol 281:31–48.
5. Adzhubei IA, Adzhubei AA, Neidle S (1998) An integrated sequence-structure database incorporating matching mRNA sequence, amino acid sequence and protein three-dimensional structure data. Nucleic Acids Res 26:327–331.
6. Xie T, Ding DF (1998) The relationship between synonymous codon usage and protein structure. FEBS Lett 434:93–96.
7. Li XQ, Luo LF, Liu CQ (2003) Abnormal preference of synonymous codons for protein secondary structure types.??? Chinese J Biochem Mol Biol 19(4):441–444 (in Chinese)
8. Jia MW, Luo LF, Liu CQ (2004) Statistical correlation between protein secondary structure and messenger RNA stem-loop structure. Biopolymers 73:16–26.
9. Luo LF, Jia MW, Li XQ (2004) Protein structure preference, tRNA copy number and mRNA stem/loop content. Biopolymers 74:432–447.
10. Luo LF (2004) Theoretic-Physical Approach to Molecular Biology. Shanghai Science Technical Publishers, Shanghai.
11. Tukey JW (1949) One degree of freedom for non-additivity. Biometrics 5:232–242.
12. Mathews DH, Sabina J, Zucker M, Turner H (1999) Expanded sequence dependence of thermodynamic parameters improves prediction of RNA secondary structure. J Mol Biol 288:911–940.
13. Kabsch W, Sander C (1983) Dictionary of protein secondary structure: Pattern recognition of hydrogen-bonded and geometrical features. Biopolymers 22:2577–2637.
14. Hofacker IL, Fontana W, Stadler PF, Bonhoeffer S, Tacker M, Schuster P (1994) Fast folding and comparison of RNA secondary structures. Monatsh Chem 125:167–188.
15. Zuker M, Stiegler P (1981) Optimal computer folding of large RNA sequences using thermodynamics and auxiliary information. Nucleic Acid Res 9:133–148.
16. Katz L, Burge CB (2003) Widespread selection for local RNA secondary structure in coding regions of bacterial genes. Genome Res 13:2042–2051.
17. Luo LF, Lee WJ, Jia LJ, Ji FM, Tsai L (1998) Statistical correlation of nucleotides in a DNA sequence. Phys Rev E 58:861–871.
18. Seffens W, Digby D (1999) mRNAs have greater negative folding free energies than shuffled or codon choice randomized sequences. Nucleic Acid Res 27:1578–1584.
19. Workman C, Krogh A (1999) No evidence that mRNAs have lower folding free energies than random sequences with the same dinucleotide distribution. Nucleic Acid Res 27:4816–4822.
20. Li WH (1997) Concerted evolution of multigene families. In: Li, W H (ed) Molecular evolution, pp. 309–334. Sinauer Associates, Sunderland, Massachusetts, and references cited therein.

Conclusion

After over a century of neurophysiological and many years of molecular biological research, we still do not understand the principle by which a stimulus such as an odour, an image, or a sound is represented by distributed neural ensembles within the brain, how the development is controlled by gene networks, or how a cell's function is governed by spatially distributed protein networks. While large numbers of studies have made detailed analyses of response profiles of single cells, single genes, and single proteins in isolation, such techniques cannot address holistic issues of how large ensembles of neurons, genes, and proteins can integrate information and interact both spatially and temporally. There is little doubt that much of the information-processing power of the brain and the control of the development of a gene network or a protein network resides in the activities of cooperating and competing networks of neurons, genes, and proteins, and that if we can unlock the principles whereby information is encoded within these networks as a whole, rather than within single neurons, genes, or proteins in isolation, we may actually be able to understand how the brain, gene networks, and protein networks work.

While some progress toward understanding how this is achieved at a gross structural level is being achieved, the only way to provide an understanding at the level of multiple cell-cell, gene-gene, and protein-protein interactions is to record from large numbers of cells, genes, and proteins within a defined system simultaneously. The three main challenges for achieving this step have been first to develop appropriate tools to record simultaneously the activity of ensembles of neurons, genes, and proteins, second to be able to analyse the huge amounts of multivariate, high frequency sampling data that would be generated as a result, and third to use these data to mathematically model the system and to then use these models to predict how different changes in the system's environment affect the system. To overcome these three difficulties, the chapters in the this book were written by computational biologists, neuroscientists, statisticians, mathematicians, computer scientists, engineers, and physicists, to improve our ability to make sense of these high-throughput measurements, and create, refine, and revalidate the models.

Network theory is fundamental for the development of systems biology. We therefore expect this book to be of interest to biologists, neuroscientists, computer scientists, mathematicians, statisticians, and physicists.

Index

Printed in the United States of America